1,000,000 Books
are available to read at

Forgotten Books

www.ForgottenBooks.com

Read online
Download PDF
Purchase in print

ISBN 978-1-5278-3959-5
PIBN 10891942

This book is a reproduction of an important historical work. Forgotten Books uses state-of-the-art technology to digitally reconstruct the work, preserving the original format whilst repairing imperfections present in the aged copy. In rare cases, an imperfection in the original, such as a blemish or missing page, may be replicated in our edition. We do, however, repair the vast majority of imperfections successfully; any imperfections that remain are intentionally left to preserve the state of such historical works.

Forgotten Books is a registered trademark of FB &c Ltd.
Copyright © 2018 FB &c Ltd.
FB &c Ltd, Dalton House, 60 Windsor Avenue, London, SW19 2RR.
Company number 08720141. Registered in England and Wales.

For support please visit www.forgottenbooks.com

1 MONTH OF
FREE
READING

at
www.ForgottenBooks.com

By purchasing this book you are eligible for one month membership to ForgottenBooks.com, giving you unlimited access to our entire collection of over 1,000,000 titles via our web site and mobile apps.

To claim your free month visit: www.forgottenbooks.com/free891942

* Offer is valid for 45 days from date of purchase. Terms and conditions apply.

English
Français
Deutsche
Italiano
Español
Português

www.forgottenbooks.com

Mythology Photography **Fiction**
Fishing Christianity **Art** Cooking
Essays Buddhism Freemasonry
Medicine **Biology** Music **Ancient
Egypt** Evolution Carpentry Physics
Dance Geology **Mathematics** Fitness
Shakespeare **Folklore** Yoga Marketing
Confidence Immortality Biographies
Poetry **Psychology** Witchcraft
Electronics Chemistry History **Law**
Accounting **Philosophy** Anthropology
Alchemy Drama Quantum Mechanics
Atheism Sexual Health **Ancient History**
Entrepreneurship Languages Sport
Paleontology Needlework Islam
Metaphysics Investment Archaeology
Parenting Statistics Criminology
Motivational

FORESTS COMMISSION
VICTORIA

. . . . . preserving the beauty of our forests for your enjoyment.

The Victorian Naturalist

Editor: G. M. Ward

Assistant Editor: E. King

CONTENTS

Articles:

Notes on Tasmanian Rat Kangaroos.
By Ron C. Kershaw 4

The Last Madimadi Man.
By Luise A. Hercus and Isobel M. White 11

The Discovery of the Lizard Genus *Carlia* in Victoria.
By A. J. Coventry 20

Winter Collection.
By J. C. Le Souëf 23

Field Naturalists Club of Victoria: 25

Diary of Coming Events. 26

Front Cover:

John Wallis took this photograph of the Little Northern Native Cat *(Dasyurus hallucatus)*.

The colour treatment of the cover has been introduced to improve the appearance of the Victorian Naturalist. Whether the objective has been successful or not depends on readers' comments.

Notes on Tasmanian Rat Kangaroos

by

Ron C. Kershaw

(Honorary Associate, Queen Victoria Museum, Launceston)

Introduction

There are two species of "Rat Kangaroo" in Tasmania. They are *Bettongia cuniculus* Gray and *Potorous tridactylous* Kerr. These animals will be referred to below as the Bettong and the Potoroo respectively. These notes are a record of observations made between the years 1950 and 1956 in an area near the Tamar River, North Tasmania. Comparisons between the animals and the types of habitat in which they are found have been made.

The terrain in Tasmania is very mountainous and an abrupt rise of several thousand feet within a few miles is not uncommon. The climate ranges from super-humid in the west to sub-humid in the east coastal area. Rat Kangaroos are found over a great deal of the State and in a wide range of conditions. The area studied herein consists of a few square miles in the vicinity of Clarence Point on the west bank of the Tamar River, near the river mouth, North Tasmania.

In this area the climate is relatively dry; the average rainfall being less than 25 inches annually with considerable fluctuations over a period. The terrain is undulating to hilly, the highest elevations being in the vicinity of 200 to 250 feet. The vegetation consists of sclerophyll forest in part associated with heath spp., grass sward, or with tea-tree scrub. Very little of the area can be said to be a natural bush environment. Burning and clearing with regrowth have been the order for many years. Since the observations were made a more vigorous clearing programme has been conducted. Cleared areas are given over to orchard or pasture. Both the Bettong and the Potoroo were common in the area during the period of observation.

Distinguishing Features

Rat Kangaroos are small kangaroo-like marsupials, which, owing to their small size and rat-like gait or visage have been compared to the rat. According to Troughton (1943b) the name "Kangaroo Rat" was used when the animal was first described and illustrated in 1789 in the "*Voyage of Governor Phillip to Botany Bay*", because of kangaroo-like limbs and rat-like visage. Another similar animal is the Brush Wallaby or Pademelon, a bulkier animal.

The Potoroo is not unlike the Bettong, but it has no crest on the tail, the white tip of which is much longer. The hair is grey-brown to brown. The hind foot is shorter than the head (Lord and Scott, 1924). It is the short hind foot which is responsible for the gait which involves the fore-limbs as well as the hind (Troughton, 1943b). This gait is rat-like and the animal scurries but does not hop. This is a useful field recognition factor as the Bettong hops like a kangaroo. The Potoroo is also a plumper animal. Pearson (1945) and Finlayson (1938) list important points of distinction. Further detail of externals is given by Guiler (1961) from examination of numerous specimens in South

Tasmania. Pearson (1946) has concluded that Rat Kangaroos are distinct from Kangaroos. He describes them as a highly specialized offshoot from the primitive phalanger. Lyne and McMahon (1951) studied the surface structure of hair in these animals.

DESCRIPTIVE

Observations were made in the field of as many animals as possible. Whenever a dead animal was encountered measurements and descriptions were made. Of the hazards to which these animals are exposed vehicles and dogs take toll. Rabbit poisoning campaigns also are responsible for deaths.

Bettongia cuniculus Gray.

The male of a pair of Bettongs killed by strychnine, still in rigor mortis, was carefully examined for the following description. It is typical of those observed.

Fur dense, smoky grey at base to pale grey at tip, dorsal interspersed hairs banded with black, with grey tips giving a darker appearance but paling laterally to greyish-white ventrally. Tail fur grey, darker above, faintly dusted brown. Interspersing hairs of tail black, forming a distinct crest for two-thirds the length. Tip of tail with a small white tuft — a somewhat variable feature. Pes grey, sole naked, granular, dark grey.

Manus grey, soles naked, granular, flesh pink. Rhinarium rounded, naked, granular, dark grey. Ears rounded, slightly brown, head faintly brown, eyes brown with blue iris, eyelashes black. Mystical bristles well developed, dark coloured, the longest 46 mm., supra-orbitals dark, few, 40 mm., genals dark, few, 25 mm.

Body length 340 mm. Hind foot 110 mm. Tail 290 mm. Ear 40 mm.

The measurements of a juvenile male, examined in July 1952, were body length 275 mm., tail 205 mm., hind foot 85 mm., ear 45 mm. This animal was paler ventrally and the dorsal fur was shining suggesting a young healthy animal. A later inspection revealed that the sheen had gone. Cause of death could not be ascertained. The small white tip of the tail was almost devoid of hair. Presumably the white tuft develops with age. The mystical bristles measured 45 mm. The animal was evidently very nearly mature.

These animals are typical of those observed and appear to represent a uniform population. However, one pale grey animal was observed, some 3 mile distant from the area. No marked colour differences were observed between the sexes although the female is of slighter build. The tail of the Bettong seems generally longer in proportion to the body

Plate L. *Bettongia cuniculus* Gray.

than is the case with the Potoroo. Guiler (1961) notes that head length is slightly greater in the male Potoroo.

Potorous tridactylous Kerr.

Fewer dead animals of the Potoroo have been encountered. It is thought that this is due to the animal living in less accessible places.

The Potoroo is similar in general appearance to the Bettong but is much more obese and hence is not so graceful. The tail has no crest and the white tip is longer enabling it to be seen at a distance. Wood Jones (1924) describes a uniform darkening to black at the tip of the tail. He seems to regard a white tip as rare. He gives eye colour as brown.

The fur is grey-brown to brown, rather variable throughout the population, but not apparently over a wide range. An old male was found dead probably within the previous twenty-four hours.

Fur dense, long, soft, grey brown on dorsal surface, paler brown laterally, smoky grey ventrally. Tail dusty brown at first passing into short black hairs for 140 mm., then dirty white to tip for 45 mm. Pes dusty brown, darker on sides; manus paler with pink soles. Interspersing hairs grey at base, then white, but often with brown or black tip. Ears dark brown, rounded; rhinarium naked, granular, black, rather elongated, nostrils lateral, slit like. Eyes and eyelashes blackish. Mystical bristles dark, developed, 40 mm. supra-orbitals dark, 25 mm., genals poor.

Head and body length 370 mm.; tail 265 mm.; hind foot 90 mm.; ear 48 mm.

The Habitat
(a) Environment:

The area under observation is a triangular patch bounded by the West Arm, the West Tamar Highway, and the Tamar River. The climate is

Plate 2. *Potorous tridactylous* Kerr.

mild, sub-humid, with relatively low rainfall reliability. There are no permanent streams and few springs. Watercourses may remain dry for long periods when rainfall is insufficient to induce run-off. There is a low lying area of marshy ground where surface water may persist for some time except under very dry conditions.

The terrain is undulating to hilly with some flat terrace surfaces. These latter have been cleared for farming activities for a long period. Orchards are frequent along the West Arm. At the time of observation rather more than half was not cleared though used for rough grazing. Orchard blocks were separated by nature strips and hedges. Several hills of dolerite rock have a thin brown-red, or black clay soil. The remainder of the area is occupied by sandy loam to fine sandy loam in part on terrace surfaces and hill slopes above the River and West Arm, or gravels and sandy gravels on ridges at 200 feet above water level, passing down to 160 feet. Heavy wet sandy loams occur in hollows and marshy areas. Podzols and podzolic soils with clay "B" horizon are present.

Areas of poor drainage, or congregated drainage support dense *Melaleuca* scrub, which has been almost impenetrable. Some gums are present here. In somewhat less dense areas ground cover is commonly of large clumps of long-broad-bladed species curving to the ground to form natural tunnels. A thick mat of leaves, sticks and litter cover the surface.

On sandy soil slopes there is a peppermint (*Eucalyptus amygdalina*)-sheoke (*Casuarina*)-blackwood (*Acacia*) association, with an understorey of shrubs rarely including *Melaleuca*. This is very dense at low levels, forming an almost closed canopy but thinning with elevation. In one part of the area the sclerophyll forest changes to a stringy-bark (*Eucalyptus obliqua*) dominated association. Here a large hedge of *Cupressus* trees fringes an orchard and forms a dense shade.

On the gravel ridges the sclerophyll forest thins, and the ground cover is dominated by heath species. Sunlight penetrates readily to the ground, and the observer may move readily. The dolerite hills support a similar forest, with scrub understorey on lower slopes. Here the open areas are covered with a native grass ground cover. In some areas the scrub understorey may be fairly dense, but it is not continuous and sunlight penetrates readily. This is in sharp contrast to the lower areas of dense cover where shade is permanent.

It is noticeable that two types of cover are available. The one of more or less low levels of dense scrub fringed by dense forest where there is virtual permanent shade, and ground cover is a foot or more deep. The second is of open forest on higher well drained areas with ground cover of heath or grasses where sunlight penetration is high. The differences are due to factors of soils and drainage with height being of minor significance. There are therefore two types of habitat available in the area. Graduations in constitution of these seem of little importance and are ignored.

1b) The habitat of the Bettong:

The Bettong is apparently less common in the area than the Potoroo, but is more strictly nocturnal in habit. This may account for the reduced number observed. Subsequent observations seem to support this. The relationship in numbers is variable. Fires have had a serious effect on the habitat. Approximately twenty animals were observed during the period 1950-1953.

The Bettong has been observed only in the open sclerophyll forest, either with heath or grass ground cover, or in nearby orchard and pasture. It may penetrate areas of light scrub cover. There seems every reason to believe that the habitat normally has a high degree of sunlight penetration. The animals have not been observed in hollows or gullies where vegetation is dense and sunlight penetration very low. It does not, however, follow that the habitat is purely of hills. Suitable flat areas are also popular; however an animal has been observed skirting a marshy area more densely vegetated.

A nest of the Bettong previously described (Kershaw, 1952) was situated on a dolerite hill slope of westerly aspect in open forest. It was fashioned of grasses in a shallow hollow in the surface, with a bark floor. Nests of these animals are always completely covered and consequently difficult to see. The site may have been chosen on the basis of suitability of material available to contrive a small tunnel. Such nests may be temporary. In some cases, although Bettongs were observed in an area, no amount of searching revealed a true nest structure. They may make use of suitable available cover in general, and confine nest building to periods of mating activity.

Nests in any case are destroyed readily by stock grazing through the area and by fires. After fires the animals repopulate an area quickly when enough cover returns. The ubiquitous bracken fern serves almost as well in providing cover for foraging expeditions. Although shy and difficult to observe the Bettong will appear in quiet places early in the evening when observations are possible. Torchlight observations have been made when the animals appear fascinated by the light or even ignore it.

Pouch young have been observed in autumn and spring. Since these observations were made, Guiler (1960) has found that breeding takes place throughout the year, but there is more activity in spring and late summer months. Observations seem to suggest also that mild, relatively dry conditions as well as the local environment may favour the Bettong.

Plate 3.
Typical habitat of the Bettong on a dolerite hill near Clarence Point, Tamar River Tasmania.

(c) The habitat of the Potoroo:

Greater numbers of Potoroo have been observed than Bettong. In daylight hours Potoroo have been observed only in areas of maximum shade, or closely nearby. At night it has been observed on roads and tracks in the vicinity of suitable areas, rarely at any distance.

The environment favoured by the Potoroo is one of dense scrub or forest where increase of moisture ensures prolific ground cover. Most favoured nesting sites occur where the ground is entirely covered and natural tunnels are formed between and under overhanging blades of grass clumps. Nests are difficult or impossible to find even when an animal has been disturbed. Those found by the writer have been found by accident.

The nest described by the writer (1952) was found in the stringy bark (*E. obliqua*) formation already described, on the bank of West Arm. The animal took advantage of overhanging shrubs, grass tufts and rubbish which formed a tunnel at the foot of a tree. Earth was banked up to close one end and a rough lining of soft grass was used. The nearby hedge of *Cupressus lambertiana* accentuates the density of cover and shade. Hence it is in keeping with other sites noted.

On foraging ventures from the nesting area the route taken may be circuitous in order to take advantage of thickly vegetated drainage ditches or "nature strips". An animal observed escaping from a rabbit trap left a considerable portion of leg, and a trail of blood. To follow the trail it was necessary to crawl for more than one hundred yards. Only once was the animal in the open when crossing a narrow track. Finally it was lost in dense scrub near a spring. A number of animals have been seen in this rarely disturbed area.

Discussion

The Potoroo has been observed more frequently in daylight than the Bettong. This may be due to the nature of its habitat which provides greater security. No animals have been seen outside this habitat in full day-light. Because of difficulty of penetration few opportunities occur for observing completely undisturbed animals. However, on one occasion two Bettong were observed for at least an hour. The animals scampered about under shrubs and tufts of grass, around logs and other cover,

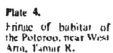

Plate 4.
Fringe of habitat of the Potoroo, near West Arm, Tamar R.

continually emitting shrill squeaks as though of excitement. The purpose of the display was not apparent.

In the area studied the Potoroo has been found to favour a closed canopy with dense ground cover habitat. The habitat in Victoria and New South Wales is recorded as similar. No evidence has been seen to suggest that the habitat elsewhere in Tasmania differs in any essential feature. The observations illustrate how these animals have adapted to an area where natural cover is greatly reduced. In a normal habitat the animals would move about with greater confidence probably over a wider range of conditions.

The Bettong has been observed in open forest where sunlight penetrates and grass or heath is the ground cover. The two types of habitat approximate and Potoroo have been disturbed in densely vegetated hollows within fifty yards of a heathy ridge where Bettongs have been seen. In the Asbestos Range, west from Frankford, Potoroo have been observed in gullies. No Bettong have been seen here. The nature of the habitat from the point of view of vegetative cover and effectiveness of drainage in relation to this appears most significant. The topography seems important only in so far as it has a bearing on these matters.

The Bettong appears able to withstand some degree of dryness. The area studied commonly has a dry summer season which may be prolonged into the autumn. Little water may be available and the question of how much water these animals may require arises. Troughton (1943a) remarked that Bettongs in the London Zoo were reported to have consumed a good deal of water. In the present area water is available in farm water troughs. Roadside excavations may also hold water for long periods.

A comparison has been made between the habitat occupied by the Bettong and that of the Potoroo in a small area near the Tamar River, Northern Tasmania.

REFERENCES

Finlayson, H. H., 1938. On a New Species of Potoroo (Marsupialia) from a Cave Deposit on Kangaroo Island, South Australia. Trans. Roy. Soc. S. Austr., 62 (1): 132-140.

Guiler, E. R., 1957. Longevity in the Wild Potoroo, Potorous tridactylous (Kerr), Aust. J. Sci. 20 (1): 26.

—————, 1960. The Breeding Season of Potorous tridactylous (Kerr), Aust. J Sci., 23 (4): 126-127.

—————, 1961. Notes on the Externals of the Potoroo, Potorous tridactylous, (Kerr), Pap. Proc. Roy. Soc. Tasm., 95; 41-48; Pl. 1, Figs. 1-9.

Kershaw, R. C., 1952. Notes on Marsupials at the Tamar River, Tasmania. Vict. Nat., 69 (8) 102-105.

Lord, E. C and Scott, H.H., 1924, A Synopsis of the Vertebrate Animals of Tasmania. Hobart.

Lyon, A. G. and McMahon, T. S., 1951. Observations on the Surface Structure of the Hairs of Tasmanian Monotremes and Marsupials. Proc. Roy. Soc. Tasm., for 1950: 71-84.

Pearson, J., 1945 The Female Urogenital System of the Marsupialia with special Reference to the Vaginal Complex. Proc. Roy. Soc. Tasm. for 1944: 71-96.

Pearson, J., 1946. The Affinities of the Rat Kangaroos (Marsupialia) as Revealed by a Comparative Study of the Female Urogenital System. Proc. Roy. Soc. Tasm. for 1945 113-125.

Troughton, E. Le G., 1943a. Furred Animals of Australia 2nd Ed. Sydney.

—————, 1943b. The Kangaroo Family. Rat Kangaroos I. Aust. Mus. Mag., 8 (5): 171-175.

Wood Jones, F., 1924. The Mammals of South Australia. Part ii. Brit. Sci. Guild Sth. Austr. Branch Govt. Print. Adelaide.

The Last Madimadi Man

by

LUISE A. HERCUS* AND ISOBEL M. WHITE**

Since 1911 or thereabouts, Jack Long[1], also known as Jack Edwards, has lived at Point Pearce Aboriginal Settlement, which is between Port Victoria and Moonta on Yorke Peninsula, South Australia. Point Pearce was established as a Mission to the Aborigines in 1868, and taken over by the South Australian Government in 1914, the reserved land having by then been increased to over 17,000 acres[2]. Farming this land now gives employment to many of the able-bodied men at the settlement, which is governed by a council of the inhabitants. These inhabitants, of whom there are about four hundred, are nearly all of part-Aboriginal, part-European descent with a handful who are pure Aboriginal. Though housing has been much improved over the last few years, it is still below the standard of an ordinary country town. Jack Long lives in one of the older, smaller houses, and a kindly woman neighbour is paid a small sum to keep his house and his clothes clean, and to provide meals for him in her own house. For his age of about a century he is remarkable for his physical activity and lively intelligence. He can recall vividly and accurately events in his life up to ninety years ago. Since he lives right in the middle of the settlement, he is not cut off from everyday happenings among his neighbours, who call him "Mate" or "Matey". This contrasts with the care commonly given to very old Australians, whether they be white or Aboriginal, which all too often involves complete isolation from the life of the community.

There are men at Point Pearce who rate as "old" and yet can remember Jack Long as a middle-aged man when they were boys, and some say he must be a hundred and six. He thinks he is about a hundred and from checking his memories we calculate that he was born no later than 1872. After learning that he has lived the last sixty years of his life at Point Pearce, among the descendants of Nararga speakers who inhabited the surrounding area at the time of European settlement, it is surprising to discover that he was born some four hundred miles away near Balranald in New South Wales. He is of pure Aboriginal descent and belongs to the Madimadi people, whose territory was on the north bank of the River Murray, and whose language is of the "Kulin" group[4]. He is of the *Maewara* moiety and his totem is kangaroo (*hugumanama*).

Jack Long's father was a Madimadi man, whose native name was *Lalugu*; his mother, called Bindul, was part Madimadi and part Dadidadi. The Dadidadi language belongs, together with Jidajida, to an isolated group with closer relationship to the languages of the Murray mouth than to those of the immediate vicinity[5]. Bindul was born on Kulkyne Station,

*Australian National University and Australian Institute of Aboriginal Studies

**Monash University and Australian Institute of Aboriginal Studies

January, 1971   11

and in his childhood Jack lived on the stations on both sides of the Murray near its confluence with the Murrumbidgee, namely Kulkyne (on the Victorian side), Canally, Yanga, Moulamein (on the New South Wales side)[5]. As a boy his main language was Madimadi, but he could also speak Daḍidaḍi, learnt from his mother's people. It is usual for Aborigines to speak more than one language, particularly where marriages occur across linguistic boundaries. In addition Jack Long learnt English,

Plate 1. Jack Long in January 1970 near Point Pearce Settlement.

and today not only speaks it fluently and with an extensive vocabulary, but can also read and write. He makes no more grammatical errors than a while Australian of equivalent education.

When L. Hercus first discovered him in 1965, on the advice of Dr. Catherine Ellis, as a valuable informant for Madimadi, he had forgotten all but a few words of Dadidadi perhaps because, as he recounts—

"I lost my mother when I was a little fellow baby; she got poisoned in them early days. They used to poison the potatoes, fruits and all that, and put them out for the rabbits; it killed a lot of rabbits too, rabbits were so bad; and she happened to eat one somehow or other and she passed away".

That the rabbits were a serious pest south of the Murray in the 'seventies and across the Murray by 1880 is a matter of record, and the method of extermination he describes was commonly used[9]. The most usual poison was strychnine and one wonders how many other unwarned Aborigines met an untimely and horrible death.

Plate 2. The Last Dadidadi Man: photograph taken about 1932 of Dinny Myers and his dog Briggo, who helped Dinny Myers and the Day family to survive the depression by his skill in getting foxes for scalp money.
(By courtesy of Hubert Day).

His father took another wife, Charlotte McDuff, who came from further west towards the Flinders Ranges. She already had a grown son, Isaac, and helped him to bring up the motherless Jack, his brother and his sister, Maria, who eventually married John Pearce¹. (Neither Jack Long nor his brother were ever married.) Later Isaac McDuff and his son, Paddy, went off to settle at Lake Condah, and Jack saw little more of them. Others of his contemporaries in the Balranald district were George Ivanhoe, Jimmy Morris and Reg Wise (Madimadi), Angus and Dinny Myers (Dadidadi), Peter Bonney (part Madimadi, part Dadidadi) and Sid Webber (part Madimadi, part Narinari). George Ivanhoe, whose totem was pelican (badaŋal), and Jimmy Morris lived to be the last of the Madimadi "clever men". Dinny Myers died at Moonacullah in the nineteen-forties, the last Dadidadi speaker, but tragically his language died with him, except for some information given to R. H. Mathews in the eighteen-nineties by Angus Myers at Cummeragunga, and fragmentary word lists published by Curr and Brough-Smyth². Of Peter Bonney and Sid Webber we shall hear more in Jack Long's story.

When Jack Long was old enough he got work on neighbouring stations as a stockman, and later became a drover. He tells us that he held a Dalgety drover's licence, and is emphatic that in those days, the eighteen-eighties, eighteen-nineties and nineteen hundreds, Aboriginal and white drovers were paid and treated equally and that "we were just as free as other men". Later they suffered under what he calls "The Act", which was "very hard on us". (As he had by then moved into South Australia, this is presumably the Act of 1911, which gave the Chief Protector the right to order an Aboriginal to stay on a particular reserve.)

He remembers the days of the Kelly Gang, and knew some of the Aboriginal police trackers, who were employed to track the bush-rangers, particularly he mentions old *Mungawida*. However, he may be echoing the opinion of Aborigines and less privileged white people of the time when he says,

"It was a very wrong affair, it was the law-people was doing the damage, the law was the foundation of all those businesses, they didn't treat people properly at all, no proper advice, no proper statement and all that. I know a lot of cases and those people, some of them called guilty, and didn't do it".

The story of Jack Long's droving days ties in closely with the building of the railways in north-western Victoria. He and his mates would drove mobs of cattle, horses and sheep from the stations on the Murray to the nearest rail-heads, their journeys getting shorter as the line extended. (The line from Kerang to Swan Hill was being built from 1882 to 1890, the Warracknabeal-Beulah line was extended to Hopetoun by 1894³.) Sometimes they drove a mob into the Western District "to Dunkeld, Casterton, Hamilton and them places", the drovers then returning to Swan Hill by train, changing at Bendigo. Once when they were on holiday, Jack Long and George Ivanhoe and Sid Webber packed their horses and rode on a visit to Cummeragunga Settlement.

Jack Long remembers that many of the stations were taken over by the big companies — he mentions "Australian Land Mortgage and Finance Corporation and the London Bank" (which may have been the London Finance Corporation). This happened to Kulkyne, Yanga and Cun

Subscriptions Now Due

The Field Naturalists Club of Victoria is administered by a relatively small number of honorary office-bearers.

The growth of the club and the expansion of its activities, particularly in connection with the production of the *Victorian Naturalist*, are continually adding to the burden of work.

It is therefore requested that fees and subscriptions be paid as promptly as possible, in order to help lighten some of this burden. The financial year commences on 1 January, 1971.

If you will not be paying your fees at one of the forthcoming general meetings, please remit them by post, using the form provided on the reverse side of this leaf.

This procedure will save office-bearers' time, and expense, in sending out reminder notices.

PLEASE ATTEND TO THIS MATTER NOW

You may help further by passing the following leaf on to an acquaintance who is not a member of the F.N.C.V. or a subscriber to the *Victorian Naturalist* but who might be interested in either.

NOTES:

1. Membership fees for the year 1971 are as follows:

Ordinary Members	$7.00
Country Members	$5.00
Joint Members	$2.00
Junior Members	$2.00
Junior Members receiving *The Victorian Naturalist*	$4.00
Subscribers to *The Victorian Naturalist*	$5.00
Affiliated Societies	$7.00
Life Membership (reduction after 20 years' membership)	$140.00

   (Ordinary, and country members receive the *Victorian Naturalist* free of any further charge.)

2. The scheme of supporting membership was introduced so that those who are able and willing to do so might help club finances. You are invited to become a supporting member by making a voluntary addition to the normal annual fee of any sum you choose, from $1 upward. Details relating to supporting members and their payments are regarded by the treasurer as confidential, and no distinction or extra privilege is bestowed on the members concerned.

(To be removed)

Field Naturalists Club of Victoria

FORM FOR RENEWAL OF MEMBERSHIP OR OF SUBSCRIPTION TO THE "VICTORIAN NATURALIST"

(To be used by existing members or subscribers for payment of fees)

Name(s) ....

Address..

....

....

(Please indicate if there is a joint member)

Mr. D. E. McInnes
Hon. Treasurer, F.N.C.V.
129 Waverley Road, East Malvern, 3145

Dear Sir,

Please find enclosed the sum of $. to cover annual membership fees
subscription to the *Victorian Naturalist* for the year 1971. Please enter this sum as follows:

Membership fees .. .. $
Supporting membership .. .. $
Subscription to *Victorian Naturalist* .. .. $

Yours faithfully.

Field Naturalists Club of Victoria

APPLICATION FORM

To be used by new members or subscribers.

(Cross out parts which are not applicable)

I wish to subscribe to the *Victorian Naturalist* for 1971. Please post it monthly to the address below.

                      Ordinary
I wish to apply for Country membership of the Field Naturalists Club
                      Junior
of Victoria.

My full name and address is:

    Mr.
    Mrs...
    Miss

    ....

    ....

I enclose the sum of $ in payment of the year's fee.

Date   /   /1971.

        Signature...

Field Naturalists Club of Victoria

AN INVITATION TO PERSONS INTERESTED IN AUSTRALIAN FAUNA, FLORA AND COUNTRYSIDE

If you have not already an affiliation with the F.N.C.V., you may apply to the club either for membership or for regular subscription to the *Victorian Naturalist*.

These are some of the club's activities:

- General meetings each month, with informative, illustrated talks by prominent naturalists. These are held on the second Monday of each month, at the National Herbarium, South Yarra.

- Meetings of study groups comprising those with specialized interests such as geology, botany, microscopy, entomology, native fauna, etc.

- Organized excursions led by nature experts, to places of interest, both near and far.

- The maintenance of a large lending library of nature books and magazines.

- The publication monthly of the *Victorian Naturalist*, a well-illustrated nature magazine produced for the general reader as well as the expert. This is issued free to all members.

Membership is available to any person interested in nature; it is not necessary to have any specialized knowledge.

Membership Fees for the year 1971 are:

Ordinary members (living within 20 miles of G.P.O. Melbourne)	$7.00
Country members (living over 20 miles from G.P.O., Melbourne)	$5.00
Junior members (under 18 years of age)	$2.00

Non-members may subscribe to the "Victorian Naturalist" for $5.00.

If you are interested in either membership of the club or subscription to the *Victorian Naturalist*, please complete the appropriate parts of the form on the reverse side of this leaf and post it to:

Honorary Secretary, F.N.C.V.,
National Herbarium,
South Yarra,
Victoria 3141.

(To be removed)

ally. "I was there when they took over — Aborigines, stock, stations and everything". This was presumably in the depression of the 'nineties when many mortgages were foreclosed and many stations were taken over by the finance companies.

He left the Balranald district finally in about 1896, and has never been back there since. Perhaps his droving work fell off with the extensions to the railways and the takeover of the stations. He left with his mate, Sid Webber, and says.

"We came to Mildura first, we worked at Mildura garden, Chaffey Brothers, and after the gardens we used to go down into them stations, stock, shearing and one thing and another, working in the woolsheds".

He also earned some money as a professional sideshow boxer. He had quite a reputation as a boxer, and for a while was a member of a well-known troup. Once he and Sid Webber took the train from Mildura to Woomelang,

"and when we go to Woomelang on that line, we leave the line and go south-west to Hopetoun, Little work in Hopetoun, come down to place called Beulah, going towards Warracknabeal then. We got work here and there through farmers and that. Eventually we went into Antwerp (Ebenezer Moravian Mission), and met some of our people and children — mixed children they were. We stayed there until we come down to Bordertown and got work there. I left Sid Webber at Bordertown and came on to Tintinara, and I eventually come down to Tailem Bend, Murray Bridge and all those places and come right into these parts."

He joined up with Sid Webber again and they decided to try for work in the Renmark gardens. They went first to Adelaide to the "company office" and then straight up to Renmark where they worked for two or three years. Sid Webber left him and he never saw him again. From Renmark, Jack Long went to the south-east of South Australia, first to Point McLeay and then to Point Pearce, and never went back to live in Victoria, though he travelled up and down the Murray as far as Boundary Bend, before finally settling to live at Point Pearce. After the Act of 1911 he says that he had to live at the reserve, though "we could still come and go and didn't bother much, never worried much as long as we had something to do, and plenty food and stuff".

About forty years ago Peter Bonney, an old childhood companion and droving mate, stayed for some time at Point Pearce, working on the dam and in the stone quarry. Peter Bonney, like Jack Long, could speak Madimadi and Dadidadi. This was the last time that Jack Long spoke his own languages with a native speaker.

Jack Long gave us the following additional information on the Madimadi language:

Additional Sentences

1. Mother said:

mada ɲindi ɲygaḍi warlbaḍi weguḍu

not   you   go   play   long-way-off
              (imperatives)

Don't go and play a long way off.

gima   wadada bermilada buŋaŋi
Straight away comes sneaks   evil-spirit
Straight away there comes and sneaks about an evil spirit.

nagadi gini buindi
see   this darkness
Look it is getting dark.

buŋaŋi   gawaŋadin gini buindi
evil-spirit follow-will this darkness
The evil spirit will follow this darkness.

gagadin ŋinan maŋgadin ŋinan giaga-minu
grab-will you   take-will you   altogether
It will grab you and take you away altogether.

2. When there was a flood:
gadini waiwulada, gewadin ŋalan, ŋurgadin ŋalan
water rises   overtake-will us swallow-will us
The water is rising, it will overtake us and drown us.

wigadin   ŋali gadinaŋ
perish-will we   water-in
'We'll perish in the water.

bai, nagadi nini daŋi gegada, jiŋgadi ŋindi gegada guragaŋ
hey! see   this ground above, go   you   up   sandhill-on
Look at this high ground up there, go up to the top of the sandhill.

daŋi   gaiu nagadi gima dirilaŋ
ground there see   here   sky-in
Look at that place up there, it's right up in the sky.

nagadi gini didi   jiŋgada gaŋuŋ gadinuŋ
see   this animal goes   that water-from
Look at the animals coming away from the water.

bambada nuni gima gadinuŋ, wuduŋi bambada
fears   that-one here water-from, man   fears
Those animals are frightened of the water, and the people too.

giadin gini wuduŋi: baigadi ŋindi, maŋgadi gujunin, leŋin, banemi
said this man   get-up you,   take spear-yours, camp-yours, food

One man said: get up, take your spear, your camp and your food.

maŋgadi gima wanabi
take   here fire

take the fire here.

ŋeŋgadin ŋali gegada, niwi-ma nagiladia banema, winmuru
sit-will we above, close-by look-may food-for, sow-thistle

We'll stop up there, and you can look for food around there, such as sow-thistles.

gadini ŋeŋgada, bai nuni buigadin
water stops   oh! this fall-will

The water has stopped rising, oh! it will drop soon.

mada-ma ŋindi bambadia, winagada ŋalan gini gadini
not-indeed you fear-may, leaves us   this water

Don't be afraid, the water is leaving us now.

bai ŋindi wegadi, waribadi, waŋiladi
Now! you laugh   play   sing

Now you can laugh and play and sing.

ADDITIONAL VOCABULARY AND MORPHOLOGY

bai
  exclamation of surprise and of encouragement. This word shows the characteristic final -ai which is common in exclamations not only in the Kulin languages, but over much of Australia. Cf. Madimadi gai "hey you".

bed-bed
  owlet nightjar. This was a bird of ill omen: it brought news of death. While discussing nocturnal birds Jack Long recalled seeing the night parrot long ago, but he had forgotten its name in Madimadi.

buindada
  at night.

buiŋdi	night. These are variant forms of the previously recorded Madimadi word buiŋgada, huiŋgi. This whole group of Madimadi forms is cognate with the Djadjala buɼunj and Wemberwemba buɼinj "night", and shows the usual loss of ɼ in Madimadi.
duni	the wood-pecker or brown tree-creeper. It too was a kind of news-bird; it could understand what people were saying and would repeat it elsewhere.
ganagal	shrimp.
gaɼuŋ	from that one. Like other adverbs of time and place gaiu "over there, not far away" must have corresponded to a demonstrative pronoun: gaɼuŋ is the ablative of such a pronoun gaiu.
gewada	to overtake.
gurugi	magpie. Cf. Djadjala gurug, Wembawemba gurulug.
jil'elilburi jilelburi (shortened form)	a bird "almost the same as the Willie Wagtail (diri-diri)." Probably the restless flycatcher.
-ma	enclitic particle, used for emphasis, probably very much like the Jodajoda -ma and the Wembawemba and Djadjala -min which generally followed the imperative and adverbs of place: Madimadi ɲiwi-ma "close by (indeed)" and madamu "not (indeed)". (Cf. also the fixed locution madawa "oh don't" which was previously attested).
waiwulada	to rise. This is a variant form for waiwilada "to rise", and it shows that Madimadi resembled Wembawemba in the optional use of a frequentative suffix -ula alongside -ila, particularly when the verbal root ended in a labial consonant (e.g., Wembawemba gubila, gubula "to drink").
wawal	night hawk.
wegada	a long way off.
winmuru	this plant with edible leaves has now been identified by Jack Long as the "sow-thistle" (a native species of Sonchus[11]).

Literature Cited

1. Hercus, Luise A. 1969. *The Languages of Victoria: A Late Survey*, Australian Institute of Aboriginal Studies, Canberra. Part I, Ch. 4, Part II, 323-358.
   ——— 1970. "A Note on Madimadi", *Vict. Nat.*, Feb., 1970, 43-47.
2. Gale, Fay, 1964. *A Study of Assimilations: Part-Aborigines in South Australia*, Libraries Board of South Australia, Adelaide, 155-156, 165-169, 171-173.
3. Hercus, L. A. 1969. op. cit., 141-142.
   Tindale, Norman B. 1940. "Distribution of Australian Aboriginal Tribes: A Field Survey", *Trans. Roy. Soc. S.A.*, Vol. 64, 192.
4. Hercus, L. A. 1969. op. cit., 141-142.
   Tindale, N. B. 1940, op. cit., 198.
5. Jervis, James. 1952. "The Western Riverina: A History of Its Development", *Journal and Proceedings of the Royal Australian Historical Society*, Vol. XXXVIII, 1-30.
   Kenyon, A. S. 1914-1915. "The Story of the Mallee", *The Victorian Historical Magazine*, Vol. IV, 23-56, 57-74, 121-150, 175-200.
6. Buxton, G. L. 1967. *The Riverina, 1861-1891: An Australian Regional Study*, Melbourne University Press, Melbourne, 208-209, 248-249.
   Fennessy, B. V. 1962. "Competitors with Sheep: Mammal and Bird Pests of the Sheep Industry", in A. Barnard (ed.), *The Simple Fleece*, Melbourne University Press, Melbourne, 228.
   Radcliffe, F. N. and J. R. Callaby, 1958. "Rabbits" in *The Australian Encyclopedia*, Vol. 7, 340-347.
7. Kenyon, A. S. 1914-1915, op. cit. 200.
8. Mathews, R. H. 1898. Unpublished manuscript.
   Brough Smyth, R. 1878. *The Aborigines of Victoria*, Government Printer, Melbourne, Vol. II, 72.
   Curr, E. M. 1887. *The Australian Race*, Government Printer, Melbourne, Vol. II, 285-289.
9. Hercus, L. A. 1969, op. cit. 255.
10. Harrigan, Leo J. 1962, *Victorian Railways to '62*, Victorian Railways, Melbourne, Chs. 14 and 15, and 283-285.
11. Cleland, J. B. 1966 in *Aboriginal Man in South and Central Australia*, Adelaide, 132.

Index to Victorian Naturalist.

An index to the volumes of the *Victorian Naturalist* is being compiled, and advice would be welcome, if given soon.

The compiler would also like help from volunteers who would be willing to either write clearly on system cards under direction; or check cards done by others; or place cards in order.

The writing of cards would be done at home: and if anyone found the work too great, he or she may give it up without causing any inconvenience.

Any person interested should ring 81-2147, or write to —

Miss K. E. Hall,

79 Kooyong Koot Road,

Hawthorn, 3111

The Discovery of the Lizard Genus *Carlia* (Scincidae: Lygosominae) in Victoria

by A. J. Coventry*

On 6 June, 1970, Mr. P. Robson of Tatong collected two specimens of a scincid lizard near Taminick Gap, in the Warby Ranges west of Wangaratta, Victoria. Mr. W. Osborne subsequently brought one of these specimens to the National Museum of Victoria (Reg'd. No. D14399), where it was identified as *Carlia maccooeyi* (Ramsay and Ogilby). Since this date, Mr. Robson has kindly donated the second specimen to the National Museum (D14616), while two additional specimens have been collected from the same area (D14563 and D14576). Another two specimens have been collected from a granite outcrop approximately 10 miles north of Albury, N.S.W. (D14636-7), while Mr. P. A. Rawlinson of Latrobe University has donated N.S.W. specimens from Tarcutta (D14522-4), 4 miles N.E. of Tarcutta (D14525-6), 10 miles N.E. of Tarcutta (D14527), 13 miles N.E. of Tarcutta (D14528), 7 miles S. of Lyndhurst (D14530-1), and Coonabarabran (D14529). All of these lizards were collected in the vicinity of rocky outcrops. The National Museum also has specimens from the type locality, Brawlin, N.S.W. (D4152-4), donated by Mr. McCooey.

The genus *Carlia* Gray, 1845, was removed from the synonymy of *Leiolopisma* Dumeril & Bibron, 1839, by Mittleman, 1952, to contain those species with four fingers and five toes which had been referred to *Leiolopisma*. He says: "Differs essentially from *Leiolopisma* as follows: A single well developed frontoparietal, interparietal very small or absent; digits 4-5. Distribution: Moluccas, New Guinea, Australia and Papuasia".

The only previously published record of a species referable to this genus in Victoria appears to be Lucas & Frost, 1893, when they listed *Leiolopisma tetradactylum* (O'Shaughnessy), with distribution (within Victoria) as Brown's Plains and Barnawartha. (The National Museum has two specimens of *maccooeyi* labelled *Leiolopisma tetradactyla* from Victoria, Reg'd. Nos. D652 and D1723, donated by Frost, and probably at least part of the material before Lucas & Frost). Both of these localities are within 25 miles of Albury, N.S.W. Lucas & Frost say "Habits: . . . is found amongst the grass and herbage on the dry open plains near the Murray". This contrasts with the habitat where recent specimens have been collected.

There is some confusion as to the relationship of *tetradactyla* and *maccooeyi*. Ramsay and Ogilby, 1890, when describing *maccooeyi* commented that the two could well be synonyms, and listed seven points in which they differed. They were that in *tetradactyla*; 1. The head is much larger, 2. The prefrontals are in contact, 3. The frontal is much shorter than the frontoparietal, 4. The scales have no trace of carination, 5. Non enlargement of the preanals, 6. Shorter tail, and 7. Different colour

*National Museum of Victoria

pattern. Mr. A. F. Stimson, of the British Museum (Nat. Hist.), has kindly compared the types of the two taxa for me, and comments (pers. comm.). "By direct comparison of the types I have checked these differences and most of them appear to break down. All three of our syntypes of *maccooeyi* have smooth scales and I can only assume that Ramsay and Ogilby were misled by the striated colour pattern on some of the scales. The preanal scales are very slightly enlarged in both species (as in your specimen), and I can find no significant difference in colouration, both agreeing with your specimen, although the type of *tetradactyla* is a little faded. The largest of our syntypes of *maccooeyi* has the frontal nearly as long as the frontoparietal compared with a frontal ⅔ the length of the frontoparietal in the type of *tetradactyla*. However, in one of the smaller syntypes of *maccooeyi* the frontal is only 70% the length of the frontoparietal, although in this specimen the minute interparietal is absent, presumably fused with the frontoparietal. This leaves only two significant differences, the proportionally larger head and shorter tail of *tetradactyla*".

All eleven complete specimens of *maccooeyi* in the National Museum have been measured, and the tail length was found to vary from 52.9 to 65.9% of the total length, with a mean of 59.3. From the measurements quoted by Boulenger, 1887, for the type of *tetradactyla*, and by Ramsay and Ogilby for the types of *maccooeyi*, the percentage of tail to total length respectively are 53 and 57.3. None of our specimens has a head as large as that reported for the type of *tetradactyla*.

From the above, it would seem that further specimens will prove that the two species are in fact synonyms, and will be known as *Carlia tetradactyla* (O'Shaughnessy).

The discovery of this genus within Victoria raises interesting points on its distribution. With the exception of the species under discussion, *Carlia* is apparently confined to humid and subhumid tropical or subtropical climates. Storr, 1964, says that it has spread south and west from the well-watered northern and eastern fringes

Plate 1
Carlia maccooeyi (Ramsay & Ogilby) from Taminick Gap, Victoria. (D14616). Total length 119 mm.

Photo: F R Rotherham

January, 1971

of the continent, into subhumid and semi arid habitats. In Western Australia (Storr, pers. comm.) this genus is rare south of the Kimberleys. Cogger, 1967, says that *Carlia* is most plentiful in northern Australia, and that *munda* is found in New South Wales and Queensland. Keast, 1962, draws attention to the distribution of four-fingered *Leiolopisma* species, and gives distribution maps of the Australian species. Mitchell, 1953, gave localities of all specimens of four-fingered *Leiolopisma* which he had seen, and listed a specimen from Dubbo, N.S.W. Apart from the record by Lucas & Frost, this is the southernmost record of the genus which I have been able to locate.

Acknowledgements:

I thank Mr. Robson for drawing attention to this species in Victoria, Mr. Rawlinson for donating his collection of *C. maccooeyi* to the National Museum, assisting with field work, and reading the manuscript, Miss J. M. Dixon of the National Museum for reading the manuscript, and Mr. E. R. Rotherham for photography.

REFERENCES

Boulenger, G. A., 1887. Cat. Lizards in the British Museum. iii. p. 288.
Cogger, H. G., 1967. Australian Reptiles in Colour. p. 66.
Keast, A., 1962. In Leeper, C.W. The Evolution of Living Organisms. Melbourne University Press. p. 402.
Lucas, A. H S. & Frost, C., 1894. The Lizards Indigenous to Victoria *Proc Roy. Soc. Vict.* 6: 79-80.
Mitchell, F. J., 1953. A Brief Revision
Rec
Mittleman, M. B., 1952. A Generic Synopsis of the Lizards of the Subfamily Lygosominae. *Smith. Misc Coll.* 117: 1-35.
Ramsay, E. P. & Ogilby, J. D., 1890 Re-description of an Australian Skink, *Rec. Aust. Mus.* 1: 8-9.
Storr, G. M., 1964. Some aspects of the geography of Australian Reptiles. *Senck. biol.* 45: 557-589.

F.N.C.V. PUBLICATIONS AVAILABLE FOR PURCHASE

FERNS OF VICTORIA AND TASMANIA, by N. A. Wakefield.
The 116 species known and described, and illustrated by line drawings, and 30 photographs. Price 75c.

VICTORIAN TOADSTOOLS AND MUSHROOMS, by J. H. Willis.
This describes 120 toadstool species and many other fungi. There are four coloured plates and 31 other illustrations. New edition. Price 90c.

THE VEGETATION OF WYPERFELD NATIONAL PARK, by J. R. Garnet.
Coloured frontispiece, 23 half-tone, 100 line drawings of plants and a map. Price $1.50.

Address orders and inquiries to Sales Officer, F.N.C.V., National Herbarium, South Yarra, Victoria.

Payments should include postage (11c on single copy).

Flowers and Plants of Victoria in Colour

Copies of this excellent book are still available, and of course would make a wonderful gift. They are obtainable from the F.N.C.V. Treasurer, Mr. D. McInnes

Winter Collection

by

J. C. Le Souëf

Although winter months are not the generally accepted time of the year for collecting in the tropical north of Australia, there are always specimens to be taken which are worth recording. With variations in seasonal conditions from year to year, it would seem that at least the insect fauna in a given locality at the same date differs each year. One of the great attractions of field entomology is that one is never quite sure what is going to turn up.

Fruit Moths

A new experience to me was the taking of a number of species of moths feeding on fruit near Cairns in June last. While waiting for insects to be attracted to the mercury vapor light at The Intake, I noticed a large moth flying to a small guava near the roadside. Closer investigation showed that it was feeding on one of the ripe fruits on the tree. Searching further with a torch, I was surprised to find a number of moths feeding on ripe fruit in the scrub. A sidelight was that there were two species of bats feeding on the moths and a large python probably preying on the bats.

The following is a list of the moths taken, the nomenclature being that of similar specimens in the National Museum, Melbourne.

Mecodina semaphora Lower
Mecodina praecipua Walker.
Corethrobela melanophaes Turner
Achaea janata Linnaeus
Ischyja porphyria Turner
Erchneia cyllaria Crampton
Cosmophila erosa Hubner
Cosmophila latimarga Walker
Phyllodes mericci Oliffe
Ophideres tyrannus Crampton
Nyctipau dentifascia Walker.

New Locality for *Trapezites macqueeni* Kerr and Sands

This recently described butterfly, originally taken from the Mareeba area of the Atherton Tableland, was recorded from Musgrave. It was taken in open eucalypt forest country with short grasses and occasional small flowering plants. Noticing *Pimelea* (akin to humilis) in flower near the road, I stopped to look for *Trapezites* skippers as one does in Victoria. I was not disappointed with the taking of this specimen feeding on a foot high blue flowering plant.

The first specimen of the Family Hesperiidae, the Skippers, was taken by Banks at Cooktown during his sojourn there. It was of great interest to take a specimen of this species, *Trapezites lacchus* (Fab.), at almost the same date and probably in the same locality, two hundreds years later. All the other species taken by Banks at this time are still to be found in the vicinity of Cooktown.

General Collecting and Observing

As a centre for the study of a wide variety of wildlife, Cooktown is without peer. Within a few miles of the town there is rain forest, the normal open eucalypt forest of the north, large tracts of mangroves, both sandy and rocky shore lines, as well as a good coral reef easily reached by foot, although it is a bit further afield being some eighteen miles from the town.

To the south towards Daintree is the magnificent rain forest of the Bloomfield River while to the north past the aerodrome is the Melvor River country with its surprising patches of "wallum" communities of *Banksias, Galmius* and *Xanthorrhoeas*, cheek by jowl with tropical vegetation. Cooktown is also the jumping off place for roads to the various northern settlements and the Cape itself.

* * *

The main object of this visit was to again look for the rare *Virachola democles* (Misk), one specimen of which I took in 1964. On that occasion there had been a "normal" wet season with forty-one species noted in a small sunny gully. These were seen in a half hour of very active flight at midday. But this year with a much drier season the number of species noted would probably not exceed thirty-five with many hours searching and no sign of *Virachola* at all. With these different conditions, two other rather rare species made their appearance. They were *Narathura wildei wildei* (Misk.) the Small Oak Blue, and *Hypochrysops polycletus rovena* Druce the Rovena Jewel, neither of which we had seen here before at this time.

* * *

During this last visit, twelve species of snakes were taken, and other reptiles seen. There were fewer beetles on the wing than usual. Several were taken south of the Archer River, a few at Coen; but there were two different emergences on the night we camped on the Hann River several days apart. Here anaplognathus, scarabs, dytiscids and a number of microcoleoptera as well as other insects, small and large, came to the light. Most of these have gone to the National Museum Collection.

* * *

Each year there is a re-enactment of the landing of Captain Cook on 18 June, but this being the bicentenary year, there were week-long celebrations with a re-enactment each morning depicting the events of that particular day two centuries ago. It is of great credit to Cooktown that with its very small population it was able to produce a cast able to take the parts of Cook, Banks, Solander, Parkinson and a number of others with the help of the Lardel Dancers from Mornington Island. This fascinating show, with able commentary, was indeed a bonus to the pleasure a visit to Cooktown always brings.

Field Naturalists Club of Victoria

General Meeting
14 December, 1970.

The final General Meeting of the year was well attended by approximately 110 members and guests, who were welcomed by the President, Mr. T. Sault.

Guest speaker for the evening was Mr. John McNally, Director of the National Museum of Victoria, whose subject, "Museology '70", proved a most interesting and unusual one. Mr. McNally showed an excellent selection of slides illustrating buildings, layouts and contents of museums which he had visited on a recent world tour. Combined with an informative commentary, the subject was covered in a most comprehensive manner. Ancient and modern concepts of museums were shown in England, Denmark, Sweden, Greece, Canada, U.S.A. and Mexico. Mr. McNally said that the role of museums varied according to type and status but all were concerned with the storage of collections, research and education. There were new ideas of what constituted a museum, for example Disneyland in Los Angeles, the Viking Ship and Marine Museum in Stockholm and the Guggenheim Museum of Modern Art in New York. Countries such as Greece and Mexico had preserved their antiquities as museums and fine examples were shown, particularly the Aztec and Mayan civilisations of Mexico.

Mr. T. Sault thanked Mr. McNally and said the club appreciated the talk, as it was a subject of much interest to all, but knowledge was not easily available.

An Honorary Member's Certificate was presented to Mr. I. Hammett, who said that he was honoured to receive it from a club which had kept him interested for over 40 years.

The President said assistance was still required with the collection of non-marine molluscs and urged members to be on the lookout for specimens while on holidays. He also spoke of a recently created reserve at Balnarring where sickle greenhoods (*Pterostylis falcata*) were flowering profusely and of the sighting of a platypus in the river at Werribee Gorge by members of the Geology excursion on 13 December. Mr. Sault said a Curator of Blocks was needed, as the storage problem at the Herbarium was becoming critical.

Correspondence tabled by the Secretary covered the following items:

A reply from Lysaghts Ltd. to a letter seeking help to conserve an area near their property at Tyabb, where the New Holland mouse was discovered. They expressed interest and requested further details of the area concerned.

As a result of a letter from Mr. N. Wakefield, a letter was sent to the Minister of Environmental Protection in West Australia requesting consideration for the conservation of the district in which the numbat is found. Proposed development is regarded as a threat to this rare animal's habitat.

From the Secretary of the Macedon Range Conservation Society, a report of the recent koala survey (201 koalas sighted) and a letter of thanks for assistance rendered by members.

Exhibits:

A. J. Swaby—Green Kangaroo Paw (*Anigozanthus flavidus*), Blue Daisy (*Lagenophora stipitata*), Pratia 2 spp., Lilli-Pilli (*Eugenia smithii* var. form), Melaleuca (*M. hypericifolia*), Q'ld Bleeding Heart Tree (*Homalanthus populifolius*), Liverwort (*Marchantia* spp.), Mazus (*M. pumilio*).

R. Riordan—Stick Insect (*Ctenomorpha marginipannis*)

Mrs. Bennett—Cup Moth Caterpillar, samples of silk and cocoons supplied by the Japanese Embassy.

A. Parker—"Bullseye" agates from Victoria River, Wave Hill, N.T.

Mrs. North—Glass Sponge (?)

Mrs. Woollard—Red Kangaroo Paw (*A. rufa*)

Mrs K. Meehan—Galls on Lilli-Pilli.

K. Strong—Galls on Black Wattle. Flower Spider under microscope.

P. Curtis—Callistemon flowers showing fasciation.

Dr. Beadnell—Case Moth, Astronomy Chart.

I. Hammett Eremaea (*E. violacea*) Melaleuca (*M. filifolia* & *M. ceratula*).

J. Ros Garnet—Collection of Cone Shells from Barrier Reef near Townsville.

I. Cameron—Clustered Everlasting (*Helichrysum semipapposum*), web of moth larvae on Boneseed.

January, 1971

F.N.C.V. DIARY OF COMING EVENTS

GENERAL MEETINGS

Notice is given of an Extra-ordinary Meeting to precede the February General Meeting. It is to consider the application from the Mid-Murray Field Naturalist Trust for affiliation.

Monday, 11 January—At National Herbarium, The Domain, South Yarra, commencing at 8 p.m.

1. Minutes, Reports, Announcements.

2. Nature Notes and Exhibits.

3. Subject for the evening — Members' Night.

4. General Business.

5. Correspondence.

Monday, 8 February — Mr. J. H. Willis: "New Guinea, Scenic and Biotic Wonderland".

GROUP MEETINGS

(8 p.m. at National Herbarium unless otherwise stated)

Wednesday, 20 January — Microscopical Group.

Wednesday, 3 February — Geology Group.

Thursday, 11 February — Botany Group.

F.N.C.V. EXCURSIONS

Sunday, 17 January — Marysville. The coach will leave Batman Avenue at 9.30 a.m. Fare $2.00, bookings with excursion secretary. Bring two meals.

GEOLOGY GROUP EXCURSIONS

Sunday, 7 February — Maribyrnong River Terraces—upstream from railway bridge, Albion-Broadmeadows Line. Leader: Mr. D. McInnes.

Sunday, 14 March — To Bulla-Deep Creek Gorge. Transport by private car. Spare seats are usually available for those without their own transport. Excursion leave from the western end of Flinders Street Station, opposite the C.T.A Building at 9.30 a.m.

No excursion is planned for January, 1971.

AVAILABLE NOW!

The Wild Flowers of the Wilson's Promontory National Park

by

J. Ros. Garnet

This is the publication which we have been eagerly awaiting ———

PRICE: $5.25

Members are entitled to discount

Contact local Secretary or write direct to Book Sales Officer:

Mr. B. FUHRER
25 Sunhill Avenue   North Ringwood, 3134

Postage: 20 cents

ENTOMOLOGICAL EQUIPMENT
Butterfly nets, pins, store-boxes, etc.
We are direct importers and manufacturers,
and specialise in Mail Orders
(write for free price list)

**AUSTRALIAN
ENTOMOLOGICAL SUPPLIES**

14 Chisholm St., Greenwich
Sydney 2065   Phone: 43 3972

**Genery's
Scientific Equipment
Supply**

183 Little Collins Street
Melbourne
(one door from Russell Street)

Phone 63 2160

Microscopical stains and mountants.

Magnifying lenses and insect nets.

Excellent student microscope with powers from 40X to 300X resolution: 20,000 lines per inch. $29.50.

Standard laboratory equipment, experimental lens set, etc.

Field Naturalists Club of Victoria

Established 1880

OBJECTS: To stimulate interest in natural history and to preserve and protect Australian fauna and flora.

Patron:
His Excellency Major-General Sir ROHAN DELACOMBE, K.B.E., C.B., D.S.O.

Key Office-Bearers, 1970/71

President:
Mr. T. SAULT

Vice-Presidents: Mr. J. H. WILLIS; Mr. P. CURLIS

Hon. Secretary: Mr. D. LEE, 15 Springvale Road, Springvale (546 7724).

Hon. Treasurer & Subscription Secretary: Mr. D. E. McINNES, 129 Waverley Road, East Malvern 3145

Hon. Editor: Mr. G. M. WARD, 54 St. James Road, Heidelberg 3084.

Hon. Librarian: Mr. P. KELLY, c/o National Herbarium, The Domain, South Yarra 3141.

Hon. Excursion Secretary: Miss M. ALLENDER, 19 Hawthorn Avenue, Caulfield 3161. (52 2749).

Magazine Sales Officer: Mr. B. FUHRER, 25 Sunhill Av., North Ringwood, 3134.

Group Secretaries:

Botany: Mrs. R. WEBB-WARE, 29 The Righi, South Yarra (26 1079).
Microscopical: Mr. M. H. MEYER, 36 Milroy Street, East Brighton (96 3268).
Mammal Survey: Mr. P. HOMAN, 40 Howard Street, Reservoir 3073.
Entomology and Marine Biology: Mr. J. W. H. STRONG, Flat 11, "Palm Court", 1160 Dandenong Rd., Murrumbeena 3163 (56 2271).
Geology: Mr. T. SAULT.

MEMBERSHIP

Membership of the F.N.C.V. is open to any person interested in natural history. The *Victorian Naturalist* is distributed free to all members, the club's reference and lending library is available, and other activities are indicated in reports set out in the several preceding pages of this magazine.

Rates of Subscriptions for 1971

Ordinary Members	$7.00
Country Members	$5.00
Joint Members	$2.00
Junior Members	$2.00
Junior Members receiving Vict. Nat.	$4.00
Subscribers to Vict. Nat.	$5.00
Affiliated Societies	$7.00
Life Membership (reducing after 20 years)	$140.00

The cost of individual copies of the Vict. Nat. will be 45 cents.

All subscriptions should be made payable to the Field Naturalists Club of Victoria, and posted to the Subscription Secretary.

victorian naturalist

Vol. 88, No. 2   FEBRUARY, 1971

F.N.C.V. DIARY OF COMING EVENTS

GENERAL MEETINGS

Monday, 8 February—At National Herbarium, The Domain, South Yarra, at 8 p.m.; to be preceded by an Extra-ordinary Meeting at 7.55 p.m. to consider the Affiliation of the Mid-Murray Field Naturalists Trust.

1. Minutes, Reports, Announcements. 2. Nature Notes and Exhibits.
3. Subject for the evening — "New Guinea, Scenic and Biotic Wonderland", Mr. J. H. Willis.
4. New Members.

Ordinary:
- Miss Caroline Davy, 246 Bluff Rd., Sandringham, 3191. (Interests—Mammal Survey and Marine Biology.)
- Mr. Geoffrey A. Christopher, 21 Madden St., Nth. Balwyn, 3104. (Interest—Spiders.)
- Mr. Andrew Haffenden, 51 Grosvenor St., Balaclava, 3183. (Interests—Reptiles, Mammals, Beetles, Geology.)
- Mr. Peter Charles Williams, 51 Sutherland Rd., Armadale, 3143. (Interests—Mammals, Birds, Reptiles, Plants.)
- Mrs. A. C. Burleigh, 7 Lucas St., E. Brighton, 3187. (Interests—Botany, Geology, Photography.)
- Miss Sarah Wetzel, 50 Williamson Rd., Maribyrnong, 3032. (Interest—Entomology.)
- Miss Joan Simmonds, Flat 1, 333 Orrong Rd., East St. Kilda, 3183. (Interests—Botany, Birds.)
- Miss June Leevers, 15 Dempster Ave., North Balwyn, 3104. (Interests—Botany, Marine Biology.)
- Mr. W. J. R. Osborne, 38 Moushall Ave., Niddrie, 1042. (Interests—Mammal Survey, Botany.)
- Mr. Alan T. McCallum, 17 Foulds Court, Montrose, 3765. (Interests—Animal Navigation and Migration.)
- Miss N. Wilson, 5/13 Linlithgow Ave., Nth. Caulfield. (Interest—Botany.)
- Mrs. Audrey Geyson, 68 Morris Rd., Upwey, 3158. (Interest—General.)
- Mr. C. Randell Champion, P.O. Box 70, East Melbourne, 3002.
- Mr. Ross Haughton, Box 34, Heidelberg, 3084.
- Miss Margaret McKenzie, 75 Bellairs Ave., Yarraville, 3013. (Interests—Birds, Geology.)
- Miss Mallon Elizabeth Upton, 5/4 Gordon Grove, South Yarra, 3141.
- Mr. Alberto Coloano, Box 2666X G.P.O., Melbourne, 3001.
- Mrs. Jean Youm, 11 Acheron Ave., Camberwell, 3124.

Country:
- Mr. Henry Salisbury, Elizabeth Ave., Portsea, 3944.

5. General Business. 6. Correspondence.

Wednesday, 10 March—Annual Meeting and election of Council Members and Officebearers.

F.N.C.V. EXCURSIONS

Sunday, 21 February—Geelong-Anglesea. Marine Biology and general. Leader Mr. J. Strong. The coach will leave Batman Avenue at 9.30 a.m. Fare $2.00. Bring two meals.

PRELIMINARY NOTICES

Easter, Friday, 9 April - Monday, 12 April—Hamilton. Accommodation has been booked and a coach chartered for this excursion. Fare $12.00 to be paid when booking. Accommodation $7.75 daily D., B. & B. to be paid individually.

GROUP MEETINGS
(8 p.m. at National Herbarium unless otherwise stated)

Thursday, 11 February—Botany Group. Mrs. M. Corrick will speak on "Plants of the Victoria Range".

Wednesday, 17 February—Microscopical Group.

Friday, 26 February—Junior meeting at 8 p.m. at Hawthorn Town Hall.

Monday, 1 March—Entomology and Marine Biology Group meeting at National Museum in small room next to theatrette at 8 p.m.

Wednesday, 3 March—Geology Group.

Thursday, 4 March—Mammal Survey Group meeting at 8 p.m. in Arthur Rylah Research Institute, Heidelberg.

Friday, 5 March—Junior meeting at 8 p.m. at Rechabite Hall, 281 High St., Preston.

General meetings in March, April and June will be on the second Wednesday instead of second Monday because of Public Holidays.

The
Victorian Naturalist

Editor: G. M. Ward

Assistant Editor: E. King

Vol. 88, No. 2 — 3 February, 1971

CONTENTS

Articles:

Studies of an Island Population of *Rattus fuscipes*.
By R. P. Hobbs ... 32

Notes on the Small Black Wallaroo *Macropus bernardus* (Rothschild, 1904) of Arnhem Land. By S. A. Parker ... 41

Naturalists in Bass Strait. By A. H. Chisholm ... 43

Victorian Non-marine Molluscs — 3. By Brian J. Smith ... 45

Obituary:

A Tribute to Noel F. Learmonth ... 39

Book Review:

A Guide to the Native Mammals of Australia ... 46

Field Naturalists Club of Victoria:

Reports ... 50
Diary of Coming Events ... 30

Front Cover:
John Wallis photographed this fine study of *Tachyglossus aculeata*, the Echidna.

Studies of an Island Population of *Rattus fuscipes*

by

R. P. HOBBS

Very little is yet known about the ecology of the small mammals on the islands of Bass Strait. Some islands maintain only a population of rabbits while others support native species. Two species of small mammals inhabit Greater Glennie Island: the murid rodent *Rattus fuscipes* and the dasyurid *Antechinus minimus maritimus*.

In December 1967, the Monash University Biological Society began a study of *R. fuscipes* which had four basic aims:—

1. To compare the morphological characteristics with those of two mainland populations of *R. fuscipes*.
2. To establish the density of the population and to compare it with that normally occurring on the mainland.
3. To determine the effect of the high population density on home range.
4. To establish the age structure of the population with a view to finding out when the breeding season was.

VEGETATION

The majority of the trapping area was covered in low scrub, varying from two to six feet in height and forming a dense canopy. The ground cover in this area was not very dense but was composed of decaying plant material. This was quite dry in the summer. Interspersed among the scrub were patches of tussock grass.

The south-western section of the trapping area was at a higher elevation. The vegetation there was *Casuarina*. In this section the ground cover was also light. Toward the lower section in the north-west of the trapping area, were exposed rocky patches with a mixed cover of tussock grass, low scrub and some pig-face succulent.

The western section was exposed to strong winds carrying salt spray. However, only the extreme western trapping stations could be affected by this as the scrub formed such a dense canopy.

METHODS

The trapping area for estimation of population density and home range was 3.6 acres (14,600 square metres). Eighty trapping stations were marked out, in ten rows of eight, each station being fifty feet distant from its nearest neighbour.

The traps used were similar to National collapsible cage traps. Twenty traps were used each trapping night and were distributed over the grid in three different ways:—

1. The traps were evenly distributed, each being two stations distant from the next. Every night, they were rotated so that after four successive nights, all stations had been trapped.
2. One quarter of the grid was trapped in a block. The traps were shifted as a block so that all stations had been trapped in four nights.
3. Random numbers were chosen each night.

Rats were transferred from the trap to a polythene bag in which they were anaesthetised with ether. While in the bag, they were weighed. They were then ear-tagged with Monel fingerling tags. After recovery, they were released.

Twenty-one rats, trapped away from the grid, were killed in order to obtain skins and skulls. Body measurements of these were also made. Twenty skulls collected at Noojee in Gippsland, by Mr. R. M. Warneke of the Fisheries and Wildlife Department, were measured and compared with the measurements of the skulls of the island population. Another ten skulls loaned by the National Museum of Victoria, collected at Portland, Western Victoria, were also measured and compared with the Glennie population.

Two live rats, one of each sex, were removed from the island to be used in breeding experiments, but in fact they never bred in captivity, for reasons unknown.

RESULTS

1. Density estimates

The trapping programme was designed for estimation of home range and population density. Trap success was calculated by dividing the total number of rats caught by the number of traps set and expressing this as a percentage. Any trap which was found empty, with the bait removed, was not counted as a set trap. Any trap found sprung, with the bait still inside was counted as half a set trap, because it would probably have been open, on the average, for half a night. Traps containing other species were not counted as set traps. Thus:

$$\text{Trap success} = \frac{100N}{T - [a + h + (c/2)]}$$

N = total individuals caught.
T = total traps set.
a = traps found without bait.
h = traps with other species.
c = empty traps, sprung, with bait inside.

By this method, the overall trap success for the 1967-1968 period was 101%. In some cases, two rats were found in a trap. The trap success was so high in each period that it could not be used as a method of determining population density changes.

The population density was calculated by a method outlined in Eberhardt (1969). The formula used is given below.

$$N = \frac{r(s-1)}{s-r}$$

where N population size
s = total captures.
r = total individuals caught.

The first estimate (See Table 1) was made using data collected over only three nights. This resulted in an inaccurate estimation because $(s-r)$ is small. Consequently, a small error in $(s-r)$ results in a large error in N. The estimates for later periods are probably more reliable.

TABLE 1.
Population density.

Date	Number captured			Estimated population size	Individuals per acre
	M	F	Juv.		
December 1967	20	18	1	216	60
January 1968	39	48	14	92	25
February 1968 (early)	18	45	28	97	27
February 1968 (late)	32	25	67	121	33
December 1968	31	28	6	92	25

2. Home range

Home range is here defined as the area encompassed by the trap sites of capture. Each trap site, as the basis of tabulating home range, was a square of sides 50 feet, or an area of 2,500 square feet. Home range was estimated by multiplying this area by the number of trap sites within the area of capture. Only those rats which were trapped nine or more times were used in this estimation. The results are shown in Table 2.

TABLE 2.
Home range.

Tag number	Sex	Number capts.	Number sites	Home range (sq. yds.)
806	F	11	3	844
812	F	13	8	4,170
813	F	15	11	6,110
817	F	9	7	2,780
825	F	12	9	4,170
838	M	11	5	3,060
844	M	11	5	1,390

3. Comparisons of morphological features.

(a) Sex differences.

Body measurements of males of the island population were compared to those of females by means of a "Student's t-test". As is shown in Table 3, a difference was found only in body weight. Skull measurements were compared in the same manner. There were no significant differences between sexes for all skull measurements taken.

TABLE 3.
Body measurements of island population.

	Mean		Result
	Female	Male	
Weight (gms.)	189.6	213.4	Difference at 5% level of significance
Head-body length (mm.)	174.4	178.0	No significant differences
Tail length (mm.)	181.2	186.9	,, ,, ,,
Hind-foot length (mm.)	37.9	37.4	,, ,, ,,
Ear length (mm.)	24.1	23.4	,, ,, ,,

(b) Comparison of skull measurements between Glennie, Gippsland and Portland populations.

The results of a statistical comparison of skull measurements are shown in Table 4. Skull measurements used are shown in the diagram. Since the Glennie population has larger skulls than the other two populations, a similar comparison of some ratios of skull measurements was made. This is summarised in Table 5.

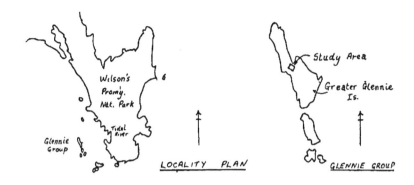

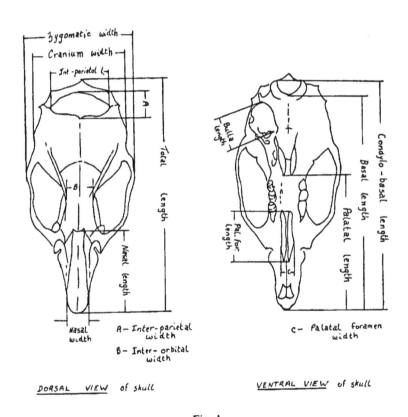

Fig. 1

TABLE 4.

Skull measurements.

	Lengths of Glennie population in mm. Range	Mean	Comparison with level of significance.	
			Gippsland population	Portland population
1. Total length	40.0 - 44.9	42.24	larger 1%	larger 1%
2. Condylo-basal length	38.7 - 43.2	40.74	,, 1%	,, 1%
3. Basal length	36.4 - 41.1	38.39	,, 1%	,, 1%
4. Zygomatic width	20.7 - 23.2	21.78	,, 1%	,, 1%
5. Inter-orbital width	6.2 - 6.9	6.49	,, 1%	,, 1%
6. Inter-parietal length	10.8 - 12.3	11.66	,, 1%	,, 1%
7. Inter-parietal width	4.7 - 5.9	5.34	,, 1%	,, 1%
8. Cranium width	16.1 - 17.5	16.73	,, 1%	,, 1%
9. Nasal length	16.3 - 18.8	17.43	,, 1%	,, 1%
10. Nasal width	4.6 - 5.2	4.94	,, 1%	,, 1%
11. Palatal length	21.7 - 24.0	22.73	,, 1%	,, 1%
12. Palatal foramen length	7.2 - 8.2	7.76	no sig. dif.	,, 1%
13. Palatal foramen width	2.7 - 3.4	3.02	larger 1%	,, 1%
14. Width inside M1-M1	4.0 - 4.9	4.36	,, 1%	,, 1%
15. Width outside M1-M1	8.3 - 9.7	8.83	,, 1%	,, 1%
16. Bulla length	6.3 - 7.0	6.67	no signif. difference	no signif. difference
17. Length of crowns M1-M3	6.7 - 7.3	6.92	larger 1%	larger 1%
18. Length of alveoli M1-M3	7.7 - 8.9	8.29	,, 1%	,, 1%
19. Length of crowns M1-M2	5.2 - 6.0	5.60	,, 1%	,, 1%
20. Incisor width (each)	1.4 - 1.6	1.46	,, 1%	,, 1%

TABLE 5.

Ratios of skull measurements.

Ratio	Means			Result	
	Glennie	Gipps	Portld.	Gl-Gi	Gl-Por
1. Zyg. wid./tot. leng.	0.516	0.510	0.511	NSD*	NSD
2. Inter-pariet. leng./tot. leng.	0.267	0.251	0.273	1%	NSD
3. Cranium wid./tot. leng.	0.396	0.402	0.420	NSD	1%
4. Nas. leng./tot. leng.	0.412	0.368	0.381	1%	1%
5. Palatal leng./tot. leng.	0.538	0.543	0.524	NSD	1%
6. Inter-par. leng./Inter-par. wid.	2.191	2.192	2.284	NSD	NSD
7. Nas. leng./nas. wid.	3.522	3.364	3.308	1%	1%
8. Pal. foram. lehg./pal. foram. wid.	2.572	3.417	2.725	1%	NSD

*no significant difference

The results tend to show that the two mainland populations are more similar to each other than to the Glennie population. Also, the Glennie population does not appear to have more affinities with one mainland population than the other.

4. *Breeding season*

Only one juvenile animal was trapped in December 1967. (See Table 1.) Many were trapped in January 1968, but only one new juvenile was trapped in late February. Mating must therefore take place from November to late December, allowing a gestation period of 21 days and a weaning period of four weeks. A more detailed study of the reproductive condition of the rats by Mr. R. M. Warneke during the period 27 September to 3 October 1968 indicated that the breeding season was about to begin.

Four animals which were mature in January 1968 were still alive in December 1968. It is not known if any animals live any longer than two years.

CONCLUSIONS AND DISCUSSION

There has recently been a revision of *Rattus* species in Australia, based on cross breeding experiments. Horner and Taylor, 1965). The population of rats in Portland was once regarded as a separate species from the Gippsland population, the former being called *R. greyi*, and the latter *R. assimilis*. These two species, and a Western Australian species called *R. fuscipes* have now been grouped together as the one species *R. fuscipes*. It is almost certain that the population on Greater Glennie Island can be ascribed to the species *R. fuscipes*, in spite of the morphological differences here described. It would be unrealistic at this stage to call the Glennie rat a new subspecies. As was seen in Tables 4 and 5, there are not many morphological differences between the Glennie, Gippsland and Portland populations. The most striking difference is the large overall size of the Glennie population. This is of interest as it seems to be a common feature of island races of small mammals to be of large size. (Corbet, 1963). The Glennie population also has relatively longer nasal bones, but few other significant differences exist.

The density of the population is very much higher than is normally found on the mainland. One source (Warneke, unpublished) quotes a figure of 4.8 individuals per acre. The high density on Glennie may be due to all or any of a number of factors. There is a general lack of predators on the island. Natural predators on the mainland, but absent on Glennie, include snakes and many species of predatory birds. There are some predatory birds from the mainland feeding on Glennie, but very few inhabit the island, due to a shortage of suitable nesting sites. Other predators absent on Glennie are feral cats, and foxes. The introduction of these to the mainland may have had considerable effect on the small mammal fauna of Australia.

Greater Glennie Island does not support a rabbit population. Although it is not suggested that direct competition takes place, rabbits have seriously affected the natural habitat on other Bass Strait islands.

It appears, from the little information on home range obtained, that there is no appreciable difference from that occurring in less dense populations.

The skulls and skins which were collected from the island in December 1967, have been deposited in the museum of the Department of Zoology and Comparative Physiology, Monash University, Melbourne.

SUMMARY

The density of the population of *R. fuscipes* on Greater Glennie Island is approximately 30 rats/acre, which is unusually high compared with populations elsewhere. Size of home range does not seem to be affected by this.

By comparing skull measurements with those of two separate populations on the mainland, it was found that the Glennie population is very probably *R. fuscipes*. Tables of skull and body measurements of the island population are given.

The breeding season is early summer; and individuals were observed to live for two years.

Acknowledgements

The author wishes to thank Dr. D. F. Dorward of the Department of Zoology and Comparative Physiology, Monash University, for his encouragement, assistance, and helpful criticism. Thanks are also due to the many members of the Monash University Biological Society, who collected the field data, and to Mr. R. M. Warneke of the Fisheries and Wildlife Department, Victoria for most of the mainland skulls, and for his reporting of some of our marked animals trapped.

REFERENCES

Corbet, G. B. (1961). Origin of the British insular races of small mammals and of the "Lusitania" fauna. *Nature, Lond.*, *191*, 1037.

Corbet, G. B. (1963). An isolated population of the bank-vole *Clethrionomys glareolus* with aberrent dental pattern. *Proc. zool. Soc. Lond.*, *140*, 316.

Corbet, G. B. (1964). Regional variation in the bank-vole *Clethrionomys glareolus* in the British Isles, *Proc. zool. Soc. Lond.*, *143*, 191.

Eberhardt, L. L. (1969). Population estimates from recapture frequencies. *J. Wildl. Mgmt.* *33* (*1*), 28-39.

Green, R. H. (1967). The murids and small dasyurids in Tasmania. Pts. 1 & 2. *Records of the Queen Victoria Museum*, *28*.

Green, R. H. (1968). The murids and small dasyurids in Tasmania. Pts. 3 & 4. *Records of the Queen Victoria Museum*, *32*.

Horner, B. Elizabeth & Taylor, J. Mary (1965). Systematic relationships among *Rattus* in southern Australia: Evidence from cross breeding experiments. *C.S.I.R.O. Wildl. Res.*, *10* 101-9.

Tate, G. H. H. (1936). Muridae of the Indo-Australia region. *Bull. Amer. Mus. Nat. Hist.*, *72*, 501-728.

Warneke, R. M. (unpublished) Life history and ecology of *Rattus assimilis*.

GEOLOGY GROUP EXCURSIONS

Sunday, 14 March—Deep Creek Gorge, Bulla. Leader: Mr. George Carlos

Transport is by private car. Spare seats are usually available for those without their own transport. Excursions leave from the western end of Flinders Street Station, opposite the C.T.A. Building, at 9.30 a.m.

A TRIBUTE TO NOEL F. LEARMONTH

The death of Noel F. Learmonth, which occurred at Portland on 9 September, 1970 at the age of 90, has left a very big gap in the ranks of Field Naturalists, Conservationists and Historians.

In those 90 years Noel established legions of friends and admirers, and he amassed a knowledge of the history and natural history of the south-west of Victoria, which is unsurpassed.

Noel Fulford Learmonth was born at Ettrick, near Portland, in the far south-west of Victoria. His family arrived shortly after the Hentys, and his grandfather was the first Mayor of Portland. Noel attended Geelong Grammar School from 1895 to 1898, and maintained his connection with his old school right to the last.

On leaving school his first work was with a Government survey team sent out to survey the last connecting rail link with Mildura. Noel was the last surviving member of this team. Their experiences in this useless and almost desert country, as it was considered then, are well told in his latest book *Four Towns and a Survey*.

Next he was private secretary to the Minister for Lands, M. K. McKenzie, and about 1905 selected land in Queensland. After a number of years he left this property, because of the prickly pear menace, and returned to Tyrundarra where he and his wife made their home at "Carramar". In 1951 they retired to live in Portland.

Noel Learmonth wrote extensively.

For many years he was a contributor to "The Bulletin", writing under the name of "Leo". He contributed nature notes to the "Portland Guardian" and several natural history journals. In the last two or three years he wrote many articles and letters on the Little Desert and Kentbruck controversy. His dry humour and ability to assemble words enabled him to strike at the very core of problems in a most telling manner.

In 1934 he wrote *The Portland Bay Settlement* (now out of print), in 1960 *The Story of a Port* (also out of print); and in 1967 *The Birds of Portland* (almost sold out). He donated this book to the Portland Field Naturalists Club. His last book *Four Towns and a Survey* has just been released. Unfortunately Noel did not see it.

Noel was interested in most sport, with cricket as his favourite. As a member of the Melbourne Cricket Club he always attended Test matches on that ground.

Outside his home and grazing interests his greatest love was the bush. From earliest days he seemed to have a strong leaning towards birds. He has made many valuable contributions to the ornithology of this far south west of the State. His recordings of sea birds, particularly beach washed specimens, sometimes in collaboration with Cliff Beauglehole, deserve particular mention. A number of rare finds have been recorded, the last one being the Broadbilled Prion (*P. vittata*), recorded in Bird Observer Notes in August 1970.

With one or two helpers he carried on the World Bird Day counts after this observation had been started. His diaries and notes of all his observations were meticulously kept.

In 1945, with a few other enthusiasts, he was instrumental in forming the Portland Field Naturalists Club, a Club which is still flourishing.

In 1946 Noel Learmonth initiated a move for an 80,000 acre Forest Park on the lower Glenelg river. After many years of effort and frustration a 22,000 acre National Park was finally created by act of Parliament in late 1969. Learmonth Creek, one of the creeks of the Moleside system, within the Park, bears his name.

In 1955 he started another move to have Mount Richmond reserved as a National Park. This 1,500 acre (now 2,000) National Park containing 450 species of flowering plants was created by act of Parliament in late 1960. A lookout tower with a round-the-compass view has been erected on the summit and named Learmonth Lookout in his honour. He was one of the prime movers in having small areas reserved at Heathmere and near Cape Nelson. More recently he did an extensive survey for a south west coast planning scheme, extending from Tyrandarra to Nelson, for the Town and Country Planning Authority.

Noel has faithfully recorded much of the early history of the Portland area. This is to be found in his books Portland Bay Settlement. The Story of a Port, and Four Towns and a Survey. To travel in the Portland district with Noel, whether it be for an hour or a day, was an experience and an education. He had a running commentary for the properties, the old homesteads, an endless number of features with historical associations, the wrecks along the coast where the rugged coastline had taken a heavy toll in sailing boat days, and last but not least, the personalities who have contributed so much to the history and the humour of this part of Victoria.

Noel Learmonth was a perfect gentleman and a perfect host. He was at home in any company. An hour or two spent in his company was always a refreshing and rewarding experience. His home at 45 Must Street, both before his wife passed away in 1964, and since, has been open to friends and strangers alike. As would be expected, his home was placed in a bush setting with every encouragement to visiting "feathered friends".

About eighteen months ago, my wife and I gave Noel a lift to Melbourne, and in the evening were his guests for dinner. During the dinner Sir James Darling, former head master of Geelong Grammar, walked passed; and on seeing Noel his quip "What! are you still walking around" illustrates perfectly the timeless institution which Noel Learmonth had become.

Sorrow was no stranger to the house of Learmonth. One son died in infancy, and the other two sons gave their lives in the Second World War. John as a prisoner of war after Crete, and Charles, who was in the Air Force, in rectifying a technical fault in a particular make of aircraft, near Perth, in Western Australia. Charles was awarded the D.F.C. and Bar, and the air base of Learmonth in the north west of Western Australia bears his name.

Noel Learmonth is survived by a daughter, Mrs. Don Baulch, five grandchildren, and a brother, Cecil

FRED DAVIES

Notes on the Small Black Wallaroo *Macropus bernardus* (Rothschild, 1904) of Arnhem Land.

by S. A. PARKER*

Introduction

Ride (1970: 198) writes of this little-known species:

"... last collected in 1922 when three specimens were obtained by Mrs. P. Cahill at Oenpelli for the National Museum of Victoria. Earlier, in 1918, Mr. Cahill presented live specimens to Taronga Park Zoo. Previously it had been collected in Arnhem Land by J. T. Tunney in 1903 and K. Dahl in 1895".

The following observations are preliminary to the treatment of this species in a forthcoming checklist of the native land-mammals of the Northern Territory (Parker, in press).

Taxonomic status

Frith & Calaby (1969: 32) write: "*M. bernardus* is sympatric with *M. antilopinus* [Antilopine Wallaroo] over probably all of the former's range, but its relationship with nearby *M. robustus* [Euro] has not been satisfactorily cleared up. *M. robustus* is known to occur in the South Alligator River area near the range of *bernardus* and collections and observations should be made in the critical area to determine the possibilities of overlap or inter-breeding".

During September-October 1969, B. I. Bolton, D. Howe and I undertook a five-week faunal survey of the sandstone escarpment country of the Deaf Adder Creek valley 55 miles little south of Oenpelli. Here we observed and collected both *M. bernardus* and *M. robustus* in the same rocky areas; on one occasion a female of each with pouch-young was obtained on the same hillside within a few minutes. Wilkins (1928: 168) observed *bernardus* in sandstone ranges fifteen miles up the King River north-east of Oenpelli in 1924; he collected a specimen of *robustus* in this area (BMNH 26.3.11.69) and another further east on the upper Goyder River (BMNH 26.3.11.68) (see Map 1). This evidence of sympatry, together with the clear and constant morphological differences between *bernardus* and *robustus* (see Table 1), justifies the recognition of *bernardus* as a separate species.

Field observations

Although *M. bernardus* has been found so far only in the escarpment country of western and north-western Arnhem Land, it is by no means rare there. In the sandstone gorges of Deaf Adder Creek valley we encountered this small, blackish, thickset wallaroo several times, always on boulder-strewn hillsides, singly and in pairs (once in a group of three). In the daytime it was wary, but a female (NTM 4748) located by eyeshine at night allowed a close approach. Our observations were too limited to detect any ecological differences between this species and *M. robustus*, which occurred in the same rocky habitat. In the same area *M. antilopinus* was recorded only on the flat valley floor.

In the National Museum of Victoria there is a note from Mrs. Marie

*Arid Zone Research Institute, Alice Springs, N.T. 5750.

February, 1971

Cahill to Professor W. Baldwin Spencer dated 9 April, 1923, pertaining to live specimens purchased of Mrs. Cahill the previous year:

"Kakadu name — Barr-ark. Food — grass and herbs. Lives in caves on Ranges. When frightened goes into dark caves. Killed by natives for food thus — spinifex set on fire causes dense smoke when animal can be approached quite easily, as smoke partly blinds animal, which stands rubbing its eyes".

Specimens examined

National Museum of Victoria: C6380-4, labelled "Oenpelli" (4) and "near East Alligator River" (1), collected in 1912-1914 by P. Cahill and purchased of his wife, Marie Cahill, in 1922 (not *collected* in 1922 as stated by Ride).

South Australian Museum: M282-4, collected on the [upper] Mary River in December 1913 by W. D. Dodd.

Northern Territory Museum: NTM 4737, 4748, 5413, collected 55 miles due south of Oenpelli on Deaf Adder Creek in September 1969 by B. L. Bolton, D. Howe and S. A. Parker, the first specimens taken since 1918, a lapse of 51 years.

Acknowledgements

I am indebted to Miss Joan Dixon, Curator of Vertebrates, National Museum of Victoria, and Mr. Peter Aitken, Curator of Mammals, South Australian Museum, for the loan of specimens and for helpful comments; also to Mr. John Edwards Hill, Mammal Section, British Museum (Natural History) for information on the two hitherto unreported specimens of *M. robustus* collected by G. H. Wilkins on the King River and the Goyder River.

REFERENCES

Frith, H. J., & Calaby, J. H., 1969. Kangaroos. Melbourne: F. W. Cheshire.

Ride, W. D. L., 1970, A Guide to the Native Mammals of Australia. Melbourne: Oxford University Press.

Wilkins, G. H., 1928. Undiscovered Australia, being an account of an expedition to tropical Australia to collect specimens of the rarer native fauna for the British Museum. 1923-1925. London.

TABLE 1.

	M. bernardus	*M. robustus* (Arnhem Land)
Vertical groove in outer face of i³	Pronounced	Faint or absent
Sides of rostrum seen from above	Straight or v. slightly concave	Slightly convex
Length of nasal overhang	14-17 mm (6)	9-15 mm (6)
Length of nasal overhang / Length of entire nasals	0.24-0.29 (6)	0.18-0.20 (6)
Angle subtended by outer edges of nasals at tip	± 40°	± 60°

Some cranial and dental differences between *Macropus bernardus* and *M. robustus*.

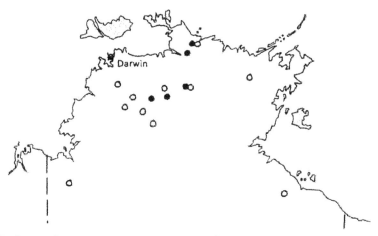

Distribution of *Macropus bernardus* and *M. robustus* in the northern N.T.

Naturalists in Bass Strait

by A. H. CHISHOLM

Although *The Literature of Australian Birds*, by H. M. Whittell (1954), is an extremely useful bibliography, and although its compiler uncovered numerous worthwhile items that had fallen into obscurity, it has at least one entry that is remarkably inadequate. This relates to the work of Donald Macdonald, who for many years was a distinguished nature and general writer on the staff of the Melbourne *Argus*.

Very surprisingly on the part of such a diligent and conscientious worker, Major Whittell not only overlooked Macdonald's books *Gum Boughs and Wattle Bloom* (1887) and the posthumous volume *The Brooks of Morning* (1933), each of which contains significant references to birds; but he failed also to record the many ornithological articles which Mac wrote—in addition to the highly useful columns of Nature Notes—from year to year in the *Argus*.

I am reminded of this matter by the fact that, as in the case of Whittell, the authors of a recent substantial paper on certain islands of Bass Strait (*Vict. Nat.*, Dec. 1970), also appear to be unacquainted with relevant articles written by Macdonald in the *Argus* during 1908. Their author obtained his material as a member of the Ornithologists' Union party, which from 24 November (1908), spent a rewarding fortnight touring Bass Strait in a special steamer.

The titles of the Macdonald articles are: "The Life of the Sea", "Gannets of the Bass", and "An Island Race". All three are lengthy and, as a matter of course, all are thoroughly readable and informative. Indeed, they fired my youthful enthusiasm so strongly that I carefully pasted them in a scrapbook of natural history I was then assembling, and here they are to this day.

Regrettably, I did not attach dates to the articles, but references in them suggest that they appeared in December, soon after the party returned to Melbourne, and therefore some three months before S. A. White's report on the outing appeared in the *Emu*.

In recommending attention to these writings by anyone adding to the records of visits to Bass Strait islands, I should mention that two further articles in point, "Bass Strait Islands Excursion" and "Cat Island", appeared in the *Argus* soon afterwards. Their author was J. W. (later Sir James) Barrett, who had led a party of 15, including 10 medical men, on a trip that began on 1 Jan., 1909, and which was carried out in the *Manawatu*, the vessel used for the Ornithologists' Union expedition.

There are many informative observations in both of Barrett's articles, notably in relation to the gannets of Cat Island, then estimated at about 4000 birds—a sharp contrast with the unhappy situation now obtaining.

Moreover, alongside the five articles mentioned above, my scrapbook presents another informative *Argus* feature; one dealing with the then newly established reservation on Wilson's Promontory. Entitled "The Promontory Park: a National Sanctuary", it carries the initials "T.S.H" (probably indicating the geologist T. S. Hall), and in the course of a general review of the area it says roundly, "There is no part of the Victorian coast that approaches the Promontory in beauty".

Perhaps I should add, too, that yet another cutting in this section of my venerable scrapbook, one dated 29 October, 1909, is a Melbourne *Herald* report that opens with this sentence:

"Bird Day was celebrated for the first time in Victoria by the children attending the State schools today (Friday)". Sponsors of the observance were the Ornithologists' Union, the Bird Observers' Club, the Field Naturalists' Club, and the Gould League of Bird Lovers.

Further details regarding that historic celebration may be found in an article entitled "First Bird Day in the Commonwealth", by H. W Wilson, published in the *Emu* in Jan 1911.

F.N.C.V. PUBLICATIONS AVAILABLE FOR PURCHASE

FERNS OF VICTORIA AND TASMANIA, by N. A. Wakefield.
   The 116 species known and described, and illustrated by line drawings, and 30 photographs. Price 75c.

VICTORIAN TOADSTOOLS AND MUSHROOMS, by J. H. Willis.
   This describes 120 toadstool species and many other fungi. There are four coloured plates and 31 other illustrations. New edition. Price 90c.

THE VEGETATION OF WYPERFELD NATIONAL PARK, by J. R. Garnet.
   Coloured frontispiece, 23 half-tone, 100 line drawings of plants and a map. Price $1.50.

   Address orders and inquiries to Sales Officer, F.N.C.V., National Herbarium, South Yarra, Victoria.

   Payments should include postage (11c on single copy).

Victorian Non-Marine Molluscs — 3

by Brian J. Smith

There are four or five species of introduced snails of the family Helicidae which look very much alike and which are often confused. They are all sandhill snails and, where they occur, usually reach near plague proportions.

Theba pisana (Muller)

This is a medium sized globular shell about 15 to 20 mm in diameter usually bearing numerous dark brown spiral bands on a white background. The inside of the aperture is often pink but the most obvious character to separate this species from the other similar species is nearly closed umbilicus or hole in the centre of the spiral under the shell. In all the species described below the umbilicus is open and obvious to a greater or lesser extent. This snail is found very commonly on coastal sand-dunes along the entire Victorian coast-line where it can occur in very great numbers. However recently this species has also been recorded along the Murray, so may occur in other dry sandy areas.

Genus *Helicella*

Species of this genus have white shells with concentric brown bands of varying thickness. These are globular to flattened discoid in shape and all have a prominent wide umbilicus. They live in sandy, fairly dry situations along the coastal fringe and inland, associated with places frequented by man. *Helicella virgata* and *Helicella neglecta* are two very closely similar species inhabiting mainly the drier areas in central Western Victoria. They can be seen clinging in clusters to the fence poles in many of the drier areas of this part of the State. These seem largely confined to inland areas. They can be separated with difficulty chiefly by their spire shape, that of *H. neglecta* being much flatter.

A small species of this genus, *Helicella caperata* is found living with *Theba pisana* in coastal dunes, and is often confused with the juveniles of the latter species. However, it can be readily separated by the presence of a large umbilicus. There may be other *Helicella* species introduced into the State but only more collecting can elucidate this.

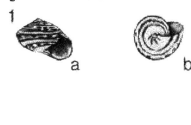

Figures. Aperture (a) and Ventral (b) views of
1. *Theba pisana* (note lack of umbilicus).
2. *Helicella virgata*.
3. *Helicella caperata* (small size).

(Drawings by Miss Rhyllis Plant.)

book review

by N. A. WAKEFIELD

A Guide to the Native Mammals of Australia
by W. D. L. Ride, with drawings
by Ella Fry.
Published by Oxford University Press, Melbourne; 1970. $7.50.

"This guide to the mammals of Australia, about the mammals of an altered and altering continent, is as much an introduction to the problems of conserving the mammals as it is a guide to the kinds of mammals themselves".

This introductory statement may indicate an important theme in Dr. Ride's book, but the major value of the work lies in its presentation of an up-to-date summary of the Australian mammal fauna.

For the past 30 years, naturalist and student have had to rely on books such as Troughton's *Furred Animals of Australia*, a work which included 161 "species" of marsupials and 86 native members of the mouse family. In Ride's book, marsupials number 124 species and native murids only 48. This 30 percent reduction in species is not indicative of omissions but represents a more rational system of classification arising from work of many mammalogists during recent decades. It is an interesting thought that this rationalization is by no means complete, and further reduction in recognized species may be anticipated in the future. In the meantime we have a much-needed up-to-date list of species names, as a standard both for those who write and for those who read about Australian mammals.

The book begins with discussion of patterns of natural distribution and habitat of our mammals, and of changes brought about by the intrusion of man into their realm. This is followed by an intensely interesting chapter which outlines the history of rediscovery or rehabilitation of several species, such as Koala and Parma Wallaby, and, as the author states, "fascinating human stories and romance lie behind each one of these". There is brief outline of aspects such as evolutionary history, zoogeography, adaptive radiation and convergence, that have brought about patterns which we see now in the Australian mammal fauna; and problems of the scientific and popular naming of our species are discussed.

It is gratifying to note that "many familar, and much quoted, statements have been omitted . . . because many of these observations are anecdotal and . . . are found to be based upon unreliable and unsubstantiated identifications, or even first appear in the literature at secondhand". This principle cannot be recommended too strongly to writers of those colorful natural history books which are now being published in ever-increasing numbers.

Dr. Ride avoids the problem of standard popular names by simply giving a series of alternatives for each species, and he has "tried to avoid introducing new ones". A notable departure from the last principle, however, is the innovation of the rather clumsy 5-syllable "antechinuses" for what eastern Australians know as phascogales. (As a popular name, phascogale is short and euphonious, and it is as appropriate for members of the genus *Antechinus* as "wallaby" is for species of *Macropus*.) With growing public interest in the conservation of our rarest species, it is becoming imperative that popular names should be standardized, for meaningful use by the mass news media. Ornithologists succeeded quite well in this field, with their *Official Checklist of Australian Birds*, in 1926.

Two-thirds of the book (Chapters 5 to 12) are devoted to the treatment of the 229 currently recognized species of native Australian mammals. These are arranged into 53 natural groups, and there is a full-page plate to illustrate each group. General information is given about each group, and a list is appended of all species of the group together with precise data about each species. The subject is thus covered exhaustively and a most informative census of our mammals is presented.

The plates comprise drawings in black and white executed by Ella Fry. They have aimed to depict details of form, pose and habitat of the species chosen, and in most cases the aim is achieved admirably. However, details in some portrayals are not true to life. Toes often appear to be somewhat enlarged for example, the head of the Echidna does not have correct proportions, and the softness of the tail of the Fairy (or Leadbeater's) Possum has eluded the artist's pen.

In the preparation of the book, there has been strong bias towards Western Australia. Forty-one of the sixty-two plates depict Western Australian specimens. The Western Australian distributions are mostly set out in detail, but in a number of cases data given for distributions in other States are vague or inaccurate.

In the main body of the book (Chapters 5-12), the distributions given for mammal species are those which applied to the early years of European settlement, and, with few exceptions, modified present-day distributions are not indicated. In a number of cases this matter is somewhat rectified by data in Chapters 13 and 14, in which there is discussion of rare or extinct species and of certain others with changed status. The fact that the distributions given are original ones could well have been explained in the introduction to the book, or the word "formerly" could have been inserted in appropriate places in the statements of distribution.

There is lack of definition and uniformity with regard to the use of terms such as "southern" and "eastern" in connection with mammal distributions in Victoria. For example, for the Brown Bandicoot (*Isoodon obesulus*) "southern" may be interpreted, correctly, as a general distribution from far SW. Victoria to the extreme east; but the term means something else in the "southern and south-eastern" given for the White-footed Dunnart (*Sminthopsis leucopus*).

Chapter 13 (The Rare Ones) sets out data of species which have become very rare or extinct, and Chapter 14 (Those with Changed Status) deals with additional species which have had their distributions reduced substantially in modern times. However, further species, such as the Red-bellied Pademelon (*Thylogale billardieri*), which have equal claim to inclusion in Chapter 14, have been omitted.

Chapter 15 is a plea for conservation, and it should be read carefully by the lay conservationist as well as the specialist. The author sums up his philosophy with these comments.

"Each species is the product of over six hundred million years of evolution, and we ourselves are one of these. In destroying any one of these species, we are destroying a facet of our potential understanding of the world about us—we owe it responsibility to each other to see that this does not happen".

The book has three appendixes. The first (Suggestions for Further Reading) lists 202 references, set out under a sequence of headings which comprise groups of chapters of the book. Secondly, there are 125 references "for the student and professional reader", pointing to detailed descriptions of the morphological characters of each Australian mammal species. Thirdly, there is information about the animals and backgrounds comprising the book's 62 plates.

DISTRIBUTION DATA OF VICTORIAN MAMMALS

As a supplement to the foregoing review, the following details of distribution are presented for Victorian users of the *Guide to the Native Mammals of Australia*. These details amplify and amend distributional data given in the book for certain Victorian mammals. References are cited when the details are based on published articles. In most other cases, the data are based on specimens in the mammal collections of the National Museum of Victoria and the Victorian Fisheries and Wildlife Department. In a few cases, the data are from the present writer's observational records.

p. 44 Black-faced Kangaroo, *Macropus fuliginosus*. Areas of Vic adjoining S. Aust., from far NW. to southern Grampians.

Wallaroo, *Macropus robustus*. For Vic., only in NE. Gippsland (upper Snowy River area).

Red Kangaroo, *Megaleia rufa*. For Vic., only in far NW.

p. 46. Toolache, *Macropus greyi*. No confirmed living record for Vic. Finlayson (1927) provided an unsubstantiated secondhand report of its one-time occurrence in Vic.

p. 47. Swamp Wallaby, *Wallabia bicolor*. Not recorded living, anywhere west of Otway Ranges area, SW. Vic., or in S. Aust.

p. 48. Red-bellied Pademelon, *Thylogale billardieri*. No longer in Vic. The former Aust. mainland population is now extinct.

p. 54. Crescent Wallaby, *Onychogalea lunata*. Not recorded living in Vic. The sole Murray-Darling record was in N.S.W. (Wakefield 1966).

Bridled Wallaby, *Onychogalea fraenata*. NW. and N. Vic. (Mildura area and near Mt. Hope) should be added to the original distribution both here and on p. 207. Not now in Vic.: possibly quite extinct. (Wakefield 1966).

p. 58. Brown Hare-wallaby, *Lagorchestes leporides*. No longer in Vic. probably quite extinct. (See p. 1981.

p. 61. Brush-tailed Rock-wallaby, *Petrogale penicillata*. In Vic., recorded only from NE. and E. (upper Murray R. area to Buchan district (Wakefield 1963), and from Grampians in SW. (Audas 1925, p. 127).

p. 67. Rufous Rat-kangaroo, *Aepyprymnus rufescens*. Formerly in N. Vic. (Violet Town, Gunbower); no longer extant in state (see Wakefield 1966).

Brush-tailed Bettong, *Bettongia penicillata*. See also p. 207, but the SE. Aust population (formerly extending to NE. N.S.W.) is presumably now extinct.

Eastern Bettong, *Bettongia gaimardi*. No longer in Vic.; the Aust mainland population is presumably extinct (see p. 207).

p 68 Potoroo, *Potorous tridactylus*. Authentic Vic. records are north-central (near Kinglake) and far-eastern (SW. of Bonang).

p. 80. Squirrel Glider, *Petaurus norfolcensis*. Scattered from Dadswell's Bridge (W. Vic.) across N. Vic. to near Benalla.

p. 82. Fairy Possum, *Gymnobelideus leadbeateri*. Originally in east-central and NE. Vic. (Wilkinson 1961).

p. 94. Common Wombat, *Vombatus ursinus*. As well as "southern" (from far SW. to extreme E.), common in north-central and NE Vic.

p. 100. Long-nosed Bandicoot. *Perameles nasuta*. Extends west to Otway Ranges, SW. Vic.

Tasmanian Barred Bandicoot. *Perameles gunnii*. Former Vic. distribution was SW. only; survives now near Hamilton.

Marl. *Perameles bougainville*. The E. Aust. population is presumably now extinct. The statement on p. 207, that "it has only been recorded in recent times from small pockets in Victoria", arises from other authors' misidentifications of *P. gunnii*—for example, the "*P. fasciata*" of Brazenor (1950).

p. 102. Pig-footed Bandicoot, *Chaeropus ecaudatus*. Not now in Vic.; possibly wholly extinct (see p. 200).

p. 108. Chuditch. *Dasyurus geoffroii*. See p. 206; but the E. Aust. population is presumably now quite extinct.

p. 110. Tasmanian Devil. *Sarcophilus harrisi*. The "possibly southern Vic." is evidently based on the capture of a specimen, considered to be an escapee from captivity, near Toohorac, north-central Vic., in 1912 (see Troughton 1941, p. 47).

p. 112. Tuan. *Phascogale tapoatafa*. Besides "southern Vic." (far SW. to far E.), it extends across N. and NE. Vic.

Red-tailed Wambenger. *Phascogale calura*. Not now in Vic.; the E. Aust. population is presumably now extinct (Wakefield 1966).

p. 119. Dusky Phascogale, *Antechinus swainsonii*. Distribution includes north-central and NE. Vic. Not recorded living in S. Aust.; the "coastal south-eastern S.A." arises from confusion with *A. minimus*. (See Wakefield and Warneke 1963).

p. 122. Common Dunnart, *Sminthopsis murina*. All Vic records are western.

p. 124. Fat-tailed Dunnart, *Sminthopsis crassicaudata*. Also central, north-central (east to Alexandra) and N. Vic.

p. 126. Kultarr. *Antechinomys laniger*. Recorded in Vic. only from far NW., in 1857; not now in the state. (Wakefield 1966).

p. 136. Swamp-rat, *Rattus lutreolus*. As well as "coastal Vic.", the species extends up to 100 miles inland (Grampians, west of Maryborough, north of Nooice).

p. 140. Water-rat, *Hydromys chrysogaster*. Also frequents fresh-water lakes and salt-water estuaries and lakes in Vic.

p. 142. White-footed Rabbit-rat, *Conilurus albipes*. Original range included Vic. (as indicated on p. 202), but not now in state and probably quite extinct.

p. 148-52. Broad-toothed Rat, *Mastacomys fuscus*. Besides occurring at comparatively low elevations (Otway Ranges, Dandenong Ranges, W. Gippsland), the species is found near Mt. St. Bernard (NE. Vic.) at 5,000 ft.

p. 154. Pebble-mound Mouse. *Pseudomys hermannsburgensis*. Not now extant in or near Vic.

New Holland Mouse. *Pseudomys novaehollandiae*. Discovered recently at Tyabb, central Vic. (Seebeck and Beste 1970).

p. 166. Eastern Horseshoe Bat, *Rhinolophus megaphyllus*. In Vic., restricted to Buchan-Nowa Nowa area, E. Gippsland.

p. 168. Little Flat Bat. *Tadarida planiceps*. In Vic., restricted to west (Wakefield 1966).

p. 176. Tasmanian Pipistrelle. *Pipistrellus tasmaniensis*. Not recorded living in Vic. There are subfossil specimens from E. Vic. (Wakefield 1967).

p. 177. Large-footed Bat, *Myotis adversus*. In Vic., two areas only Warrandyte (central) and Buchan (E. Vic.).

Little Broad-nosed Bat. *Nycticeius greyi*. For Vic., western only (Wakefield 1966).

p. 178. Red Fruit-bat, *Pteropus scapulatus*. Not resident in Vic.; the two state records represent sporadic visits.

Grey-headed Fruit bat, *Pteropus poliocephalus*. Not resident in Vic., but regular summer visitor to E. and rare visitor to central areas.

February, 1971

p. 186. Dingo, *Canis familiaris dingo*. Extant in north-central, NE and E Vic.; absent now from northern and western regions of state.

p. 190. Australian Sea-lion. *Neophoca cinerea*. Sporadic visitor to W. Vic. seas and Lady Julia Percy Is.

New Zealand Fur Seal. *Arctocephalus forsteri*. Sporadic visitor to Vic. seas.

p. 203. Smoky Mouse. *Pseudomys fumeus*. Only one colony is known in Grampians. The report of "two localities" was due to an error in museum records.

REFERENCES

Audas, J. W. (1925): *One of Nature's Wonderlands. The Victorian Grampians*. Ramsay Publishing Pty. Ltd., Melbourne.

Brazenor, C. W. (1950): *The Mammals of Victoria*. National Museum of Victoria, Melbourne.

Finlayson, H. H. (1927): Observations on the South Australian species of the subgenus "Wallabia". *Trans. R. Soc. S. Aust.*, **51**: 363-77.

Seebeck, J. H., and Beste, H. J., (1970): First Record of the New Holland Mouse (*Pseudomys novaehollandiae* (Waterhouse, 1843)) in Victoria. *Vict. Nat.*, **87** (10): 280-87.

Troughton, E. LeG. (1941): *Furred Animals of Australia*. Angus & Robertson, Sydney.

Wakefield, N A. (1963). Notes on Rock-wallabies. *Vict. Nat.* **77** (11): 322-32.

——— (1966): Mammals Recorded for the Mallee, Victoria. *Proc. Roy. Soc. Vict.*, **79** (2): 627-36.

——— (1967): Mammal Bones in the Buchan District. *Vict. Nat.*, **84** (7): 211-14.

———, and Warneke, R. M (1963) Some Revision in *Antechinus* (Marsupialia) — I. *Vict. Nat.*, **80** (7): 194-219.

Wilkinson, E. H. (1961): The Rediscovery of Leadbeater's Possum, *Gymnobelideus leadbeateri* McCoy. *Vict. Nat.*, **78** (4): 97-102.

Field Naturalists Club of Victoria

Due to unforeseen circumstances, the report of the January General Meeting could not be included in this issue, but will appear in the March *Victorian Naturalist*.

Marine Biology and Entomology Group

7 December, 1970

The meeting was chaired by Mr. R. Condron, fifteen members being in attendance.

Mr. McInnes spoke of the photos taken by Mr. P. Kelly of the proposed new site for the National Museum, and said that he had forwarded it to Mr. Messer, Environment writer for the Age newspaper.

Exhibits:

Mr J. Selford—some pond life from the Little Desert area which included an unidentified aquatic larva.

Mrs McInnes—a small species of snail, and a species of spider both as yet unidentified.

Master Stephen Condron showed a live lizard. Mr. R. Condron gave a short talk on its anatomy, explaining that it was in process of losing its legs.

Mr. R. Condron — a small case of butterflies and other insects collected on a recent trip to the Little Desert. Included in the Lepidoptera were pupae of a species of the Lycaenidae family of butterflies, *Ogyris amaryllis*, which he bred out. Also he showed beetles from Salt Lake, Dimboola, a cicada, and some species of bugs.

Mr. P. Kelly gave a short talk on a species of beetle, family Chrysomelidae Genus—*Paropsis*. This species has a very short life cycle, in all two weeks.

Mr. K. Strong, under a stereoscopic microscope, showed a species of wasp which emerged from a gall taken from the Black Wattle. Its body length was 3½ mm; also a flower spider taken on the club outing to Toolebewong in the Healesville area.

Mr. J. Strong—a moth, family Anchelidae — species *Anthela ocellata* taken at Murrumbeena, Victoria.

Mr. D. McInnes, under his binocular microscope, showed a species of noise and smells, but not the aromatic protozoan *Loxodes*. He gave a short talk on this.

Magnificent stand of White Mountain Ash, *Eucalyptus regnans*, in the Marysville State Forest

FORESTS COMMISSION VICTORIA

. . . . . preserving the beauty of our forests for your enjoyment.

Field Naturalists Club of Victoria

Established 1880

OBJECTS: To stimulate interest in natural history and to preserve and protect Australian fauna and flora.

Patron:
His Excellency Major-General Sir ROHAN DELACOMBE, K.B.E., C.B., D.S.O.

Key Office-Bearers, 1970/71

President:
Mr. T. SAULT

Vice-Presidents: Mr. J. H. WILLIS; Mr. P. CURLIS

Hon. Secretary: Mr. D. LEE, 15 Springvale Road, Springvale (546 7724).

Hon. Treasurer & Subscription Secretary: Mr. D. E. McINNES, 129 Waverley Road, East Malvern 3145

Hon. Editor: Mr. G. M. WARD, 54 St. James Road, Heidelberg 3084.

Hon. Librarian: Mr. P. KELLY, c/o National Herbarium, The Domain, South Yarra 3141.

Hon. Excursion Secretary: Miss M. ALLENDER, 19 Hawthorn Avenue, Caulfield 3161. (52 2749).

Magazine Sales Officer: Mr. B. FUHRER, 25 Sunhill Av., North Ringwood, 3134

Group Secretaries:

Botany: Mr. J. A. BAINES, 45 Eastgate Street, Oakleigh 3166 (57 6206).
Microscopical: Mr. M. H. MEYER, 36 Milroy Street, East Brighton (96 3268)
Mammal Survey: Mr. P. HOMAN, 40 Howard Street, Reservoir 3073.
Entomology and Marine Biology: Mr. J. W. H. STRONG, Flat 11, "Palm Court", 1160 Dandenong Rd., Murrumbeena 3163 (56 2271).
Geology: Mr. T. SAULT.

MEMBERSHIP

Membership of the F.N.C.V. is open to any person interested in natural history. The *Victorian Naturalist* is distributed free to all members, the club's reference and lending library is available, and other activities are indicated in reports set out in the several preceding pages of this magazine.

Rates of Subscriptions for 1971

Ordinary Members	$7.00
Country Members	$5.00
Joint Members	$2.00
Junior Members	$2.00
Junior Members receiving Vict. Nat.	$4.00
Subscribers to Vict. Nat.	$5.00
Affiliated Societies	$7.00
Life Membership (reducing after 20 years)	$140.00

The cost of individual copies of the Vict. Nat. will be 45 cents.

All subscriptions should be made payable to the Field Naturalists Club of Victoria, and posted to the Subscription Secretary.

F.N.C.V. DIARY OF COMING EVENTS

GENERAL MEETINGS

Wednesday, 10 March — Annual Meeting at National Herbarium, The Domain, South Yarra, at 8 p.m.
1. Minutes, Reports, Announcements.   2. Nature Notes and Exhibits.
3. Subject for the evening — "Beetles": Mr. P. Kelly.
4. New Members.
   Ordinary:
   Miss Mary Griffith, 15 Shoobra Road, Elsternwick, 3185.
   Mr P. T. Lewis, Flat 25, 34-50 Neil St., Carlton, 3053.
   Mrs Gwynneth I. Holmes, 75 Centre Road, Brighton East, 3187.
   Mr. Ian R. McLeod, 83 Railway Crescent, Williamstown, 3016.
   Mrs H. Davey, 7 Bolton St., Box Hill, 3128.
   Mrs. D. C. Long, 68 Abbeyate St., Oakleigh, 3166.
   Joint:
   Mr. and Mrs. D. M. Barham, Flat 2, 94 Gillies St., Fairfield, 3078.
   Country:
   Mr. Allan M. Hurse, 18 Railway St., Seymour, 3660
   Mr. A. T. Marlio, 1 Harbour Drive, Sebastopol, 3356
   Junior:
   Mr. Leslie Cross, 2A Maud St., Nth Balwyn, 3104

5. General Business.   6. Correspondence.

Wednesday, 14 April — "Wilson's Promontory": Mr. J. Ros Garnet.

F.N.C.V. EXCURSIONS

Sunday, 21 March — Brisbane Ranges. Subject: Entomology. Leader: Mr. P. Kelly. The coach will leave Batman Avenue at 9.30 a.m. Fare $1.60. Bring two meals.

Friday, 9 April to Monday, 12 April (Easter) — Hamilton, led by members of the Hamilton F.N.C. A coach has been chartered for this excursion and accommodation booked at $7.75 per day for dinner, bed and breakfast. The coach will leave Flinders St. from outside the Gas and Fuel Corp. at 8.45 a.m. Bring a picnic lunch. Fare $12.00 including day trips, to be paid to the Excursion Secretary when booking, cheques to be made out to Excursion Trust.

Saturday, 21 August to Sunday, 5 September — Flinders Ranges. The coach will leave Melbourne Saturday morning and travel via Bordertown, Adelaide, Quorn, Wilpena (2 nights), Arkaroola (4 nights), Wilpena, Port Augusta (3 nights), Renmark, Swan Hill, Melbourne. Most of the accommodation will be on a dinner, bed and breakfast basis but at Arkaroola there will be bunks in huts, bed only, but meals are available in the dining or food can be obtained at store, with a certain amount of motel accommodation as well. Cost will depend upon numbers going but should be in the vicinity of $150 with bunks at Arkaroola of $195 with motel accommodation. Deposit of $50.00 to be paid when booking, cheques being made out to Excursion Trust.

Easter Campout to Mt. Eccles and Glenelg — Anyone desiring further information please contact Mr. D. McInnes or Mr. B. Cooper.

GROUP MEETINGS

(8 p.m. at National Herbarium unless otherwise stated)

Thursday, 11 March — Botany Group — "Uncommon Native Plants" by Mr. H. Alan Morrison.
Wednesday, 17 March — Microscopical Group.
Friday, 26 March — Junior meeting at 8 p.m. at Hawthorn Town Hall.
Thursday, 1 April — Mammal Survey Group Meeting in Research Institute buildings, cnr Brown St. and Stradbroke Ave., Heidelberg, at 8 p.m
Friday, 2 April — Junior meeting at 8 p.m. at Rechabite Hall, 281 High Street, Preston.
Monday, 5 April — Entomology and Marine Biology Group meeting at 8 p.m. at National Museum in small room next to theatrette.
Wednesday, 7 April — Geology Group.
Thursday, 8 April — Botany Group — Mr. F. Woodman will speak on "Some aspects of Natural History in South Africa".

The Victorian Naturalist

Editor: G. M. Ward

Assistant Editor: E. King

Vol. 88, No. 3   3 March, 1971

CONTENTS

Articles:

    Fossil Podocarp Root from Telangatuk East, Victoria.
        By G. Blackburn   56

    The Grinding Rocks at Stratford.   By Aldo Massola   58

    The Mammals of the Brisbane Ranges.   By J. W. F. Hampton   62

    Northern Strzelecki Heathlands.   By Jean Galbraith   71

Feature:

    Readers' Nature Notes and Queries   61

Book Review:

    Birds of Victoria 2 (The Ranges)   73

Field Naturalists Club of Victoria:

    General and Group Meeting Reports   74
    Diary of Coming Events   54

Front Cover:

Last month, the Echidna was featured on the cover. This month it is featured again to show variation in colour and density of hair between spines. This photograph is of a Central Australian form. Last month's was of a Victorian form.   Photo: John Wallis.

Fossil Podocarp Root from Telangatuk East, Victoria

by G. Blackburn[*]

During percussion drilling for water in the Telangatuk East district, pieces of wood were found in sandy spoil brought up from a depth of 65 ft (20 m). They were retained by the landholder, Mr. F. W. Dunstan, who told me of this in May 1963 when I was at his farm; but he was unable to show me the specimens. My first response was to mention the possibility of establishing their age by radiocarbon dating. Weeks later, one specimen several centimetres long, approximately 1 sq. cm. in cross-section and woody, was brought to me in Adelaide. Larger pieces had accidentally been broken up after discovery — the fragments were not sent.

There were then two considerations apart from expense, to discourage an attempt at radiocarbon dating. Firstly, the specimen sent to me weighed less than 10 gm, the minimum quantity of wood then required for dating. (There was a possibility of acquiring the other fragments and I wrote to Mr. Dunstan to warn him against discarding them).

Secondly, the wood was collected from sediments which possibly were as old as Pliocene, judging by the latest geological information then available, (Victorian Resources Survey for the Wimmera Region (1961)). Considering this indication and the depth of the specimens, radiocarbon dating was considered unlikely to establish an age within the limit—for this method—of approximately 35,000 years. It also appeared that if dating was undertaken, the largest remaining specimen might be needed for the requisite combustion, along with other pieces of the material.

Due to these circumstances, an alternative was to find out if an examination of the specimen would indicate the plant species it represented. The wood was accepted for examination by the Wood and Fibre Structure Section of the CSIRO Division of Forest Products, Melbourne. The report on the sample (H. D. Ingle, personal communication 1963) stated that it was coniferous, belonged to the Podocarpaceae, could not be matched with any other Australian species, came from root rather than stem wood, and was regarded as reminiscent of *Podocarpus amara* from Queensland.

The failure to match the specimen with another Australian species indicates that it is fossil material. The possibility of the root being part of a tree grown on the present surface requires that there was a rooting depth far greater than usual. A maximum depth of 5 to 6 metres is indicated for roots of *Pinus* growing in deep sand in South Australia (J. W. Holmes, personal communication). the maximum depth reported by Kramer and Kozlowski (1960) for tree root penetration is "over 30 feet" (9 m). The podocarp specimen therefore appears to represent a tree grown on soil now buried.

[*]CSIRO Division of Soils, Glen Osmond, S.A.

Fossil podocarps occur in Tertiary brown coals of Victoria (Cookson and Pike 1953) and in buried soil near Hamilton, as recorded by Gill (1964). He found stumps of *Phyllocladus*, identified by H. D. Ingle, in soil developed on Lower Pliocene marine rocks and sealed off by basalt. The Telangatuk site (Hamilton military map 1:250,000 (SJ 54-7) ref. 499436) is approximately 50 miles (80 km) north of the site near Hamilton, but it shows no evidence of basalt flows or of marine fossils in its sandy sediments, and no buried soil has been recorded there. The sedimentary rocks at the Telangatuk site were mapped, provisionally in 1960, by Spencer-Jones (1965) as "laterites on Palaeozoic rocks, sometimes including gravel and sand", and were allotted to Pliocene-Miocene (?) age. The geological survey map of Victoria (1:1,000,000, 1963) shows Pliocene sedimentary rocks at the locality, but does not distinguish non-marine from marine sediments. The podocarp specimen is terrestrial in origin and appears to be the sole fossil recorded for the Cainozoic deposits near Telangatuk. In view of the geological information, it is not older than Pliocene.

The Telangatuk site lies in a region containing widespread evidence of north-south-trending ridges consisting of sandy sediments which near the surface occur as ferruginous sandstone. Features of these ridges suggest that they were formed at coast lines (Blackburn, Bond, and Clarke 1967). The pattern of ridges in the Telangatuk East district is not distinctive—there are gaps associated with valleys of the Glenelg River system—and the podocarp specimen was not embedded in an unmistakable portion of a ridge. The specimen therefore fails to provide evidence on the origin of the ridges—one of the main reasons for interest in it—but knowledge of its discovery and botanical features may be useful to others with opportunities for collecting in the region. The podocarp root is one of the terrestrial fossils found in western Victoria in materials which generally show no organic remains—others include a fossil tortoise and a feather found in lateritic ironstone from Carapook and Redruth (Gill 1965).

REFERENCES

Blackburn, G., Bond, R. D., and Clarke, A. R. P. (1967). Soil development in relation to stranded beach ridges in County Iowan, Victoria. CSIRO Aust. Soil Publ. No. 24.

Cookson, I. C., and Pike, K. M. (1953). The Tertiary occurrence and distribution of *Podocarpus* (section Dacrycarpus) in Australia and Tasmania. Aust. J. Bot. 1: 71-82.

Gill, E. D. (1964). Rocks contiguous with the basaltic cuirass of western Victoria. Proc. Roy. Soc. Vic. 77: 331-355.

Gill, E. D. (1965) Palaeontology of Victoria. pp. 6-24 in Victorian Year Book No. 79.

Kramer, P. J., and Kozlowski, T. T. (1960). "Physiology of Trees" (McGraw-Hill: N.Y.)

Spencer-Jones, D. (1965). The geology and structure of the Grampians area, western Victoria. Mem. Geol. Surv. Vic No. 25.

CHANGE OF MEETING NIGHTS

The General Meeting nights for the months of April and June, 1971 have been changed from the Monday to Wednesday.

The Grinding Rocks at Stratford

by Aldo Massola*

This paper reports the discovery of rocks bearing grinding grooves at the Knob Reserve, which is located on a prominent bend of the Avon River, to the south-east of and close to Stratford, Gippsland.

On its way to Lake Wellington, the Avon River meanders through the Great Plain slowly cutting downwards into the alluvial filling of the plain. Its erratic course caused both by the presence of patches of more resistant terrain and by the scouring action of the strong current during floods; a process most active on the outer banks of the bends. This shifting of the stream's bed gives rise to alluvial flats and swampy terraces; the former generally growing a good crop of grass which in the past attracted kangaroos and other herbivores, and the latter soon becoming the home of a variety of amphibious animals and birds. Both conditions were taken advantage of by the Aborigines. To them a stream of this nature becomes a highway, since, besides supplying food and water, it is also a well defined path.

In a former paper, and in my book *Journey to Aboriginal Victoria*, I have stated that one of the reasons for the Aborigines of the lake's country leaving their comfortable camps by the lake-side and braving the hidden terrors and hidden dangers of the inland was the need of raw materials of a kind not obtainable at their usual haunts. In this case this commodity was the ready-made axe-shaped pebbles lying in their thousands on the banks of the upper reaches of the Avon River, only wanting to have one edge ground to be ready for use—or barter.

Barter was most important in the sociology of the tribes, since it enabled contact to be maintained with distant groups; and it was also the means by which new ideas and techniques were circulated between one tribe and the next. For this reason certain raw materials or manufactures were always obtained by barter, even if they could be obtained locally with much less trouble.

It is almost certain that a local group inhabited the country about Stratford. It will be recalled that soon after Angus McMillan, the explorer, formed the Nuntin Station in the early part of 1840, it had to be abandoned because of fierce and continual attacks by the Aborigines. It was not until December of that year that he was able to re-occupy the station, and even then the attacks upon it did not cease for some months. This points to the Aborigines having permanently occupied this country, and not just to a group being there on walk-about.

Also, the Moravian Missionary, Rev. F. A. Hagenauer's first choice of site for his Mission Station was near Maffra, which means that the Aborigines must have been numerous there; and although he was not able to establish himself at that locality because of opposition from local squatters, his second choice, which became Ramahyuck Aboriginal

*4 IR Wolseley Str., Mont Albert Vic. 3127

Station, was still on the Avon, although close to its mouth. The official reason given for this second choice was that "the Aborigines were in the habit of congregating there".

All this points to a fairly large Aboriginal population, not only on the shores of Lake Victoria, but also along the Avon River and its tributaries.

In a former paper I have attempted to reconstruct the scene in the Munro District, but the same activity could have taken place anywhere along the upper Avon before the coming of the white men. Groups of Aborigines, having collected suitable pebbles from the river bank, dispersed through the comparatively open forest of red gum, stringy bark and red box. There were large mobs of kangaroos and many emus about and this meant plenty of food for them and for the visiting groups who would soon come up river to barter for axes; and away from the river surface-water was available everywhere. Scattered over this land of plenty were many smallish ferruginous sandstone outcrops, the more fine-grained of which made perfect whetstones for grinding the river pebbles into axes.

The grinding process however, leaves an impression or groove, roughly the width of the edge of the pebble being ground upon the rock, and by this means we are now able to tell which rock was so used.

To the south-east of and quite close to Stratford the current of the Avon River met a resistant obstacle in a rocky hillock, which caused the river to go around it, thus forming a prominent bend. This hillock, although in itself not much higher than 300 feet, is never-the-less high enough above the Great Plain to be a very good observation point. This is the Knob Lookout, which the Elders of Stratford, in their wisdom, have long since declared a Reserve.

Plate 1

Showing some of the grooves discovered at Stratford

Photo: Author

From it there is an uninterrupted view over the surrounding plains; to the west Ben Cruachan, 22 miles; and Mt. Erica 44 miles away, are plainly seen with the naked eye. Mt. Wellington, 35 miles away, dominates the north-west horizon.

Naturally, because of its elevation and nearness to water the Knob must have been an Aboriginal camp-site, but the Reserve has become a picnic ground, visited annually by perhaps thousands of people. Over the years the already flattish top of the Knob was bulldozed, a gravel road, lately asphalted over, was laid to give access to it, and a wooden railing to prevent cars from falling over the steep edges into the river below was erected. Signs of an Aboriginal camp cannot remain under these conditions.

It was to this spot, that a few weeks ago, Mr. and Mrs. Campbell Fletcher, of Munro, (to whom we are already indebted for their discovery of the Munro grinding rock site) took a visitor to show him the points of interest in the district. Leaning against the rail while talking to his friend, Mr. Fletcher's eyes wandered upon the rocky outcrop marking the edges of the Knob's top. What he saw were grinding grooves.

I should have explained that the Knob is edged by sandstone outcrops, which lay almost buried around its north, west and south sides, consolidating its edges. Grinding grooves appear to be well scattered upon these outcrops, but have a larger concentration on the south side, where the sandstone is finer grained. At least 18 grooves were counted, most of which were completely buried by the soil and had to be dug out. The majority of the grooves measure either 6" long by 2" wide or 8" long by 3" wide.

The depth is fairly uniform at between $\frac{1}{4}$" and 1". The grinding grooves are in such a prominent position of those rocks that it is a wonder they have not been noticed before.

Grinding rock sites are rare in Victoria, and nowhere else do we meet with the wide flung distribution found on the Avon River watershed. The whole district was obviously an axe factory, and aside from pinpointing a new site, the importance of Mr. Fletcher's discovery lay in its establishing this fact. Eleven sites have been reported in the vicinity of Munro, and the present discovery connects them with the Boisdale site, which is of a similar nature and has the same geological history. Boisdale site is about 8 miles along the river from Stratford, and it is about 8 miles from Stratford to the Munro sites. There is every possibility that intervening sites between these three points will be found. Local Field Naturalist and Historical Societies should, in fact, organise field days with this object in view.

Once again I wish to thank Mr. and Mrs. Campbell Fletcher for their companionship and hospitality, and for making it possible for me to record these new additions to the list of Aboriginal antiquities in Victoria

REFERENCES
Massola Aldo, The Grinding Rocks at Boisdale. *Vict. Nat.* 82, 1965.
———— The Grinding Rocks at Munro *Vict. Nat.* 84, 1967.
———— Letter to the Editor *Vict. Nat.* 86, 1969.
———— More Grinding Rocks at Munro. *Vict. Nat.* 86, 1969.
———— Journey to Aboriginal Victoria. Rigby. Adelaide 1969.
———— Aboriginal Mission Stations in Victoria. Hawthorn Press Melbourne, 1970.
West, Alan L.: Axe-grinding Rock near Munro, Central Gippsland. Victoria, *Vict. Nat.* 86, 1969.

Readers' Nature Notes and Queries

Maternal Magpies?

This interesting note comes from Mrs. Hilary Reid, of Somers in Victoria.

A fortnight ago a young man I know well was driving to work along a country road. He saw a young magpie in trouble in the grass. Its parents were watching from a pinetree where he concluded there must be a nest. He stopped and took it home for his children who were not yet about. He told his wife to keep an eye on it and hurried away. Very soon after, two adult magpies appeared, and perching on a shed and fence, considered the situation. Through the window of the kitchen the following scene was quite visible.

One bird flew down. It pecked and pecked at the back of the baby's neck as though attacking it. The other bird watched until the first one got a satisfactory hold on the scruff of the neck, trying to lift it into the air. Of course the squawking little one shot out its wings, whereupon the other parent seized the nearest wing and they lifted off the ground and carried it away. The nest was about a mile away by air. I wish I had seen it.

Have any other "Field Nats" had such luck?

Anteater's Diet?

Ellen Lyndon of Leongatha in Victoria, supplied these two nature notes.

In a patch of low scrub by the roadside near Kongwak, in South Gippsland, last autumn, we disturbed an Echidna, a big one. It had been poking in loose soil under a dark object in the undergrowth, a furry patch that presently resolved itself into the carcase of a very defunct domestic cat, lying half out of the carton in which it had been hurled into the bush. The Anteater's rootings had scattered big white blowfly larvae all around. Although it did not dig in it ceased operations and did not resume them during the time we were watching, so we did not actually see it pick up anything. The question is, did Tachyglossus mistake those white grubs for termites or does he enjoy a little variation in his diet when the opportunity offers?

Bird Behavior

Walking home from the shops one bright sunny afternoon recently, a faint cheeping from the sky made me look up to see two white birds resembling seagulls coming in from the south, a little higher than the power lines. The leader appeared to be nearly exhausted and somewhat battered, with feathers missing here and there. It called regularly and weakly. As they passed over I saw they were a pair of Black-shouldered Kites.

They proceeded over the town, the rearmost bird diving often at the leading one which managed to evade it. Finally they grappled and fell, locked together, and rolled over and over out of the sky just as if they had been shot down, and disappeared from view among the rooftops and shrubbery.

Field Naturalists Club of Victoria
ANNUAL MEETING
The Annual Meeting will be held in the National Herbarium on
WEDNESDAY, 10 MARCH, 1971

Mammal Survey Group Contributions—VI.

The Mammals of the Brisbane Ranges

by

J. W. F. HAMPTON*

INTRODUCTION

The Brisbane Ranges, which are the broken and eroded edge of a plateau of Ordovician rocks, clearly mark the western boundary of the plain which extends eastwards to the coast of Port Phillip Bay in Victoria. At this boundary the hills rise sharply, in places to 1200 ft., then gradually merge westwards with the tablelands of the Meredith district. This range is bordered in the south and west by the Moorabool River and in the north by the Parwan Creek.

As little seemed to be known of the mammal fauna of these hills the Mammal Survey Group carried out three sample surveys during 1968 and 1969, the results of which are presented in this paper. The names of the members of the Group whose work provided the data upon which this report is based are given in the author's acknowledgements.

DESCRIPTION OF THE AREA

History

The history of European penetration into these hills is essentially the history of the gold mining that went on from 1853 to about 1900. This activity was centred on Steiglitz (formerly Stieglitz, which accounts for its present pronunciation) where the population exceeded 1000 in 1856 when there were six schools, four hotels and four churches in use. It again reached this level in 1894 when over forty reefs were being worked. (Anon. 1951; McKinlay, 1969). While very few signs of such a town remain at the sight of Steiglitz, the surrounding bush conceals the remains of extensive mining activity. The dependence of such a population upon wood for fuel and power, mine-timbering and building for over forty years, and the prospecting of almost every yard of the district in the search for new reefs, must have had a devastating effect on the natural habitat.

Unfortunately no records have been found which describe the natural surroundings or the animal life at the time of the first discoveries of gold on the property belonging to Charles von Stieglitz. The lack of agricultural development, in marked contrast to that in the adjoining areas along the eastern boundary of the Ranges, around Anakie and Staughton Vale, indicates the poor value of the soils. The results of the surveys presented in this paper tend to confirm this. Part of the area is used as a water catchment and, while there are some fenced properties, its only other uses appear to be as a picnic area, as a source of firewood and for shooting.

Rainfall

Official rainfall figures for Durdidwarrah (1928-1968 average) and for Anakie (1889-1968 average) are given in Table 1, together with the average number of days on which more than 0.01" of rain fell. No data is available for Steiglitz but it appeared to be a drier area than Durdidwarrah.

*Mammal Survey Group, F.N.C.V., c/- Secretary, 40 Howard St., Reservoir, Victoria, 3073

Vegetation

In the seventy years since the major reefs were worked out, the bush has partly recovered from the ravages of those times but comparison with the forest within the water reserve (which was created in 1875) show how far short of full recovery this has been. In comparable areas within the reserve, trees are mature and dense, often 70-80 ft. high with a well-established shrub layer. Outside the reserve the forest is largely open and immature and has suffered severely from felling and burning. Nevertheless, a considerable variety of trees is present and, in addition, the natural habitat varies from a dense to an open type of forest with considerable variation in shrub and ground cover.

Areas surveyed

Three sites were chosen for surveys which represented, in density and variety of vegetation, the average (Survey 1), the richest (Survey 2), and the poorest (Survey 3) that was found in these hills. The first survey was centred on Sutherland's Creek, north of Steiglitz, in August 1968; the second on Aeroplane Road, in February 1969 and the third at the junction of Sutherland's Creek and Mariner's Gully, south of Steiglitz, in September, 1969. These are shown in Figure 1.

Survey 1: Trapping was carried out in three distinct types of habitat. On the plateau there was an open forest of Messmate (*Eucalyptus obliqua*), Broad-leaved Peppermint (*E. dives*), Red Stringybark (*E. macrorrhyncha*) and Scent-bark (*E. aromaphloia*) in areas of varying density, in which the canopy at 20-35 ft. gave from 0 to 50% cover. The shrub layer was equally uneven and consisted mainly of *Hakea*, *Banksia* and Grass-tree (*Xanthorrhoea australis*). Ground cover was fairly dense in some swampy areas, being made up of Flat Pea (*Platylobium obtusangulum*), Guinea-flower (*Hibbertia stricta*) and grasses, but was elsewhere very thin, showing signs of the recent severe drought.

On the slopes there was an open tree cover of 20 ft.-high regrowth Messmate and Peppermint with less than 1% mature trees. The only other component present was Grass-tree which was generally sparse but also occurred in dense patches. There was virtually no ground cover (the survey notes written at the time describe ". . . a thin sheen of young grass, moss and ground plants a few mm. high, no cover for anything taller than ants").

The creek bed was devoid of a shrub layer and had a ground cover of *Poa australis* up to 12" tall, giving an 80-100% cover downstream and about a 20% cover upstream from the survey camp. There was an open canopy of eucalypts not differing from that on the slopes.

TABLE I

Official rainfall figures (in inches) for Durdidwarrah (40-year average) and Anakie (79-year average) together with the average number of days of rain.

	Jan	Feb	Mar	Apr	May	June	July	Aug	Sept	Oct	Nov	Dec	Year
Durdidwarrah	1.55	1.88	2.05	2.30	2.31	2.28	2.15	2.35	2.72	2.65	2.23	2.09	26.56
Days of rain	9	9	11	16	17	18	20	20	17	17	15	13	182
Anakie	1.34	1.71	1.83	2.09	1.98	1.92	1.89	2.15	2.43	2.48	2.04	2.05	23.91
Days of rain	7	8	8	13	14	16	17	17	15	15	12	10	152

Part of the water reserve (due west of Dundidwarrah), which was surveyed in daylight, consisted of a flat, fairly uniform, mid-dense to open forest of Swamp Gum (*E. ovata*), Manna Gum (*E. viminalis*) and Snow Gum (or White Sallee) (*E. pauciflora*) with isolated dense clumps of Red Stringybark, Messmate and Narrow-leaved Peppermint (*E. radiata*) and one or two single Wild Cherries (*Exocarpus cupressiformis*). There was virtually no shrub layer, except near a water channel, and the ground layer consisted of a close cropped cover of grass with a few small herbs. This gave a park-like appearance with long views through the trees. There were many signs of pig-rooting, though no feral pigs were seen.

Survey 2: The survey camp was situated on the main plateau which at this point carried a fairly uniform cover of mid-dense to dense forest with the canopy at 40 to 60 ft. This was dominated by Red and Brown Stringybark (*E. macrorrhyncha* and *E. baxteri*) with scattered Messmate, Red Ironbark (*E. sideroxylon*) and Long-leaved Box (*E. goniocalyx*). A mixed tussock grassland-sclerophyll shrub stratum occurred below the tree canopy, dominated by Grasstree and Silver Tussock (*Danthonia pallida*), with *Daviesia corymbosa*, *Acrotriche serrulata*, *Platylobium obtusangulum*, *Epacris impressa* and *Pultenaea gunnii*.

There were a few thickets of Silver Wattle saplings (*Acacia dealbata*) with a few mature trees. Ground cover was of small herbs and grasses, not giving a very dense cover. The vegetation of the gully slopes was complex and differed on the north and south facing slopes. In general, this was a grassy woodland to open forest co-dominated by Red Stringybark and Long-leaved Box with an understorey of *Danthonia pallida*, *Poa australis*, *Haloragis tetragyna* and *Senecio quadridentatus*, mosses and lichens. On the southern aspects this differed in that Messmate and Long-leaved Box occurred with a mixture of Red Ironbark, Brown Stringybark and Red Box (*E. polyanthemos*), covering a dense ground layer dominated by *Poa australis*. Gully floors were covered by a mid-dense to open canopy of Messmate and Long-leaved Box with occasional Manna Gum and Red Box, with a dense herbaceous sward dominated by *Poa australis* (broad leaf form) with patches of Wiregrass and some Sword-grass. A detailed account of the vegetation of this area is available in the Group's files. There was no running water within the area at the time of the survey but recent rains had filled all depressions.

Survey 3: This Survey was centred at the junction of Mariner's Gully and Sutherland's Creek (named after Robert Sutherland who camped here in 1836) in the driest part of the Brisbane Ranges where there was a very thin cover of sandy gravel on the plateau and rocky slopes leading down to the narrow creek bed and to the open bed of the gully. The area was much mined and pitted and there were a number of brick and stone house foundations. On the plateau Red Stringybark was dominant, growing to about 35 ft., with White Ironbark (*E. leucoxylon*), Red Box, Narrow-leaved Peppermint, an occasional Wild Cherry and Sheoke (*Casuarina littoralis*), together forming a mid-dense to open canopy of immature and stunted trees. The shrub layer was composed almost entirely of Grass-tree which was very dense in places but was limited to the flat top of the plateau. Ground

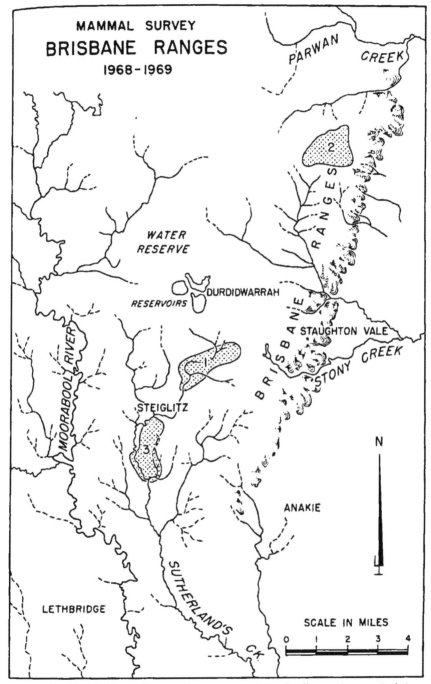

Fig. 1. The Brisbane Ranges. Areas surveyed in detail are shown stippled.

cover was extremely sparse, there being some species of heaths, lilies and orchids and some legumes. On the slopes the tree cover was similar except that some Red Ironbark occurred and the shrub layer consisted only of some patches of spindly Golden Wattle (*Acacia pycnantha*) with an occasional Gold-dust Wattle (*A. acinacea*). The ground cover consisted solely of dry forest debris, with some mosses on the lower slopes. In the creek bed the forest included an occasional Manna Gum and there was a sparse shrub layer of Black Wattle saplings (*A. mearnsii*), a few Hazel (*Pomaderris sp.*) and some patches of Bracken (*Pteridium esculentum*) with a thin ground cover of mosses and tufts of grass. Sutherland's Creek was running well at the time.

METHODS

Trapping and spotlighting methods were the same as those previously described (Hampton and Seebeck, 1970). While trapping was confined to the areas that are described in some detail above, spotlighting covered areas further afield, up to 2 miles from the survey camps, the distance travelled usually varying inversely with the number of animals encountered.

RESULTS

The total survey effort, together with the total number of animals seen and caught is summarised in Table II and a systematic list of the seven native and two introduced species that were recorded is given in Table III. An analysis of the results is given in Table IV in which the numbers of specimens recorded are given in relation to the survey effort, i.e., the numbers of animals caught per 100 trap-nights and the numbers seen per spotlight-hour. Results are presented in this way so that data obtained in different surveys may be compared.

The apparent population of native species, when compared with that found in surveys of other regions using the same methods, was low. The comparative figures are given in Table V.

NOTES ON THE SPECIES RECORDED

(Reference specimens that were retained are designated by their registration numbers in the collection of the Fisheries and Wildlife Department, Victoria).

1 Brown Antechinus, *Antechinus stuartii*.

This was the only small ground-living native species recorded. Surveys 1 and 3 yielded one specimen each and 8 were caught in Survey 2. This number represents about an average catch for this species compared with similar surveys elsewhere. The specimens caught in February were all juvenile or immature and obviously born in that

TABLE II
Survey effort, with total trapping and spotlighting results.

Survey	Date	Trap-nights	Spotlight-hours	Specimens recorded	Number of species identified
1	Aug. 1968	74	9.0	29	4
2	Feb. 1969	70	20.5	23	6
3	Sep. 1969	69	12.5	11	4
Total		213	42.0	63	8

TABLE III
Systematic list of mammals recorded in the Brisbane Ranges

Order MARSUPIALIA

Family Dasyuridae
 (1) Brown Antechinus — *Antechinus stuartii* Macleay

Family Phalangeridae
 (2) Brush-tailed Possum — *Trichosurus vulpecula* (Kerr)
 (3) Ringtail Possum — *Pseudocheirus peregrinus* (Boddaert)
 (4) Sugar Glider — *Petaurus breviceps* Waterhouse
 (5) Koala — *Phascolarctos cinereus* (Goldfuss)

Family Macropodidae
 (6) Black Wallaby — *Wallabia bicolor* (Desmarest)
 (7) Grey Kangaroo — *Macropus giganteus* Shaw

Order RODENTIA

Family Muridae
 (8) Black-rat — *Rattus rattus* (Linnaeus)

Order LAGOMORPHA

Family Leporidae
 (9) Rabbit — *Oryctolagus cuniculus* (Linnaeus)

TABLE IV
Apparent abundance of species.
(a) Numbers of animals caught per 100 trap-nights.
(b) numbers seen per spotlight-hour and
(c) animals seen in daylight.

Survey	1	2	3	Total number of animals
(a) Trapping				
Antechinus stuartii	1.35	11.4	1.45	10
Rattus rattus	—	—	2.90	2
(b) Spotlighting				
Pseudocheirus peregrinus	—	0.05	—	1
Petaurus breviceps	—	0.05	—	1
Phascolarctos cinereus	0.11	0.29	0.56	14
Wallabia bicolor	—	0.15	—	3
Oryctolagus cuniculus	0.33	0.05	0.08	5
(c) Daylight sightings				
Phascolarctos cinereus	5	2	—	7
Wallabia bicolor	—	1	—	1
Macropus giganteus	17	—	—	17
Oryctolagus cuniculus	2	—	—	2

season. The one caught in September was an adult female with young. The specimen caught in August, however, was a young female, about half-grown, its age not fitting what is believed to be the normal breeding season for this species (Woolley, 1966). Specimens were caught where leaf-litter and trash was the only ground cover; on dry slopes amongst heaths; under a mid-dense, low shrub layer of Grass-tree and Hakea; on dry, bare, rocky slopes and in damp moss and Bracken near a creek.

Specimen: D.880♀, 22 Feb, 1969.

2. Brush-tailed Possum, *Trichosurus vulpecula*.

No specimens were seen during these surveys though Brush-tail scats were found during Surveys 2 and 3. However, both live and road-killed specimens were reported by Group members near Anakie and near Steiglitz at other times during 1968 and 1969.

3. Ringtail Possum, *Pseudocheirus peregrinus*.

One specimen only was recorded, in Stringybark, during Survey 2, when a Ringtail mandible was also found.

Specimen: P.801, mandible.

4. Sugar Glider, *Petaurus breviceps*.

One specimen only was recorded in Messmate, during Survey 2. This species is generally considered to be widespread and common in Victoria, though in our experience it is seldom seen. A single sighting is therefore assumed to indicate that this species is established in this area.

5. Koala, *Phascolarctos cinereus*.

This appeared to be the most plentiful of the species recorded and was seen at all survey sites. Animals were seen at night in Long-leaved Box, Messmate, Red Stringybark and Peppermint. Within the water reserve they were seen in daylight in Manna Gum, Swamp Gum and feeding in Snow Gum. A total of 21 specimens were seen and a further 8 were heard at night.

It is not known whether Koala was present in the Brisbane Ranges prior to 1944 when large numbers were introduced by the Fisheries and Wildlife Department. In 1944-5, 155 animals from Quail Island and 258 from Phillip Island were released in the Durdidwarrah water reserve. A further 171 from French Island were released in 1957. (Records of the Fisheries and Wildlife Department, Victoria). The apparent abundance of animals at the time of these surveys showed that the species had become well established.

TABLE V

Comparative abundance of animals (native species only)

*This data is taken from Group records and includes both published and unpublished observations

	Specimens caught per 100 trap-nights	Specimens seen per spotlight hour
Brisbane Ranges	4.69	0.45
Mean, 11 other areas *	18.63	1.23

Specimen: K.190, skull, 25 Aug. 1968.

6. Black Wallaby, *Wallabia bicolor*.
Four animals were seen during Survey 2 and what were possibly wallaby were heard on a number of occasions at night. Some skeletal material was collected.
Specimens: M.2045, mandible; M.2046, femur; 22 Feb. 1969.

7. Grey Kangaroo, *Macropus giganteus*.
The water reserve has become a refuge for these animals and seems to be well suited to them. Twelve, of all ages, were seen in daylight in this reserve, two other pairs and one juvenile were seen in the area covered by Survey 2, all in daylight.
Specimens: M.1992-1998, skulls, 25 Aug. 1968.

8. Black-rat, *Rattus rattus*.
Two specimens were caught on very bare ground along Mariner's Gully in Survey 3. Ground cover was such that both could be seen in the traps from a distance of about 50 yards. There were extensive remains of mining activity and foundations of old huts in this particular area and it is possible that this species has survived here since its probable introduction during the gold mining days.
Specimens: R.40419, R.40433; 20 Sept. 1969.

9. Rabbit, *Oryctolagus cuniculus*.
A few rabbits were seen, 5 in Survey 1 and one each in Surveys 2 and 3, so that even this species was not common.

DISCUSSION

These sample surveys detected the presence of relatively few native species in the Brisbane Ranges and, insofar as our survey methods yield comparable quantitative data, the numbers of animals recorded was very low compared with similar surveys in other districts. While specific reasons for this could not readily be identified, the lack of variety in the shrub layer and the general lack of physical cover in the vegetation, especially at ground level, appeared to be the main reasons. Whether this was due to the poor soils or to exploitation, or both, is not known. The use of a large part of the area for uncontrolled, illegal shooting, must also have had an effect on the population of Black Wallaby, Grey Kangaroo and possibly Koala, for which the area should provide adequate support. The flourishing population of the latter two species within the relative security of the water reserve showed what the whole area might support under similar protection.

The presence of other species, notably Echidna, Tuan, Pigmy Possum and Water Rat was reported to us by the Geelong Field Naturalists Club (Pescott, T. personal communication) but, except that what were thought to be signs of Echidna were seen, we were unable to confirm the presence of the others in these surveys.

Many years ago (Mattingley, 1911) *Antechinus flavipes* was caught "near Parwan" but neither this species nor *A. swainsonii* were recorded in this series of surveys.

One interesting observation was the apparent complete absence of the Southern Bush-rat (*Rattus fuscipes*). In our experience this is the most common native mammal in Victoria and in only five out of 40 individual surveys carried out in other districts have no Bush-rats been caught. In three of these the Black-rat (*Rattus rattus*) was present.

In seven different regions similarly surveyed by the Group (Seebeck, Frankenberg and Hampton, 1968; Fryer and Temby, 1969; Hampton and Seebeck, 1970 and unpublished data) *R. fuscipes* has occupied 12.83% of all traps set and has accounted for 75.5% of all native animals caught. In the Brisbane Ranges the Black-rat was found only in one, very inhospitable area and as there were no indications of severe predation, (no Foxes and only one Boobook and one Tawny Frogmouth were recorded) it is supposed that some key ecological factor has prevented colonisation by *R. fuscipes*. This might be the absence of an essential nutrient, the presence of toxic plants or merely the lack of sufficient damp and well-covered habitat.

Thus, in general, the Brisbane Ranges appear to be more hospitable to kangaroo, wallaby and Koala than they are to most other species of native mammal.

Acknowledgments

The following were the members of the Mammal Survey Group whose work provided the data upon which this paper is based:

R. Dale, G. Douglas, R. Fryer, D. Hackett, J. Hampton, P. Homan, A. Howard, C. Hutchinson, A. Lewis, D. Munro, D. Poison, D. Reeves, A. Simon, M. Taylor, I. Temby, and J. Wolfenden.

From the Geelong Field Naturalists Club, G. Carr, G. Gayner and T. Pescott also assisted with these surveys and considerable help with analysis of the vegetation was given by N. Scarlett of the School of Botany of the University of Melbourne.

The Group wishes to thank the Geelong Sewerage and Water Trust for permission to enter the water reserve. Equipment used in these surveys was purchased from a grant made by the M. A. Ingram Trust. Native mammals were handled under the provisions of a permit issued by the Fisheries and Wildlife Department of Victoria and our thanks are expressed to the staff of that department for their assistance and co-operation in this work. Liki Muceniekas drew the map and J. K. Dempster, J. H. Seebeck and D. Munro read the manuscript.

REFERENCES

Anon. (1951) "The History of Steiglitz, 1835-1951", The St. Arnaud "Mercury", N.D.

Fryer, R. and Temby, I. (1969), "A Mammal Survey of Stockman's Reward, *Vict. Nat.* 86, 48.

Hampton, J. W. F. and Seebeck, J. K. (1970), "Mammals of the Riddell District" *Vict. Nat.* 87, 192-204.

Mattingley, A. H. E. (1911) Nor. *Vict. Nat.* 28, 49-50.

McKinlay, B. J (1969). "Steiglitz — Geelong's Eldorado", The Geelong "Advertiser", 27/12/1969, p. 19.

Seebeck, J. H., Frankenberg, J. and Hampton, J W. F. (1968), "The Mammal Fauna of Darlimurla". *Vict. Nat.* 85, 184-193.

Woolley, P. (1966), "Reproduction in *Antechinus* spp. and other Dasyurid marsupials", Sympos. Zool. Soc. Lon. No 15. pp. 281-294.

Flowers and Plants of Victoria in Colour

Copies of this excellent book are still available, and of course would make a wonderful gift. They are obtainable from the F.N.C.V. Treasurer, Mr. D. McInnes

Northern Strzelecki Heathlands

by JEAN GALBRAITH

Between Longford and Willung the heathlands of the Gippsland Lakes meet the first low thrust of the eastern foothills of the Strzelecki Ranges. The heathlands are divided like a stream by the increasing height of the ridge. Westward from that point it becomes two streams, one flowing south along the coast, and one flowing along the base of the northern foothills until it fades out in the plains west of Traralgon.

What we might call the southern stream is fairly well known, although I know it only here and there. The northern stream which we in the Latrobe Valley know well, seems to be little known to others, although none of our heathlands is more colourful or interesting. Only fragments remain unspoiled, and although there are four small wildflower sanctuaries there is no national park or other large reserve to protect a sample of what is there. Thousands of acres of pines have blotted out the wildflowers except in broken patches and one large area near Chesnum Road (Rosedale South), the best part of which is also threatened. The Latrobe Valley F.N.C. is now hoping to obtain a large reserve in what remains. Most, possibly all, the species known in the Rosedale South heathlands, grow there, although not in such colourful abundance.

This northern stream is bounded on the north by the red gum plains of the Latrobe Valley, and on the south it washes up the lower foothills of the Strzeleckis. Without giving a complete record of all plants, the following notes give some idea of the flora.

The commonest eucalypts are *E. nitida* (east of Rosedale South only, so far as I know), *E. radiata, E. viminalis, E. dives, E. obliqua, E. globoidea, E. bridgesiana, E. consideniana* and *E. cephalocarpa* (mainly to the east), while the commonest Acacias are *A. oxycedrus, A. diffusa, A. verticillata, A. botrycephala, A. mucronata,* with *A. mitchellii* at Providence Ponds.

Leptospermum myrsinoides forms a lower storey over most of these heathlands, with patches of *L. phylicoides* and *L. juniperinum.* Through these is a bright interweaving of *Leucopogon ericoides, L. virgatus, Epacris impressa, Acrotriche serrulata, A. prostrata, Tetratheca pilosa* and *Hibbertia stricta,* all abundant and widespread. There are large, often pure, stands of *Banksia serrata,* and *B. marginata* grows over the whole area, Wedding Bush (*Ricinocarpos pinifolius*) and Red Correa (*C. reflexa*) are also widespread though commoner in some parts than others.

There are patches of Dotted Heathmyrtle (*Thriptomene micrantha*) as far west as Dutson Downs although it is most abundant on Sperm Whale Head (Lakes National Park). The flora of the Lakes Park has been fully recorded by Mr. Fred Barton, for years the park ranger, and almost all heathland species known there extend westward through Loch Sport at least as far as the Dutson Downs area, where however *Boronia ane-*

March, 1971

monifolia, *Leucopogon ericoides*, and often *Dampiera stricta* are much more abundant and colourful than I have ever seen them in the Lakes Park, while *Bossiaea obcordata* and *Astroloma pinifolium* do not seem to occur west of Loch Sport and Providence Ponds. The dwarf pink-flowered form of *Calytrix tetragona* with downy grey foliage does not appear to grow west of Dotson Downs, but is replaced near Rosedale South by the taller pale flowered form with smooth, bright green leaves. Also at Rosedale South is the only occurrence I have seen in the heathland of a beautiful glabrous form of *Prostanthera denticulata* with long terminal leafless racemes of large wide-open purple flowers. The goldfields form of this Rough Mint-bush is hairy with leaves 1 to 1 in. long and small deep purple two-lipped flowers, but the Rosedale form is nearer to that Tasmanian form with 1 to 1 in. leaves and lighter purple 1 in. flowers, very wide open and only slightly two-lipped.

The 3 to 6 ft. bushes amongst 2 to 3 ft. bushes of bright pink *Boronia anemonifolia* and a mosaic of other flowers make the Rosedale South heathland outstandingly colourful.

Here too is, apart from a small "island" north of Moe, the most western occurrence of *Sowerbaea juncea* and it is also the western limit of *Bossiaea heterophylla*, and (in these heathlands) of *Brachyloma daphnoides* and *Calytrix*. There is a patch of the chiefly Mallee *Cryptandra tomentosa* at Rosedale South, another Mallee species, *Zieria veronicea* near Longford and *Hibbertia virgata* (Mallee and desert species) grows between Dutson and Rosedale where it is abundant. The *Zieria* is also a form characteristic of the area, with larger leaves, and smaller paler flowers than in the north, with flowers cupped or almost hooded instead of wide open.

In the North Strzelecki heathlands *Hibbertia fasciculata* also reaches its western limit near Merriman's Creek, south-west of Rosedale, but *H. stricta* and *H. acicularis* grow throughout the whole length, *Hovea heterophylla* I have noticed only at Traralgon South but it is probably widespread. *Apira ericoides* is common east of Longford, but farther west I know it only on one hill near Willung South, where there is also an isolated patch of *Leucopogon collinus*.

Correa reflexa var. *cardinalis* grows throughout these heathlands, with an especially fine form at Traralgon South in the north and Hedley in the south. *Scaevola ramosissima* and *Gompholobium latifolium* grow near Gormandale, as western "islands" of eastern species, comparable to the *Sowerbaea* north of Moe, *Grevillea chrysophaea* (var. *canescens*) grows at Sperm Whale Head, but the typical, much less hairy form is fairly common at Longford, Rosedale, and Traralgon South. *Tetratheca pilosa* is common, but *T. ciliata* and *T. ericifolia* var. *rubaeoides*, although scattered through the heaths are commoner in the forests to the north and south.

Other species abundant in this heathland strip are *Pimelea linifolia*, *Persoonia juniperina*, *Bossiaea cinerea*, *Pultenaea gunnii*, *Dillwynia sericea*, *D. glaberrima*, *Kennedya prostrata*, *Hardenbergia violacea*, *Conosperma volubile* (with *C. ericinea* and *C. calymega* less common), *Astroloma humifusum*, *Monotoca scoparia*, *Stackhousia monogyna*, *Marianthus procumbens* and dainty little *Poranthera microphylla*.

Four Hakea species, *H. sericea*, *H. ulicina*, *H. pugeoniformis* and *H. nodosa* grow in scattered patches and seven of the nine Victorian species of *Drosera* are found in one part or another. *D. arcturi*, being alpine, does not grow here, and although *D. spathulata* grows in the slightly different heaths of Yallourn North, I have not seen it in the northern Strzelecki heaths. *Platysace lanceolata* also is scattered through these heaths and *Schizea asperula*, the Rough Comb-fern grows in two places south of Rosedale, once with Bushy Club-moss (*Lycopodium deutrodendrum*), and there are many small species like *Centrolepis* and *Crassula*, as well as lilies, irids, and orchids which I have not mentioned. I have listed sixty orchid species but many more have been recorded. These will be published by others in a wider survey so are not mentioned here.

Composites are rather less common than in most places, although *Olearia ramulosa*, *Helichrysum obtusifolium*, *H. scorpioides*, and *Senecio lautus* are common and widespread (the *Olearia* mainly where heath and forest flora meet), and there is one lowland occurrence of *Helichrysum thyrsoideum* at Traralgon South.

Goodenia humilis, common near Yallourn, is uncommon here, and I have seen *G. paniculata* only near Providence Ponds. Many grassland plants, like *Stylidium graminifolium* grow also in the heaths, but not more commonly than in the surrounding country.

book review

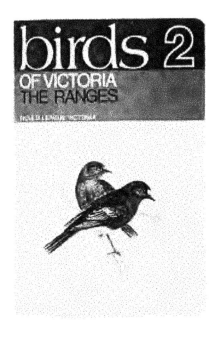

Birds of Victoria 2
(The Ranges)

Published by the Gould League.
Price: $1.50.

The second of these excellent publications is available.

To those who have already seen Book 1, anything said here would be trite: but to others, interested in bird study, this series of books should not be overlooked on any account.

In Book 2, five pages, including a colour plate of Victorian Forest Areas are given over to some valuable information on Bird-Forest association.

Margo Kröyer-Pedersen's paintings are again of a high standard, although the Grey Shrike-Thrush does perhaps lack the dark bead-like eyes of the species.

Every bird observer must have this book!

Field Naturalists Club of Victoria

Mrs. Edna King

Another tragic loss to the F.N.C.V. occurred on Tuesday, 23 February, when Mrs. Edna King, the Assistant Editor, died suddenly.

Heartfelt sympathy goes out to her husband and relatives, from all members.

Mrs. King was active in a number of fields connected with the Club, not the least of these being Assistant Editor, in which she was of tremendous help.

Edna's bright and cheerful personality will be greatly missed from Club meetings and excursions alike. —EDITOR.

General Meetings

It is regretted that we are unable to publish the report of the January General Meeting; and can only give a brief report of the February meeting. Mrs. King, our late Assistant Editor, was preparing this report at the time of her death.

General Meeting — 8 February

Prior to the meeting, an Extraordinary Meeting was held to consider the application for affiliation of the Mid-Murray Field Naturalists Trust. A motion approving of this was carried. Mr. and Mrs. McFarlane, and Miss Irvine, three members of the Club were present.

The area represented by the Trust was between Swan Hill and Robinvale on both sides of the river; and they were concentrating on conservation.

The General Meeting was chaired by the President, Mr. T. Sault, and approximately 160 members were present, including three new members, and Mr. E. Allan (the immediate past President) who has recently returned from overseas.

It was with regret that the deaths of two members were announced — Mr. William Hunter, and Mr. P. J. Hemmie. Mr. Willis spoke of the vast amount of botanical work done in E. Gippsland by Mr. Hunter who was a surveyor in the area, and discovered many new species of plants.

Mr. Hemmie was well known for his knowledge of place names, and was an authority on building stones. His work mainly concerned the Grampians area, and he led many excursions in geology.

Correspondence was received from Mr. J. Baines seeking nominations for the Natural History Medallion award to be in by May next.

Also, the recently formed Warringa and District Conservation Society requested support in their effort to conserve the marshland area near the Banyule Homestead, which is being threatened by over-development.

Notice was received from the P.M.G. that the *Victorian Naturalist* had been reclassified into Category A, instead of B.

A motion was carried to the effect that the Secretary be notified before a meeting, of any other business not listed on the Agenda, and that such business be placed on the Agenda.

Exhibits:

N. McFarlane—*Scaevola* sp. (affinities with *S. littoralis*). The Herbarium states that it is a new record for Victoria, collected in sand dune country near Annuello, south west of Robinvale.

A. J. Swaby—Cushion Bush (*Calocephalus brownii*), *Leptospermum* "Hobetta".

Mrs. R. Taylor—Two summer flowering wattles i.e. *retinodes* and *A. terminalis*).

T. Sault—Fairies' Aprons (*Utricularia dichotoma*).

Mrs. Bennett—Early photographs of Tower Hill and Warrnambool areas.

A. Morrison—*Olearia ramulosa* and *Nertera depressa*.

R. Condron—*Hibiscus*.

The subject of the evening was titled "Scenic and Biotic Wonderland" — an address on an excursion through New Guinea, given by Mr. J. H. Willis.

Marine Biology and Entomology Group
1 February, 1971

The meeting was chaired by Mr. R. Condron; 18 members being in attendance.

Mrs. Zillah Lee announced that the Preston Junior Club would be staging a nature show at Northlands Centre Court shopping centre at some date in the near future, and asked for help with exhibits from the Group. Help was offered by several members, and Mrs. Lee said that, provided the show was a success this year, it will be a regular feature in the years to come.

Dr. Brian Smith announced that "Museum Memoirs" as from the next publication, would go on sale. A report of the Westernport Survey will be published shortly (Part 2 No. 32); the price to be announced. Dr. Brian Smith also announced that the first Museum "work-in" for this year would be on Saturday, 13 February at 10.30 a.m. Subsequent meetings to be held on 13 March, 24 April, 22 May, 19 June, and 17 July.

Exhibits:

Miss McLaren—larva of a moth of the Anthelidae family from her garden in East Malvern.

Miss White—Some slugs taken from Hickory Wattle at Lake Mountain—Crusader Bugs on Acacia from Edithvale and some leaf hoppers also from Edithvale.

Mr. Kelly—some drowned beetles from his swimming pool at Ivanhoe on which some small aquatic larvae appeared to be feeding.

Mr. McInnes—what he thought might be a species of Ostracod taken from water at Cann River. Mr. McInnes said that these, if they were Ostracods, were larger than those he knew of.

Mr. K. Strong—microscope slide of a small wasp which emerged from a gall taken from a Silver Wattle tree.

Mr. Godward—a tiger moth, family Arctiidae, probably *Ardices Glatignyi*, found resting on concrete at Parkville.

Mr. Condron—some snails which Mr. Setford had given him from Dimboola. Mr. Lung identified these as *Helicella Virgata*; an introduced snail from Europe - found in West Victoria and South Australia. Also two species of butterflies—Macleay's Swallowtail (*Graphium Macleayanus*) taken at Mt. Donna Buang, and *Anaphaeis Java Teutonia* (Caper White) taken at Box Hill.

Mrs. Lee—a Holothurian washed up on the beach at Somers in early November. Dr. Brian Smith thought it was a deep water species.

Mr. Kelly showed a series of colour transparencies of the beetle family *Chrysomelidae*—leaf-eating beetles. Mr. Kelly gave an interesting talk on these, explaining that there were approximately 1,000 species in the family. Slides shown were of the Genus *Peropsis*. Close ups of several beautiful species were shown, and also its life cycle.

Dr. Smith showed slides of a Genus of shells. He explained that 16 species of these were originally listed, but that they had now been re-classified leaving only two species. This is the Triton family which is predaceous on starfish, and as such is important as it helps to reduce the numbers of the Crown of Thorns starfish. Taking of these shells is now banned.

Geology Group Excursions

Sunday, 14 March — Deep Creek Gorge, Bulla. Leader: Mr. George Carlos.

Sunday, 10 April — Eildon District. Leader: Mr. Tom Sault.

Transport is by private car. Spare seats are usually available for those without their own transport. Excursions leave from the western end of Flinders Street Station (opposite the C.T.A. Building) at 9.30 a.m.

FIELD NATURALISTS CLUB OF VICTORIA — BALANCE SHEET AT 31st DECEMBER, 1970

Year 1969	Liabilities				Year 1969	Assets		
$404	Subs paid in advance		$422		$3,038	Cash at Bank		$976
—	Sundry Creditors		—		957	Commonwealth Bonds at cost		2,000
385	M. A. Ingram Trust grant in hand				34	Sundry Debtors		1,094
	Special Funds and Accounts—				151	Badges & Car Stickers at cost		32
3,103	Building Fund	$3,100			635	Microscope Project Stock at cost		197
916	Publications Fund	4,770			100	Books for Sale Stock at cost		575
100	Library Fund	100				Flower Book Stock at cost		53
1,037	Club Account	730						4,927
200	Excursion Account	200			4,739	Library Furniture & Equipment at cost		$5,184
5,217	Estate Marion Wright Legacy	5,217				Investment of funds—		
418	Estate Ruby A. Lewis Legacy	418				Publications Fund —		
200	Estate Miss I. F. Knox Legacy	200			1,800	Commonwealth Bonds at Cost		$1,800
1,060	Microscope Project A/c.	1,090				Stocks valued at cost—		
186	Flower Book Account	2,212			139	Victorian Ferns		52
			18,037		738	Victorian Toadstools		666
5,857	Surplus of Assets over Liabilities		7,013		505	Wyperfeld National Park		403
					54	Sundry Debtors		55
					1,280	Cash at Bank		1,794
								4,770
						Building Fund —		
					2,100	Commonwealth Bonds at cost		$2,100
					1,000	S.E.C. Stock at Par		1,000
					3	Cash at Bank		1
								3,101
						Library Fund—		
					100	Commonwealth Bonds at cost		100
					5,200	Legacy Estate Marion Wright Commonwealth Bonds at cost		5,200
					140	Costick Reserve — Maryborough at cost		140
						Flower Book Account —		
					1,750	Commonwealth Bonds at cost		2,050
$24,463			$25,472		$24,463			$25,472

We report that in our opinion the accompanying Balance Sheet and Accounts of the Field Naturalists Club of Victoria are properly drawn up in accordance with the provisions of the Companies Act 1961 and so as to give a true and fair view of the state of the Club's affairs at 31st December 1970, and of its operations for the year ended off that date, and that the accounting and other records examined by us have been properly kept in accordance with the provisions of the Act.

Melbourne
3rd February, 1971

Signed
Danby Bland & Co.
Chartered Accountants
Auditors

FIELD NATURALISTS CLUB OF VICTORIA
GENERAL ACCOUNT
STATEMENT OF RECEIPTS & PAYMENTS FOR THE YEAR ENDED 31st DECEMBER, 1970.

Year 1969	Receipts				Year 1969	Payments			
	Subscriptions Received—					Victorian Naturalist—			
$162	Arrears	$155			$1,445	Printing		$4,315	
5,327	Current	5,441			1,015	Illustrating		727	
130	Supporting	143	$5,739		319	Dispatching		371	
					20	Editorial		50	
313	Sales of Victorian Naturalist		237					5,463	
113	Advertising in Victorian Naturalist		124		(808)	Less Ingram Trust Grant		957	
	Interest Received—								$4,506
	Library Fund	$5				Working Expenses —			
	Bank Account	87			177	Postage & Telephone		$179	
	Investment on M. Wright Legacy	247			279	Printing & Stationery		106	
	Premium on Redemption of Bonds	156	495		40	Rent of Room for Storage		40	
399	Sundry Income		9		51	Gel Expenses		68	
64	Amount transferred from Building Fund				141	Affiliation Fees, Subs and Donations		71	
155	Part Payment of Rent		170					22	
	Surplus — Stock of Badges		20		19	Preston Junior Club Rent		72	
694	Deficit for year		—		144	Natural History Mon Expenses		663	
						Typing & Clerical Assistance		50	
					1,169	Audit			1,271
					50				
					85	Mammal Survey Group Expenses		406	
					(85)	Less Ingram Trust Grant		406	—
					244	Rent of Hall and Library			255
					55	Insurance			51
						Surplus for year			711
$7,357			$6,794		$7,357				$6,794

March, 1971   77

FIELD NATURALISTS CLUB OF VICTORIA

BUILDING FUND

Amount of Fund at 31st December, 1969	$3,103
Interest on Investment and from Bank Account	168
	3,271
Less Amount transferred to General Account for payment of rent	170
Amount of Fund at 31st December, 1970	$3,101

PUBLICATIONS FUND

Amount of Fund at 31st December, 1969		$4,516
Surplus for the year from—		
Ferns of Victoria and Tasmania	$34	
Victorian Toadstools & Mushrooms	37	
Vegetation of Wyperfeld National Park	63	
		134
Interest on Special Bonds and Bank Account		120
Amount of Fund at 31st December, 1970		$4,770

CLUB IMPROVEMENT ACCOUNT

Amount of Account at 31st December, 1969	$1,037
Nature Show Profit	108
Profit on Booksales	30
	$1,175
Less Purchase of Library Books, Binding, Furniture and Equipment	445
Amount of Fund at 31st December, 1970	$730

A Wide Range of Microscopes
including
Biological Microscope 1500 X
Oil immersion, Condenser and
Mechanical Stage $95.00
and
Stereo-binocular Microscope
$61.33

* * * *

"A Guide to Australian Spiders"
by Densey Clyne $4.95

* * * *

Also a full range of Astronomical
and Terrestrial Telescopes
from
Genery's
Scientific Equipment
Supply
183 Little Collins St., Melbourne
PHONE: 63-2160

ENTOMOLOGICAL EQUIPMENT
Butterfly nets, pins, store-boxes, etc.
We are direct importers and
manufacturers,
and specialise in Mail Orders
(write for free price list)

AUSTRALIAN
ENTOMOLOGICAL SUPPLIES
14 Chisholm St., Greenwich
Sydney 2065   Phone: 43 3972

F.N.C.V. PUBLICATIONS AVAILABLE FOR PURCHASE

FERNS OF VICTORIA AND TASMANIA, by N. A. Wakefield.

The 116 species known and described, and illustrated by line drawings, and 30 photographs. Price 75c.

VICTORIAN TOADSTOOLS AND MUSHROOMS, by J. H. Willis.

This describes 120 toadstool species and many other fungi. There are four coloured plates and 31 other illustrations. New edition. Price 90c.

THE VEGETATION OF WYPERFELD NATIONAL PARK, by J. R. Garnet.

Coloured frontispiece. 23 half-tone, 100 line drawings of plants and a map. Price $1.50.

Address orders and inquiries to Sales Officer, F.N.C.V., National Herbarium, South Yarra, Victoria.

Payments should include postage (11c on single copy).

Field Naturalists Club of Victoria

Established 1880

OBJECTS: To stimulate interest in natural history and to preserve and protect Australian fauna and flora.

Patron:
His Excellency Major-General Sir ROHAN DELACOMBE, K.B.E., C.B., D.S.O.

Key Office-Bearers, 1970/71

President:
Mr. T. SAULT

Vice-Presidents: Mr. J. H. WILLIS; Mr. P. CURLIS

Hon. Secretary: Mr. D. LEE, 15 Springvale Road, Springvale (546 7724).

Hon. Treasurer & Subscription Secretary: Mr. D. E. McINNES, 129 Waverley Road, East Malvern 3145

Hon. Editor: Mr. G. M. WARD, 54 St. James Road, Heidelberg 3084.

Hon. Librarian: Mr. P. KELLY, c/o National Herbarium, The Domain, South Yarra 3141.

Hon. Excursion Secretary: Miss M. ALLENDER, 19 Hawthorn Avenue, Caulfield 3161. (52 2749).

Magazine Sales Officer: Mr. B. FUHRER, 25 Sunhill Av., North Ringwood. 3134

Group Secretaries:

Botany: Mr. J. A. BAINES, 45 Eastgate Street, Oakleigh 3166 (57 6206).
Microscopical: Mr. M. H. MEYER, 36 Milroy Street, East Brighton (96 3268)
Mammal Survey: Mr. P. HOMAN, 40 Howard Street, Reservoir 3073.
Entomology and Marine Biology: Mr. J. W. H. STRONG, Flat 11, "Palm Court", 1160 Dandenong Rd., Murrumbeena 3163 (56 2271).
Geology: Mr. T. SAULT.

MEMBERSHIP

Membership of the F.N.C.V. is open to any person interested in natural history. The *Victorian Naturalist* is distributed free to all members, the club's reference and lending library is available, and other activities are indicated in reports set out in the several preceding pages of this magazine.

Rates of Subscriptions for 1971

Ordinary Members	$7.00
Country Members	$5.00
Joint Members	$2.00
Junior Members	$2.00
Junior Members receiving Vict. Nat.	$4.00
Subscribers to Vict. Nat.	$5.00
Affiliated Societies	$7.00
Life Membership (reducing after 20 years)	$140.00

The cost of individual copies of the Vict. Nat. will be 45 cents.

All subscriptions should be made payable to the Field Naturalists Club of Victoria, and posted to the Subscription Secretary.

JENKIN BUXTON & CO. PTY. LTD., PRINTERS, WEST MELBOURNE

F.N.C.V. DIARY OF COMING EVENTS
GENERAL MEETINGS

Wednesday, 14 April—At National Herbarium, The Domain, South Yarra, commencing at 8 p.m.
  1. Minutes, Reports, Announcements.
  2. Nature Notes and Exhibits.
  3. Subject for the evening — "Wilson's Promontory": Mr. J. Ros Garnet.
  4. New Members.
     Ordinary:
       Miss Linda Barraclough, Flat 4, 104 Riversdale Rd., Hawthorn, 3122
       Miss Susan Mena, 73 Grey St., East Melbourne, 3002.
       Dr P. Bridgewater, Howitt Hall, Monash University, Clayton, 3168
       Mr. Graham William Wycherley, Flat 6, 492 Barkers Rd., East Hawthorn, 3123
       Mrs Edith Ann Pedrana, 25 McGregor St., Middle Park, 3206.
     Joint:
       Mr Alan Goss and Mrs Josephine Goss, 63 Chapel Rd., Moorabbin, 3189
     Country:
       Mr Geoff W Carr, 19 Highview Pde., Hamlyn Heights, Geelong. 3215.
     Joint:
       Messrs Robin and Ivan Spargo, Spargo Ski Lodge, Falls Creek, 3699
  5. General Business.
  6. Correspondence.

Monday, 10 May — "New Guinea Highlands": Mr. J. H. Willis.

F.N.C.V. EXCURSIONS

Sunday, 18 April—Barfold Columns, Leader: Mr. F. Robbins. The coach will leave Batman Avenue at 9.30 a.m. Fare $1.80. Bring two meals.

Saturday, 21 August to Sunday, 5 September — Flinders Ranges. The coach will leave Melbourne Saturday morning and travel via Bordertown, Adelaide, Quorn, Wilpena (2 nights), Arkaroola (4 nights), Wilpena, Port Augusta (3 nights), Renmark, Swan Hill, Melbourne. Most of the accommodation will be on a dinner, bed and breakfast basis but at Arkaroola there will be bunks in huts, bed only, but meals are available in the dining or food can be obtained at store, with a certain amount of motel accommodation as well. Cost will depend upon numbers going but should be in the vicinity of $150 with bunks at Arkaroola of $195 with motel accommodation. Deposit of $50.00 to be paid when booking, cheques being made out to Excursion Trust.

GROUP MEETINGS
(8 p.m. at National Herbarium unless otherwise stated)

Thursday, 8 April—Botany Group. Mr. F. Woodman will speak on "Some Aspects of Natural History in South Africa".

Wednesday, 21 April—Microscopical Group.

Friday, 30 April—Junior Meeting at Hawthorn Town Hall at 8 p.m.

Monday, 3 May—Entomology and Marine Biology Group meeting at 8 p.m. at National Museum in small room next to theatrette.

Wednesday, 5 May—Geology Group.

Thursday, 6 May—Mammal Survey Group Meeting in Arthur Rylah Research Institute building; corner of Stradbroke and Brown Sts., Heidelberg, at 8 p.m.

Thursday, 13 May—Botany Group.

GEOLOGY GROUP EXCURSIONS

11 April — CANCELLED.

8 and 9 May—A weekend excursion to the deep lead mines in the Creswick area is being arranged in conjunction with the Creswick Field Naturalists. Camping facilities and hotel accommodation are available.

6 June — To Waurn Ponds in search of sharks' teeth.

The Victorian Naturalist

Editor: G. M. Ward

Vol. 88, No. 4   7 April, 1971

CONTENTS

Articles:

    The Aborigines and Lady Julia Percy Island, Victoria. By Edmund D. Gill and Alan L. West   84

    The Brush-tailed Rock Wallaby in Western Victoria. By N. A. Wakefield   92

    Rediscovery of the Large Desert Sminthopsis on Eyre Peninsula, South Australia. By Peter F. Aitken   103

Personal:

    William Hunter (1893-1971), Doyen of East Gippsland Botanists   88

Field Naturalists Club of Victoria:

    Diary of Coming Events   82

Front Cover: Thorny Devil, Mountain Devil, *Moloch horridus*—these names all belong to this, one of Australia's most bizarre lizards. Inhabiting the desert areas of most States of the mainland, it feeds solely on small black ants; eating many thousands in a meal.

Photo: John Wallis.

The Aborigines and Lady Julia Percy Island, Victoria

by Edmund D. Gill* and Alan L. West*

From Cape Reamur, The Craigs, or the top of the dunes at Yambuk, west of Port Fairy, Western Victoria, an island can be seen five miles off the coast. It is roughly triangular in outline, over a mile long and over half a mile wide. With vertical cliffs, it stands 100 to 150 feet out of the sea (See Plate 1). When seen from a distance, it intrigues people by reason of its high steep walls and flat top. The usual reaction is to gaze in surprise and ask "What on earth is that?" Just as our curiosity is aroused by the sight of the island, so was that of the Aborigines. It held an important place in their mythology, and we now have evidence in the form of a flint scraper that Aborigines may have visited the island.

Lady Julia Percy is the name of this uninhabited island. Only seals and seabirds (and more recently rabbits) live there. The seas are commonly rough in the vicinity, which, combined with the lack of water and sandy beaches, and the presence of steep cliffs, makes it rather inhospitable. Lieutenant James Grant named the island in 1800, Flinders charted it in 1802 (commenting on its cliffs and flat top), and the Frenchman Baudin sighted it in the same year (referring to its treeless condition, being "covered only with low heath"). Sixty years later, Surveyor Allan described the thick scrub as an impediment to his work, but this has all gone now due to the rabbits. In the 1800's the island had a large colony of seals. However, the sealers who worked this coast almost wiped out the seal populations, and had to leave when the industry became unprofitable. They established no settlement on Julia Percy, but it is recorded that two were buried there in the 1820's. A small amount of guano was collected there, and the island was used briefly for pig farming. Probably the longest period for which the island was inhabited was in 1935, when for three months the McCoy Society carried out scientific research there (Wood Jones et al. 1937).

Site of a Volcano

The McCoy Society report describes the island as a volcanic complex of boulder tuff and some six lava flows. The headland called Pinnacle Point is the neck of a volcano where the lava congealed to form solid basalt. The flatness of the island is due to lava flows, while the steep cliffs are due to marine erosion of the boulder tuff causing vertical breakaways of the overlying basalt. The tuffs and included ejected rocks are evidence of an explosive volcano, while the lava flows are evidence of a subsequent milder phase of activity in the form of an effusive vent. It is interesting to enquire whether Aborigines saw the island volcano in eruption. Quite possibly they did. The McCoy researchers noted the shallow soil on the island and the lack of a drainage pattern, concluding from this that the island is young. The lava flow on which Port Fairy is built is Penultimate Glacial in age (Gill 1967), while the Tower Hill volcano northeast of Port Fairy is about 7300 years old on the latest dating. To the

*National Museum of Victoria

west, a peat immediately over a lava flow from Mt. Napier dated about 6235 years old, so probably is of the same order of age as Tower Hill. Perhaps the volcano at Lady Julia Percy Island belongs to this group; if so, it erupted out of the sea. If it is older, it probably erupted when sea level was much lower, in which case the Aborigines would have been able to walk out to it.

Aboriginal Mythology

James Dawson was a land holder in the Western District in the nineteenth century. From the pen of this keen recorder of Aboriginal life (Dawson 1881) we learn that the Gunditjmara who lived on the mainland opposite the island regarded it as a place of considerable importance. Dawson states that when the bodies of the dead were buried they were wrapped in grass. Should grass be found in the mouth of a mainland cave facing the island, following a death, this was regarded as a good omen. It signified that the body of the deceased had been removed through the cave to the island by a good spirit. From the island the spirit of the dead was conveyed to the clouds. Should a meteor be seen at about the same time, this was interpreted as fire, being taken up with the spirit of the dead. Dawson records the name of the island as Deen Maar.

R. H. Mathews (1904) spelt the name slightly differently as Dhinmar but his account of the significance of Julia Percy Island in Aboriginal myth is largely in agreement with that of Dawson. He added the interesting comment that the dead were buried with their heads pointing towards the island. As far as we know this proposition has, to date, not been tested. Contrary to Dawson, it was Mathews' view that in Aboriginal belief human spirits awaited re-incarnation on the island.

Plate 1. Lady Julia Percy Island, Dinghy Cove. Photo: R Warneke

A centre associated in the minds of Aborigines with departed spirits, would hold few attractions. For this reason the absence, in the past, of signs of Aboriginal visitation has aroused little comment.

Artefacts

In January, 1968, Mr. Trevor Pescott of Geelong forwarded several pieces of flint to the National Museum of Victoria. He reported having found them on Julia Percy Island and added that they looked rather like Aboriginal artefacts. One of the authors (A. L. W.) examined the flakes and agreed that they appeared to be the waste products of an Aboriginal craftsman. The view was expressed that the flakes could more assuredly be associated with Aborigines if an undoubted artefact were found on the island. This discovery has now taken place.

When an Australian Broadcasting Commission party visited the island in July 1970, Mr. Robin Hill picked up a well-formed Aboriginal flint scraper on the southern part of the island (Figure 1). At the same time, Mr. Robert Warneke found at Dinghy Cove on the NE shore, a small flint core and a flake. The latter fitted back perfectly on to the core. It is possible that the flake could have been struck by a geologist, but the scraper is an indubitable Aboriginal implement.

It was made on a disc-shaped flake roughly 3.4 cm. in diameter, and is secondarily trimmed all around the circumference. As the drawing shows, there is a clear colour demarcation through the material, indicating a difference in composition. The stippled area in the drawing represents that part of the scraper which is a light grey-fawn in colour and which in comparison with the other area appears to be of coarser texture. The remaining two-thirds consists of fine-grained slightly mottled flint ranging in colour from dark grey to light grey with a yellowish tint. No bulb of percussion is clearly visible on the inner surface and this surface is also typical in having two different curves at right angles to one another. In the drawing the vertical cross-section shows the convex curve while the concave surface is apparent in the horizontal cross-section. It is not suggested that this opposed curvature is other than fortuitous, but it is an unusual feature.

Aboriginal Visits?

The core and flakes, probably made by Aborigines, found in recent years on the island, and in particular, the Aboriginal scraper, present a challenge to the generally held view that Aborigines did not visit Julia Percy Island.

There are at least two reasons for Aborigines wanting to visit Julia Percy and one, the island's place in mythology, to explain their possible reluctance to do so. We have no knowledge of the antiquity of the legend which explains the island as a spirit centre. It could be that the stone material was left by Aborigines when they visited the island before the creation of the legend. On the other hand even at the time of first European contact the Gunditjmara may occasionally have gone to the island even though they regarded it as a spirit centre.

The Aborigines' curiosity and the promise of food in the form of seals, eggs and seabirds would have provided a strong incentive for them to brave the rough waters. On the island are small pebble beaches where landings can be made when the sea is not too rough. The pebbles are mostly of basalt (as one would expect); but of other kinds, those of flint being the most numerous (Stach 1937). The Aborigines found flint an ideal material for their purposes, and it may be that if they visited Julia Percy they did so as much for the supply of flint as for food.

Most of the flint found in Western and Northern Victoria came from the Mount Gambier Limestone, as the fossils bear witness. The Mount Gambier area was outside the tribal territory of the Gunditjmara, so the stone would have to be traded. Its wide distribution (Gill 1957) shows how strongly it was favoured, so the trade "price" could have been appreciable. There would therefore be real point in having their own supply from the island. The Mount Gambier Limestone passes out under the sea, so probably during times of lower sea level nodules were eroded from the formation and were moved about by currents and the surf.

The European sealers who hunted on the island in the early decades of last century were often accompanied by Aboriginal (mostly Tasmanian) women. It is possible that one of these may have dropped the flint scraper but the weight of evidence, we think, indicates that some Aborigines visited Lady Julia Percy Island possibly searching for food, flint or both.

Figure 1.

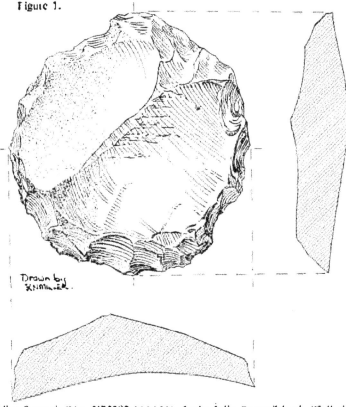

Flint Scraper (No. X75383 N.M.V.), Lady Julia Percy Island. (Full size).

REFERENCES

Dawson, I., 1881. *Australian Aborigines* pp. 51-52.

Gill, E. D., 1957. The Australian Aborigines and fossils. *Vict. Nat.* 74: 93-97.

Gill, E. D., 1967. Evolution of the Warrnambool-Port Fairy coast and the Tower Hill Eruption, Western Victoria. Chap. 15 in *Landform Studies in Australia and New Guinea*, Ed. Jennings and Mabbutt, A.N.U.

Gill, E. D., and Gibbons, F. R., 1969. Radiocarbon date related to vulcanism and lake deposits in Western Victoria. *Aust. J. Sci.* 32: 109.

Jones, F., Wood, et al., 1937. Lady Julia Percy Island. Reports of the Expedition ... *Proc. Roy. Soc. Vict.* 49: 327-437.

Mathew, R. H., 1904. Ethnological Notes on the Aboriginal tribes of New South Wales and Victoria. *J. Roy. Soc. N.S.W.*, p. 297.

Stitch, I. W., 1937. General geology (McCoy Soc Rept.). Ibid. 49: 322-343.

William Hunter (1893-1971)
Doyen of East Gippsland Botanists

On 21 January, at the Gippsland Home and Hospital in Bairnsdale, life ended for a fine, accomplished, helpful and lovable old naturalist in his 78th year. William Hunter was born at Tatura in the Goulburn Valley, Vic., on 16 July, 1893. His Presbyterian father, also William Hunter who owned a drapery and general store there, had migrated with a younger brother from Millport on the Firth of Clyde, Scotland, during the 1860's.

One of six children, young Hunter was educated at the local Tatura State School until 12 years old, when he continued at Mr. Zercho's Deaconesses' High School (Anglican) for another three years. Wishing for him to go into business at the family store, his father was eventually persuaded—by Mr. Campbell at the Working Men's College—to let William come to Melbourne and take a Junior Technical Course. After his first year at College, the boy won a scholarship for senior technical training over the next three years; but in his second year he became a surveying apprentice to W. P. Steen (who was a lecturer on the College staff), at the age of 18.

When 22, and with apprenticeship completed, he enlisted in the A.I.F. during November, 1915, as an infantryman — although he would much have preferred the engineers. He joined a mining and tunnelling company, and went abroad in September 1916 in the Second Pioneer Battalion. While in camp on Salisbury Plain he was able to visit Stonehenge and Salisbury Cathedral. In March 1917 he went on service to France until the end of World War I.

Within a month of his return to Victoria, Hunter applied for a surveying position and worked under Mr. J. D. Hope in Gippsland, but early in 1921 he went surveying in Tatura district so as to be near his family. Mr. Mott, district surveyor in East Gippsland, recommended him for feature surveying in that district and he went there in 1923. Except for occasional jobs in the Goulburn Valley, Bill Hunter's activities henceforth were virtually restricted to East Gippsland for the remainder of his life. As a licensed surveyor he worked variously for government instrumentalities and shire councils. His plans were models of neatness and thoroughness, while his faultless copper-plate lettering became almost legendary—even long letters and reports were meticulously scripted!

During the great depression of the early 1930's his father was forced out of business, and both parents came to live with their son at Bairnsdale. Hunter senior died in 1937 and his wife in 1949—at a private hospital in Melbourne where she had been invalided for two years. Mrs. Hunter was formerly a keen gardener and croquet player. Bill never married.

Soon after his parents retired in Bairnsdale, W. Hunter met Mr. Thom, a former major in the British Army who had been educated at Heidelberg (Germany) and had travelled widely. Thom gave Hunter an interest in the local flora, having himself acquired Ewart's *Flora of Victoria* (1931) to help him identify the vegetation around his selection on the Cann Valley Highway, near the N.S.W. border. Mrs. Agnes Thom became a country member of the Field Naturalists Club of Victoria (elected 13/5/1935), and she soon proposed Hunter for membership also; he was duly elected on 20 April, 1936, but his name was inadvertently omitted from the list of new members published next month in the *Victorian Naturalist*. The Thoms persuaded Hunter to get his own copy of Ewart's *Flora*, then selling for $3, and they encouraged many bush animals to frequent their property.

About this time, Hunter also came to know Frank Robbins who was teaching science at Bairnsdale Higher Elementary School, and the two went on several botanical excursions. Bill Hunter subsequently sent a long series of specimens for identification to T. S. Hart (which, on Mr. Hart's death in June, 1960, came to the Melbourne Herbarium) and orchid specimens to the late W. H. Nicholls.

Later in 1936 he compiled an annotated census of all the plants known to date for East Gippsland, followed over succeeding years by several supplements as new records came to light. These beautifully written lists are now in the writer's library, as is also his "List of the Suggan Buggan Flora", presented by the late George N. Hyam in April, 1944. By 1937 systematic botany was almost Hunter's consuming obsession. His enthusiasm heightened through acquisition of a good set of Bentham's *Flora Australiensis* (purchased for £17/17.-).

While surveying along the Deddick River for the Tubbut to Wulgulmerang road, in December, 1936, W. Hunter found a solitary bush (8 ft.) of the very rare *Myoporum floribundum*, thus substantiating F. Mueller's premature record in the *Key Syst. Vict. Plants* (1888)—Mueller himself had collected this plant in Feb. 1854, but along the Snowy River just inside New South Wales. This is the most graceful and beautiful of all Boobialla species. Hunter's surveying companion at the Deddick River was Clem Heather who, for amusement, would often carve human faces on conspicuous trunks and tree stumps. The name "Old Joe's Creek" was bestowed by Bill Hunter after a black-painted effigy that Clem. had cut nearby; they christened the carving "Old Black Joe", but it has long since disappeared.

During his surveys of potential settlement blocks at Suggan Buggan and Ingeegoodbee, for about two years in 1938 and 1939, opportunity was taken to make extensive botanical collections. As a result, many old recordings of F. von Mueller were confirmed and a number of new records made for Victoria. It was through the finding of *Cryptostylis hunterana* Nicholls (March 1938) at Marlo that Bill first came into contact with N. A. Wakefield. Orchidologist W. H. Nicholls had telegrammed Marlo, saying "Send more *Cryptostylis leptochila*" (which Hunter had called it), but the latter failed to rediscover the unusual orchid in that district. Nicholls then commissioned the Wakefield family of Orbost to make an intensive search. Norman Wakefield and his father succeeded in locating more material which became the type for this new and remarkable species. Many years later Wakefield named *Acacia hunterana* (Snowy River Wattle) in honour of his friend, unaware that the same species had previously been described in New South Wales under the name *Acacia boormanii*.

With all his immense store of field knowledge, Bill Hunter published astonishingly little. He seems to have made only three contributions to the *Victorian Naturalist*, viz.:

1. "List of orchids recorded for East Gippsland" (Dec. 1938) —with valuable notes on distribution.
2. "The Mountain Plum Pine of Goonmirk, East Gippsland" (May 1941) — record dimensions are given for this usually sprawling conifer which at Goonmirk (3000 ft. alt.) attained 20-25 ft., with trunks up to 1 ft. in diameter at the base.†
3. "Habits of East Gippsland Currawongs" (Jan. 1954).

Early in 1950 he moved from Bairnsdale to Marlo (at the Snowy River mouth), and during 1954 he built a cottage on the northern outskirts of Mallacoota; there he resided until impoverished circumstances and failing health demanded his removal to Bairnsdale Home and Hospital in 1969. The local bush nurse at Mallacoota, Sister Dickenson, latterly kept a kindly eye on his welfare; but nothing delighted the old surveyor more than a call from some visiting botanist — with, perhaps, a ramble in the nearby bush that he knew so well.

The writer's friendship and correspondence with Bill Hunter cover more than 26 years, from early in 1943, when he became very enthusiastic in collecting and identifying East Gippsland fungi, to his last letter (still in the neatest script) of 16 March, 1969. One remembers him as a stocky man with round face, pale blue dancing eyes, white bristly "prison-cut" hair and a loud penetrating voice. I well recall travelling with him once on a Melbourne train, when he began describing one of his recent botanical trophies, and how distinct it seemed from species so-and-so. Phrases like "calyx split to the bottom", "anthers with a hairy appendage" and "ovary quite inferior" —shouted in a crescendo of excitement above the rattle of the train — soon had the heads of curious and startled fellow-passengers craned in our direction!

Bill was most precise, punctilious and at times a little stubborn. Some amusing stories are told about his exploits: the time his laden packhorse, not very successfully strapped, bolted up Wulgulmerang way and left a trail of camping supplies distributed for many chains along the mountain track; the time he reached gingerly for an apple while steering his old car and ran off the road into the bush. Once he brushed with a policeman about his failure to have windscreen wipers on his car. In vain Bill explained that, whenever visibility was impaired by mud or insect splodges, he simply wiped the glass clean with a handkerchief — surely that was sufficient? "Mr. Hunter", insisted the officer, "the Law *requires* that you install windscreen wipers". All Bill could snort was "Then *I* definitely consider the Law is an ass".

On another occasion, when breakfasting with a camp-mate, he suddenly lapsed into silence and refused to speak for several days. At last his unhappy companion could hear the tension no longer and said: "Look here, Bill, we can't go on like this. *What's* the trouble? If I've offended you in any way, then out with it".

Reluctantly, Bill confessed "Well, . . . what would *you* do if someone spilt tobacco ash in your porridge?" Such foibles only serve to throw in contrast the sterling qualities of a character that is becoming annually scarcer. Our Club is grateful for one of Victoria's most distinguished naturalists and amateur botanists, a man who did so much in a busy decade to elucidate the floristics of East Gippsland. He is fittingly commemorated by a small wildflower reserve that he helped to establish near Marlo Racecourse, to protect rare orchids including his own *Cryptostylis hunterana*.

Vale, William Hunter!

—J. H. WILLIS

F.N.C.V. PUBLICATIONS AVAILABLE FOR PURCHASE

FERNS OF VICTORIA AND TASMANIA, by N. A. Wakefield.
The 116 species known and described, and illustrated by line drawings, and 30 photographs. Price 75c.

VICTORIAN TOADSTOOLS AND MUSHROOMS, by J. H. Willis.
This describes 120 toadstool species and many other fungi. There are four coloured plates and 31 other illustrations. New edition. Price 90c.

THE VEGETATION OF WYPERFELD NATIONAL PARK, by J. R. Garnet.
Coloured frontispiece, 23 half-tone, 100 line drawings of plants and a map. Price $1.50.

Address orders and inquiries to Sales Officer, F.N.C.V., National Herbarium, South Yarra, Victoria.

Payments should include postage (11c on single copy).

THE WILD FLOWERS OF THE WILSON'S PROMONTORY NATIONAL PARK

by

J. Ros Garnet

PRICE: $5.25

Contact local Secretary or write direct to:
F.N.C.V. Magazine Sales Officer
Postage 20 cents

The Brush-tailed Rock-wallaby (Petrogale penicillata) in Western Victoria

by N. A. Wakefield[1]

Abstract

Evidence is presented of past occurrences of *Petrogale penicillata* in W. Vic. — fossil remains at Byaduk Caves, faeces at Mount Arapiles, and old literature references to living animals in the Grampians. Details are given of the discovery of a surviving colony in the Grampians, and of observations of three members of the colony. The habitat is described, and results are given of an assessment of food of the local rock-wallabies, based on study of samples of faeces. Other mammals occurring in the colony area are noted, and assessment is made of the extent and numerical strength of the rock-wallaby occurrence. Suggestions are made for conservation of the Grampians rock-wallabies.

Introduction

The Brush-tailed Rock-wallaby was thought to have disappeared from Victoria over 50 years ago, and it was rediscovered in the Snowy River valley in NE. Gippsland by Mr. Keith Rogers of Wulgulmerang in about 1938 (Wakefield 1953). Subsequently it was ascertained that the species survived in about nine small colony areas in that region, and that it was apparently limited to a tract of country approximately 25 miles long and 20 miles wide (Wakefield 1961, 1963b).

An additional one-time occurrence of the species is documented by Lingard (date unknown), who describes a rock-wallaby colony which he visited in the Currawong area, near the Snowy River in SE. New South Wales, in 1842 (see Fig. 1).

During the past several years the NE. Gippsland colonies have been reduced in number. The one at Wallaby Rocks, Wulgulmerang, has died out, and the colony near Mount Seldom-seen appears to have been eliminated in the severe bushfire of 1965. Other of the smaller occurrences may also have succumbed, but the species should survive indefinitely both in the Little River Gorge, Wulgulmerang, and in the Snowy River Gorge to the east of Butchers Ridge.

Subfossil remains of a single individual of *Petrogale penicillata* were found in one of the Byaduk Caves, SW. Vic., by Mr. Lionel Elmore of Hamilton, in 1961. This material was considered to be prehistoric (Wakefield 1964a), and to represent "the sole indication that the species ever occurred in W. Victoria" (Wakefield 1964b). However, the latter opinion overlooked reports by Thomas (1868) and Audas (1925, 1950) that the species lived in the Grampians.

Figure 1 shows past and present occurrences of *Petrogale penicillata* in and near Victoria. The localities are those discussed in previous reports (Wakefield 1953, 1961, 1963b) and in the present article.

The Byaduk Specimen

This was found in the Bridge Cave, Byaduk, "on rocks in cavern behind bat chamber", and it comprises most of a skull, with adult dentition on one side, and large pieces of postcranial bones. It is lodged in the

[1] Biology Department, Monash Teachers' College, Clayton, Victoria.

National Museum of Victoria, registered number P21009. Following are some dimensions of skull and teeth:

Basal length	93.6 mm.
P⁴ length	6.8 mm.
P³ length	5.6 mm.
M¹⁻³ length	20.0 mm.

No feature of the Byaduk specimen differentiates it from the *Petrogale penicillata* of eastern Victoria.

Occurrence at Mount Arapiles

On 6 November, 1969, a small deposit of bones and debris was found on a sheltered ledge of a cliff at Mount Arapiles in far-western Victoria. In the deposit were several typical rock-wallaby faecal pellets. These were light grey in colour and appeared to have been preserved in the deposit for at least several decades. Associated bone material included specimens of *Conilurus albipes* and *Pseudomys* spp., which evidently disappeared from W. Vic. some time last century, as well as of the introduced mouse (*Mus musculus*) and rabbit (*Oryctolagus cuniculus*).

Further search brought to light about 25 similar rock-wallaby faeces, located in crevices in other parts of the cliff system. The extensive broken outcrop of Devonian sandstone which comprises much of Mount Arapiles, appears to be ideal rock-wallaby habitat, and the species probably lived there until the general reduction in status of the species about 50 years ago. There is little doubt that the species which occurred at Mount Arapiles would have been the same as that in the Grampians and the Byaduk Caves.

Occurrence in the Grampians

Early records

Thomas (1868) made the following statement in reference to the Grampians:

The light and graceful rock wallaby abound among these hills as we see by numerous traces of their presence in the caves formed by the falling together

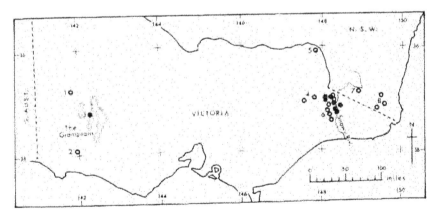

Figure 1. Occurrence of *Petrogale penicillata* in and near Victoria.

● — Extant colony   ○ — One-time occurrence, no longer extant.
1 — Mount Arapiles 2 — Byaduk Caves. 3 — Red Rock, Victoria Range 4 — Union alkd.
5 — Tintaldra 6 — Past and present occurrences in Snowy River valley, Victoria 7 — Currawong area. 8 — Scarp of East Range.

April, 1971

of detached fragments of rock. Shooting wallaby with the rifle ought to be good sport here; but we never tried it, as we always thought that we carried weight enough in our own proper person, taking into consideration the position of the rocky ascent, which in most places is perpendicular, and in some leans over on top of you in a most perplexing manner.

Audas (1925, p. 127) included this comment in a summary of wildlife in the Grampians:

> Rock Wallabies, though rare, can always be seen by anyone who wishes to do so. Inquisitiveness is their chief characteristic, and it is only necessary to seat one's self in a prominent position in some rocky section and wait. Patience is soon rewarded, for the monkey-like black face, encircled by light-coloured hair, will be quickly seen peering at the intruder from some adjacent boulder.

In the preface, Audas stated that his book was "entirely the result of personal observation . . . covering a period of years". No information was given as to when or where the rock-wallabies concerned had occurred. However, the behaviour which he describes is similar to that of the NE Gippsland rock-wallabies sixty or more years ago, before they became scarce. Today, members of the surviving Victorian colonies are extremely timid and not at all inquisitive.

Audas (1950, p. 318) repeated his 1925 statement without variation.

Rediscovery

Credit for the rediscovery of rock-wallabies in the Grampians belongs to Mr. Ellis Tucker of Brit Brit in W. Victoria, who has systematically explored sectors of those ranges over a period of years in quest of Aboriginal paintings. On 12 September, 1970, he sent me a packet containing several typical rock-wallaby faecal pellets, with these comments:

> I think I remember you saying once that there either could be, or should be, rock-wallabies in the Grampians. Until recently, I had no reason to think there were, but after collecting these specimens on ledges and shelves in caves in the Red Rock area last weekend, I cannot think of any other animal which could have been responsible. Red Rock is on the western side of the northern end of the Victoria Range, and is a very

Figure 2. Rock-wallaby tracks, Red Rock area.

Figure 3. Rock-wallaby habitat, Red Rock area.
Wallabies live under the large fallen blocks seen in the middle of the lower part of the picture, and they climb about on ledges of the cliff which overhangs their home.

rugged spot, high, with many overhangs, ledges and caves.

On 22 October, a small party, including Tucker and myself, went to the area concerned and explored several hundred yards of cliffs and rock outcrops which were certainly the focus of an extant colony of rock-wallabies. No animal was sighted, but footprints were found in patches of soft soil, faeces were present in abundance, and some living places (rock piles with crevice systems) were identified. (See Figures 2, 3, 4, 5.)

Later that day, and on 23 October, I examined the colony area further and, in particular, identified a typical basking place, where animals might be expected to be out warming themselves when the morning sun struck the spot. A concealed vantage point was selected some 60 yards away, and a route was noted by which the point could be reached without disturbing any animals which might be basking.

Observations of rock-wallaby family

From 19 to 22 December, Mr. Russell Batbard* and I carried out further work in the colony area; and on the morning of the last day, as weather conditions were suitable, we went to the vantage point selected previously. Two rock-wallabies were in the basking place—a half-grown joey and, presumably, its mother—and shortly afterwards a third specimen, presumably the male, joined them (see Fig. 6). We had the three under observation from 8.45 to 9.15 a.m., and for the latter part of this period the mother and joey had moved down to a rock shelf a few yards away while the male remained above.

On 4 January, 1971, Bathard and I observed the three rock-wallabies, in the same area, from 9 to 10.30 a.m. The female was located first, on a broad recessed ledge several yards from the original place. Later she

*Science student, Monash Teachers' College.

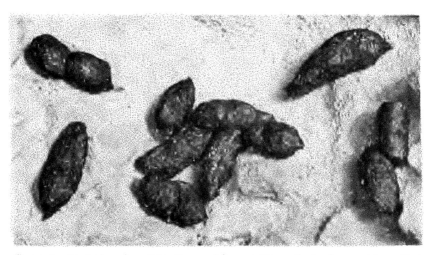

Figure 4. Typical rock-wallaby faeces, lying on the surface of a sandstone block in Red Rock area (natural size).

took cover quickly, as if alarmed. About the same time, the joey and then the male appeared at the original place, and soon afterwards the female joined them (see Fig. 7).

On 16 January, Mr. Keith Dempster, research officer of the Fisheries and Wildlife Department, accompanied Bathard and myself to the vantage point, and we observed the male and the joey at the original basking place. Later, one of us moved round and observed that the female was again on the broad recessed ledge (see Fig. 8). Eventually she took cover, as before, and at the same time the other two left their basking place. Then the three were seen moving down together through a narrow declivity between two rock masses, and finally one of the adults jumped up to the original basking place, paused a moment to look at us, then passed out of sight into the dark crevice system beyond. These observations lasted from 9.15 to 10.15 a.m.

The female and joey were bluish-grey in general body colour, with hands, feet and tail black. The male was similar but with a strong rufous infusion about the lower parts of his body. Backs of ears, longitudinal head stripe, and a patch behind each shoulder, were also black. Inside the ears was fawn, and the chest as well as the lower sides of the face were whitish. The male was larger than the female, and each of the adults appeared to be very plump, due at least partly to the long fur of the body. The general colouration, and the very long hair of the tail, identify the species as *Petrogale penicillata*.

While basking, the wallabies spent considerable time grooming themselves, both with the claws of the hands and with the small conjoined 2nd and 3rd digits on the inner side of the foot. After using one of these "twin toes" as a comb, the animal would lick the organ, evidently to clear it of accumulated fur. Precisely the same procedures are characteristic of a number of macropodid species—the Brush Wallaby, *Macropus rufogriseus*, for example.

Habitat

The rock-wallaby colony area is on a northern aspect in Devonian sandstone formation, and it extends for approximately a quarter-mile (400 m.). The elevation is from about 1300 to 1600 feet (400-500 m.) above

Figure 5. Part of crevice system of rock-wallaby home, Red Rock area.
The upper left rock is part of the cliff, while the lower rocks, and that on the right, are detached masses which have fallen from above.

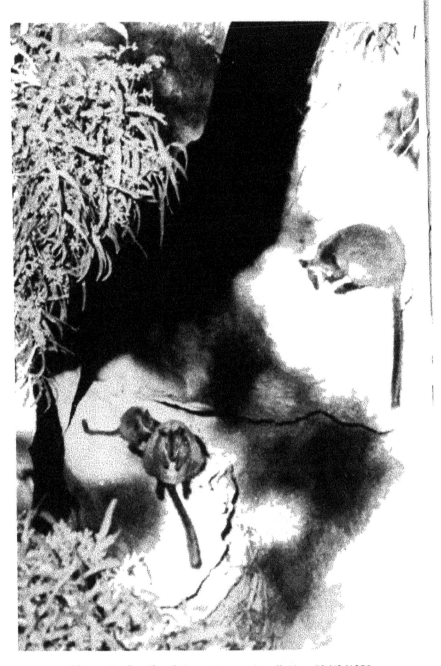

Figure 6. Family of Grampians rock-wallabies, 22/12/1970.
(Enlargement x20 from 35 mm. Ilford Pan F negative taken with 135 mm. lens at 50 m. range.)

sell level. Wallabies live in three places in the area, each home being amongst and under detached rock-masses which have fallen together against or near the bases of perpendicular or overhanging cliffs (see Figs. 3 and 5). This is precisely the kind of habitat described by Thomas in 1867.

Much of the outcropping rock about and above the actual homes is irregularly weathered to form projections and ledges. Herbage and shrubbery grow where soil has accumulated, and a sparse scattering of faeces indicates that the wallabies range widely over these outcrops to graze and browse on the vegetation. Long-leaf Box (*Eucalyptus goniocalyx*) grows amongst the rock masses, and numerous small specimens of Oyster-bay Pine (*Callitris rhomboidea*) form thickets. Myrtaceous shrubs such as Totem-poles (*Melaleuca decussata*), Thryptomene (*T. calycina*) and Fringe-myrtle (*Calytrix tetragyna*) are in association, as well as Wedge-leaf Hop-bush (*Dodonaea cuneata*) and Hop Goodenia (*G. ovata*). Three grasses grow about the rocks: Fibrous Spear-grass (*Stipa semibarbata*), Bristly Wallaby-grass (*Danthonia setacea*), and a tussock-grass of the *Poa australis* complex; but none is plentiful. On somewhat deeper soils, above and below the main outcrops, Brown Stringybark (*Eucalyptus baxteri*) predominates, and there is an abundance of heathland shrubbery, including Silky Tea-tree (*Leptospermum myrsinoides*) and Hakea species.

Analysis of food

Faeces were collected at random from throughout the colony area, both in and about living places and on open rock outcrops. Small as well as large pellets were selected, and about one-third of the sample taken were grey in colour, indicating that they were comparatively old. Forty pellets were treated, each being broken apart and a sample taken of the contents. The fragments comprising each sample were sorted out under a binocular dissector and as many fragments as possible were identified at species level by comparison with samples of vegetation found growing in the colony area.

Grass fragments comprised approximately 90 percent of the total sample of faecal material analysed. These fragments, representing both stems and leaves of grass, were mostly about 3 mm. long, and it is this material which is responsible for the flecking of lighter colour apparent on the surface of comparatively fresh faecal pellets. Most grass fragments which were specifically identified were of *Stipa semibarbata*, the most abundant grass in the colony area. A few specific identifications were made of *Danthonia setacea* but none of the *Poa* species. It was noted that the *Danthonia*, though comparatively scarce, was closely cropped in parts of the colony area.

Fragments of foliage of myrtaceous shrubs were found in two-thirds of the faecal pellets. Most of them were of Rosy Heath-myrtle (*Baeckea ramosissima*), a species quite uncommon in the colony area but heavily cropped wherever it grew. Only a few fragments each of *Leptospermum myrsinoides*, *Melaleuca decussata* and *Calytrix tetragyna* were found in the faeces, despite their all being abundant in the area.

Seeds and fragments of the capsules of Pigmy Purslane (*Calandrinia neviana*) were found in more than half the faeces examined. Evidently the wallabies are partial to this tiny succulent plant. Fragments of moss, seeds and glumes of *Scirpus antarcticus*, and seeds similar to those of *Hydrocotyle callicarpa*, each occurred in a few pellets. As these three, and the purslane, grow only about an inch in height, the indication is that the wallabies occasionally crop very short herbage.

Plants of Germander Raspwort (*Haloragis teucrioides*) and Variable Stinkweed (*Opercularia varia*) were seen to be cropped in the colony area, and several fragments of each were found in the faeces. Fragments of twigs of *Callitris rhomboidea* occurred in eight of the 40 pellets which were sampled, and floral bracts of Thread Rapier-sedge (*Lepidosperma filiformis*) occurred in several pellets.

The dioecious *Dodonaea cuneata* was represented in several pellets by either male flowers or by seeds and fragments of seed capsules. Seeds of *Goodenia ovata* were found in three pellets, and pieces of the flowerheads of unidentified species of the Compositae occurred in several of the samples. These details indicate that inflorescences of shrubs constitute a significant item in the rock-wallabies' diet.

Fragments of arthropod exoskeletons were found in almost half the samples analysed. Accidental ingestion of such animals would occur when shrub foliage and inflorescences were eaten and when very low ground herbage was cropped.

In the early days of European settlement, sheep were grazed on the Victoria Range, but now the original grassy parkland has been replaced

Figure 7. The rock-wallaby family, 4/1/1971.
The female's tail is protruding from the shadow above the joey.
(Enlargement x14 from 35 mm Pan F negative taken with 300 mm. lens at 50 m range.)

Figure 8. The female rock-wallaby basking on recessed ledge, 16/1/1971.
(Enlargement x3.3 from 35 mm negative taken with 300 mm lens at 65 m. range.)

by dense tough shrubbery (Wakefield 1963a). Presumably, that change in vegetation has been general in the Grampians, and it is probable that this has been a major factor in the almost complete elimination of rock-wallabies from the region.

Associated mammals

The Silver-grey Possum (*Trichosurus vulpecula*) is co-existent with the rock-wallabies, and it is evidently abundant. It lives in similar rock crevices and, on the evidence of its faeces, ranges over the high exposed outcrops just as the wallabies do. Also on the evidence of faeces, the Common Ringtail (*Pseudocheirus peregrinus*) is present but scarce, and an insect-eater, probably the Yellow-footed Phascogale (*Antechinus flavipes*), is present, too. Diggings of the Echidna (*Tachyglossus aculeatus*) are abundant, even on high cliff ledges, and the Brush Wallaby (*Macropus rufogriseus*) inhabits contiguous heathland tracts.

Tracks and faeces indicate that the introduced Ship Rat (*Rattus rattus*) is in the colony area, and feral goats (*Capra hircus*) have camped in two rock shelters there, though not recently. A single set of faeces indicates that the European fox (*Vulpes vulpes*) is an occasional visitor to the area. The rabbit (*Oryctolagus cuniculus*) is present in some heathland tracts, and the Red Deer (*Cervus elaphus*) visits these too, but there is not abundant evidence of either.

Extent of occurrence

Following reports by naturalists, of the observation of faeces thought to be those of rock-wallabies, a number of exploratory excursions were made to other parts of the Victoria Range. The loci were (a) near the head of Honeysuckle Creek, 8 miles south of Red Rock, (b) the

Mount Thackeray area, 7 miles SE. of Red Rock, (e) below the Coal track, 5 miles ESE. of Red Rock, and (d) above Red Rock Creek, 2 miles south of Red Rock. No evidence was found of rock-wallabies in any of these areas. The rock formations above and to the south of Red Rock Creek appeared to be ideal habitat, but, on the evidence of faeces, they contained no *Petrogale* though *Trichosurus* was obviously abundant. The faeces reported from near Mount Thackeray proved to be of the Koala (*Phascolarctos cinereus*), and at Honeysuckle Creek only faeces of *Trichosurus* were found.

At present, the rock-wallabies in the Red Rock area comprise the only extant colony known in western Victoria. There appear to be only the three living places in the colony area, and one of these is evidently occupied by no more than the family of three individuals which we observed. Thus, the population of the whole colony may be only several animals.

Conservation

Most of the Grampians is reserved forest, under the jurisdiction of the Forests Commission of Victoria, and the area is a sanctuary for wildlife. The rock-wallaby colony is within the reserved forest and sanctuary, but by only a few hundred yards. Closely adjacent areas are Crown Lands, and two small pieces of these should be added to the reserved forest so as to afford the rock-wallaby colony a greater measure of permanent protection.

Because of the probable very low numerical status of the rock-wallaby colony at Red Rock, its immediate environs should be kept as free from human intrusion as possible. Disturbance of the animals, even by a small number of zoologists or naturalists who might be genuinely interested in observing them, could upset the delicate balance of their survival there. We observed this principle so far as possible during our assessment of the rock-wallaby colony, and in particular we refrained from entering the broken outcrop which houses the three animals that we observed, though it would have been most interesting to have examined their home.

ACKNOWLEDGMENTS

Forests Commission officers at Stawell assisted in many ways, and Mr. Ian Smith in particular was most co-operative and helpful. The quantitative part of the analysis of food of the Grampians rock-wallabies was carried out by Russell Bathard, and he assisted also with general work during three field excursions to the region.

REFERENCES

Audas, J. W., 1925. *One of Nature's Wonderlands. The Victorian Grampians.* Ramsay Publishing, Melbourne.

———, 1950. *The Australian Bushland.* W. A. Hamer, Melbourne.

Linyard, J., date unknown. *Narrative of a Journey to and from New South Wales.* J. Taylor, Printer. London.

Thomas, H., 1868. *Outs. A Guide for Excursionists from Melbourne.* Edited and published by H. Thomas, Melbourne.

Wakefield, N. A., 1953. The Rediscovery of the Rock-wallaby in Victoria. *Vict. Nat.,* 70 (11): 202-6.

———, 1961. Victoria's Rock-wallabies. *Ibid.,* 77 (11): 322-32.

———, 1963a. Mammal Remains from the Grampians, Victoria. *Ibid.* 80 (5): 130-33.

———, 1963b. Notes on Rock-wallabies. *Ibid.,* 80 (6): 169-76.

———, 1964a. Mammal Sub-fossils from Basalt Caves in South-western Victoria. *Ibid.,* 80 (9): 274-87.

———, 1964b. Recent Mammalian Sub-fossils of the Basalt Plains of Victoria. *Proc. Roy. Soc. Vict.,* 77 (2): 419-25.

Rediscovery of the Large Desert Sminthopsis (*Sminthopsis psammophilus* Spencer) on Eyre Peninsula, South Australia

by PETER F. AITKEN[*]

INTRODUCTION

Sminthopsis psammophilus, The Large Desert Sminthopsis, (Plate 1) was originally described from a unique specimen taken in 1894 amongst porcupine grass tussocks on the sand-ridge desert near Lake Amadeus in the Northern Territory of Australia. The single example was captured by Mounted Trooper E. C. Cowle, then stationed at the Illamurta Police Post, and presented to members of the Horn Scientific Expedition to Central Australia, for whom he was acting as guide. Subsequently in 1895, the specimen was described as a new species by Sir W. Baldwin Spencer, then Professor of Zoology at Melbourne University. For 75 years no further example was discovered and the species was feared to be extinct.

REDISCOVERY

On 28 February, 1969, a second specimen was captured by Mr. Mervyn Andrews, a cereal farmer, on his property at Section 10, Hundred of Mamblyn, Eyre Peninsula, South Australia; approximately 650 miles south-east of the Type locality. The capture area, situated in a zone of 12-14 inch rainfall, features a topography of roughly parallel sand-dunes, between 20 and 40 feet in height, separated by wide interdune valleys. Originally the system had a fairly uniform covering of mallee-broombush vegetation with sparse ground cover. Mallees (*Eucalyptus* spp.) tended to predominate in the valley flats, but became more heavily admixed with broombush (*Melaleuca uncinata*), plus other herbaceous and ephemeral plants, on the dunes. Examples of native cherry (*Exocarpus sparteus*) and native pine (*Callitris canescens*) occurred sporadically throughout, and semi-open areas of porcupine grass (*Triodia lanata*) appeared intermittently on the dune slopes. In recent years however, considerable acreages of natural scrub have been cleared from the inter-dune valleys and the land developed for cereal growing. The sand-dune vegetation, on the other hand, has remained largely intact, or has been permitted to regenerate following initial logging and burning activities.

The Large Desert Sminthopsis was taken in daylight, while escaping from a blazing porcupine grass tussock, growing at the base of a re-vegetated dune adjacent to a cleared inter-dune valley. Clearing of both dune and valley had been completed 3 years previously, but the developed valley paddock had remained fallow during this period. The capture spot was at the extreme corner of the paddock, which itself represented the limit of clearing. Ten yards beyond, the dune vegetation continued in a virgin state and the inter-dune valley remained uncleared, although the

[*] South Australian Museum.

under-shrub had been degraded by sheep grazing.

Following presentation of the Mamblyn specimen to the South Australian Museum, the area was immediately examined by the Author with members of the South Australian Field Naturalists Society Mammal Club. The search failed to reveal either additional examples of the Large Desert Sminthopsis, or further data on the habits of the species. Numerous Mitchell's Hopping-mice (*Notomys mitchelli*) were collected nearby however, and introduced house-mice (*Mus musculus*) were found to be abundant throughout the surrounding district.

As a result of this failure, an intensive campaign for more specimens was initiated throughout northern Eyre Peninsula, and, on 23 April, 1969, Mr. Dennis Eichner, a clearing contractor of Bailey Plains, collected four further specimens while logging and burning mallee-broombush scrub on Section 45, Hundred of Boonerdoo. This area was of identical topography and vegetation to the Mamblyn locality, but situated some 50 miles to the south-east. One of Mr. Eichner's specimens subsequently escaped, but the remainder were deposited in the South Australian Museum.

Discovery of the Boonerdoo colony not only confirmed that the Large Desert Sminthopsis was a viable species of Eyre Peninsula, but provided additional evidence that mallee-broombush scrub with porcupine grass, was a preferred habitat. The discovery served also to allay fears that the species would disappear again soon after rediscovery. Such could well have been the case at Mamblyn, where clearing of the habitat is proceeding rapidly. At Boonerdoo, on the other hand, all four specimens were collected from an area immediately adjacent to the Hambidge Wildlife Reserve, where the topography and vegetation is essentially the same as that of the capture locality. It is reasonable to assume therefore, that the Large Desert Sminthopsis is represented on the reserve, where, with proper management, its chances of survival are good.

Plate 1.
Sminthopsis psammophilus, Mamblyn, South Australia.

(Photo by Konrad Kuehle).

Nomenclature

Sminthopsis psammophilus Spencer. 1895, Proc. Roy. Soc. Victoria, new ser., 7: 223.

1896. Report on the Horn Sci. Exped. to Central Australia, 2 (Zool.): 35-6, pl. 1, fig. 2, 2a-h.

Sminthopsis macrura psammophila Tate, 1947, Bull. Amer. Mus. Nat. His. 88 (3): 123.

Sminthopsis psammophila Troughton, 1964, Proc. Lin. Soc. New South Wales, 89 (3): 317.

Material

The original description of *Sminthopsis psammophilus* included no cranial details, since it was based on a single spirit preserved specimen with the skull *in situ*. An expanded account of the species is appended, based on the Holotype in the National Museum of Victoria: no. C6203, adult ♂, (spirit body with skull now removed) "Sand Hills near Lake Amadeus C.A., 1894, E. C. Cowle". Plus 4 specimens from the South Australian Museum: nos. M7662, adult ♂, (puppet skin and skull) Sec. 10, Hundred of Mamblyn, S.A., 28 Feb. 1969, M. Andrews; M7663, adult ♂, (spirit) M7971 and M7972, adult ♀ ♀, (puppet skins and skulls) Sec. 45, Hundred of Boonerdoo, 23 Apr. 1969, D. Eichner.

Pelage colour nomenclature follows the standards of Ridgeway, (1912) and all measurements are in millimetres with the terminology, unless otherwise stated, after Cockrum (1955). Skull measurements were made with Helios dial calipers under a binocular microscope. Flesh dimensions were obtained from flesh material in the case of South Australian Museum specimens and from a spirit preserved specimen in the case of the Holotype (Spencer, 1895 *op. cit.*).

Field Identification

An exceptionally large and robust *Sminthopsis*, clearly recognisable from other members of its genus by the laterally thin, dorso-ventral, feather-like crest of stiff hairs on the distal quarter of the tail (Plate 2).

External Features

Dorsally the fine, dense fur is approximately 14 mm. long, of which the basal 10 mm. are dark mouse grey, the medium 3 mm. are drab grey and the terminal 1 mm. is fuscous black. Interspersed with the fur are spines 18 mm. long, of which the basal 8mm. are dark mouse grey and the thickened, apical 10 mm. are fuscous black. The back is thus imparted with a drab, brindled appearance that is continued forward on the head in a wedge shaped patch with its apex between the eyes. Elsewhere on the dorsal surface of the head, the fur is shorter and more bristly, with less pronounced fuscous black tips and fewer spines, giving a lighter, less brindled effect. A fuscous black ring of hair is present around the eye.* Mystical vibrissae number approximately 25 on each side, are up to 35 mm. in length and are predominently fuscous black. Other vibrissae per side are: orbital, 2, fuscous black; facial, 2, fuscous black; and carpal, 2, white.

On the cheeks, flanks, shoulders, hips and thighs, the fuscous black tips of the fur are markedly reduced, the median shaft colour alters to vinaceous buff, and spines are vir-

*Spencer was in error in his 1896 account of the Holotype, in which he stated — 'There is a white line of hairs around the eye'. Similar examination of the Holotype revealed no hair of such a feature nor was it apparent in any of the fresh material to hand.

tually absent. A pronounced vinaceous buff colouration is thereby produced in these areas.

The membranous, ovate ear has a well developed, rounded tragus, 3.5 mm. long, by 3 mm. wide, an anterior fringe of vinaceous buff and white hairs, and an antero-external patch of fuscous black bristles. The remainder of each ear is sparsely covered, inside and out, with short, white bristles.

Ventrally the fur is fine, with no spines, and is approximately 10 mm. long, of which the basal 4 mm. are dark mouse grey and the apical 6 mm. are white. On the chin the fur is sparse, short and white throughout.

Both fore and hind feet are clothed with short, white hairs extending to the bases of the pads on the fore-foot and covering the sole of the hind-foot, except for the pads plus a bare, median area of fuscous black, granular skin, which extends forwards from 3 mm. behind the base of the hallux to the bases of the interdigital pads. Apart from this latter area, the soles and pads of both fore and hind feet are ivory yellow in colour. The fore-foot is approximately 4.5 mm. broad and has six semi-coalescent pads, of which three are interdigital; one is at the base of the first digit, and two are plantar. Each pad is covered with uniform, fine granulations, and is surmounted by a smooth, enlarged granule, equal to two granulations in size. The hind foot is approximately 5 mm. broad across the bases of the digits and has three interdigital pads, of which the most external is horseshoe-shaped. No hallucal pad was discerned. Each interdigital pad is covered with uniform, fine granulations and surmounted by a longitudinal row of smooth, enlarged granules, four or five in number, each granule being approximately equal to two granulations in size.

The tail is extremely muscular and tapers towards the tip, but has no sign of secondary fattening or basal incrassation. It is tri-coloured and covered with stiff hairs 3 mm. long. Dorsally the colour is brindled, drab grey with vinaceous buff toning; laterally it is pale smoke grey and ventrally it is fuscous black. On the

Plate 2.

Sminthopsis psammophilus showing details of the diagnostic tail crest.

(Photo by Roman Ruehle)

distal quarter of the tail a laterally thin, dorso-ventral, pennate crest of stiff hairs is present. These hairs are from 4-5 mm. in length and coloured fuscous black in two specimens; drab grey with fuscous black tips in two others; and vinaceous buff with fuscous black tips in the fifth specimen. Dorsally the crest is approximately 25 mm. long and is weakly developed, ventrally however, it is some 30 mm. long and strongly developed with densely packed, erect hairs.

Table 1 provides data on the flesh dimensions from all known specimens of *Sminthopsis psammophilus*, Spencer, 1895 (*op. cit.*) when tabulating the tail measurement in his original description of the Holotype, stated: "Very tip broken off". I calculate that approximately 5 mm. are missing.

TABLE 1

Flesh Dimensions of *Sminthopsis psammophilus*

Measurement	Holotype ♂ C6203 Lake Amadeus (Ex. Spencer)	♂ M7662 Mamblyn	♂ M7663 Boonerdoo	♀ M7971 Boonerdoo	♀ M7972 Boonerdoo
Body length	105	114	112	91	91
Tail length	116 (Very tip broken off)	129	114	116	126
Length of hind foot (s.u.)	25	26	25.7	25.2	25.5
Height of ear (from notch)	24.5	26.8	25.1	23.1	23.3

SKULL CHARACTERS

An elongate skull, with an exceptionally high crowned cranium, produced by the exaggerated, bulbous inflation of the anterior portions of the parietals. This results in a diagnostic cranial depth of at least 12 mm. (Plate 3).

Nasals attenuate, tapering slightly towards the front from a minimal, postero-lateral expansion. Post-orbital processes poorly defined as broad, blunt points without constrictions behind. Sagittal crest minute, but noticeable on the posterior third of the median parietal suture. Lambdoidal crests of similar appearance. A pair of supplementary palatal openings between M4-4. Alisphenoid bullae fractionally less inflated in proportion to skull length than those of *S. crassicaudata* (genotype species). In comparison with adult skulls of *S. crassicaudata* in the South Australian Museum, the ratio for greatest length of skull to greatest breadth of alisphenoid bulla of *S. psammophilus*, compared with *S. crassicaudata*, is 1.06:1. The periotic bullae appear slightly more swollen than those of *S. crassicaudata*, due to the absence of median longitudinal depressions.

Dental formula. $I\frac{4}{3} : C\frac{1}{1} : P\frac{3}{3} : M\frac{4}{4}$. Teeth, upper jaw: I^1 prominent, set apart from I^{2-4}, which are smaller and increase in size to the rear by crown length, but not crown height. C^1 more than twice as high as I^4 and separated from it by a diastema of approximately 1 mm. Minute posterior cingula are present on the canines of two out of the four specimens examined. $P^{1,3,4}$ well spaced, not separated by a diastema from the canine, and increasing evenly in size to the rear by both crown height and crown length; P^1 being approximately half the size of P^4. Small anterior and posterior cingula are normally present on all premolars, but vary considerably in development. In one specimen examined, both cingula were apparently absent from P^{1-1} and were particularly distinct on P^{3-3}. Molars high cusped.

Lower jaw: I_1 not set apart from I_2 and fractionally superior to it in both crown height and crown length,

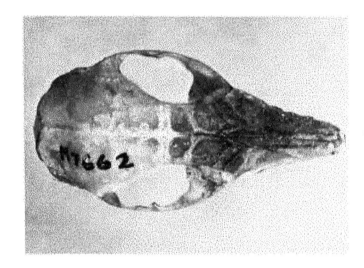

Plate 3A. Skull of *Sminthopsis psammophilus*: Dorsal view of cranium and upper jaw.

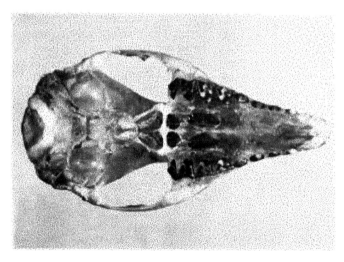

Plate 3B. Skull of *Sminthopsis psammophilus*: Ventral view of cranium and upper jaw.

(Photos by Roman Ruehle).

I_2 smaller than I_2 in crown height, but greater than either I_1 or I_2 in crown length, through the possession of a minute postero-labial cusp. C_1 approximately twice as high as I_3, with a small posterior cingulum, and not separated from I_{1-3} by a diastema. $P_{1,2,4}$ well spaced, with no diastema between P_1 and the canine. In general P_1 and P_4 are of approximately equal dimensions, although P1-1 are noticeably smaller in one specimen examined. Both P_1 and P_4 are inferior in size to P_3. All premolars possess small anterior and posterior cingula. Molars high cusped.

Table 2 provides data on the cranial and dental dimensions from the Holotype and three other skulls of *Sminthopsis psammophilus*, all hitherto undescribed.

Plate 3C
Lateral view of cranium and upper jaw.

Plate 3D
Lateral view of mandible.

(Photos by Roman Ruehle).

April, 1971

Table 2

	Holotype ♂ C6203 L., Amadeus	M7662 ♂ Mainblyn	M7971 ♂ Boonerdoo	M7972 ♂ Boonerdoo
Greatest length of skull (from anterior tip of premaxilla)	—	34.9	32.5	33.3
Basilar length	30.3	30.4	27.8	28.9
Zygomatic breadth	18.0 (cal.)	17.8	17.3	17.6
Cranial breadth	13.7	13.7	13.3	14.0
Least interorbital constriction	6.3	6.7	6.4	6.2
Rostral breadth	4.7	4.5	4.5	4.4
Greatest width across upper molars	11.0	11.2	10.6	10.6
Depth of cranium	12.3	12.2	12.0	12.0
Palatilar length (inc. spine)	18.3	18.2	17.1	17.5
Greatest breadth of alisphenoid bulla	5.3	5.0	4.9	5.0
Length of incisive foramina	4.6	4.4	4.1	4.0
Length of anterior palatine foramina	4.8	4.6	4.4	4.6
Length of posterior palatine foramina	1.4	1.6	1.5	1.4
Width of posterior palatine foramina	1.6	1.8	1.8	1.7
Length of nasals	12.7 (cal.)	13.0	12.0	13.0
Greatest width of nasals	2.9	2.6	2.8	2.6
Dorsal length of pre-maxilla	6.7 (cal.)	8.8	7.6	8.2
Length of mandible (exc. teeth)	—	25.9	24.2	24.8
Crown length of I^1	1.0	0.9	0.8	0.7
Crown length of I^{2-4} (Inc.)	2.05	2.25	2.3	2.25
Crown lengths of $P\frac{1}{1}$	1.4	1.4	1.5	1.4
$P\frac{3}{3}$	1.8	1.4	1.5	1.5
$P\frac{4}{4}$	1.9	1.8	1.8	1.7
	—	1.8	1.8	1.8
	—	1.9	1.9	1.9
	—	1.7	1.65	1.7
Crown heights of $P\frac{1}{1}$	—	1.0	1.1	1.0
$P\frac{3}{3}$	—	1.1	1.05	1.05
$P\frac{4}{4}$	—	1.4	1.4	1.2
	—	1.5	1.4	1.3
	—	1.9	1.85	1.5
	—	1.4	1.3	1.1
Crown length of M^{1-3} (Inc.)	6.2	6.4	6.5	6.5
Crown length of M^{1-4} (Inc.)	7.1	7.4	7.5	7.5

Skull Dimensions of *S. psammophilus*.

SPECIFIC RELATIONSHIP

Tate, (*op. cit.*) in the only modern attempt to evaluate phylogenetic relationships in the genus *Sminthopsis*, separated the known species into two divisions. A "primitive" division characterised by striated foot-pads and P^4 very large with P^1 (and sometimes P^3) considerably reduced; and a more "advanced" division comprising forms with granulated foot-pads, no extreme enlargement of P^4, and P^1 subequal to P^3. Tate also stated that those species in the "primitive" division possessed strongly built skulls with well formed, low sagittal crests and relatively small bullae; and that for those species in the "advanced" division, the defining characters could be accompanied variously, by partial coalescence of individual foot-pads, narrowing of the feet, caudal incrassation and enlargement of the ears and bullae.

Troughton, (*op. cit.*) pointed out that members of the *Sminthopsis murina* complex, included by Tate in his "primitive" division, do not possess striated foot-pads. Nevertheless under Tate's defined arrangement, *S. murina* could hardly be transferred to the "advanced" division, due to its enlarged P^4, relatively small bullae and low sagittal crest. A number of other anomalies are apparent in Tate's arrangement. *S. froggatti* for example, (=*S. froggatti* and *S. larapinta* of Tate) included in the "advanced" division, possesses granulated foot-pads, moderately large ears and an incrassated tail, but has relatively small bullae, a low sagittal crest and an enlarged P^4. Similar difficulties occur with *S. hirtipes*, which has large ears, specialised, coalescent, granulated foot pads, and huge bullae, but an exceptionally large P^4 and a reduced P^3.

It is thus apparent that the combination of characters defined by Tate do not separate the species of *Sminthopsis* into clear phylogenetic divisions, nor, in fact, should such divisions be considered realistic until the characters of the genus have been re-examined within a broader family and palaeontological context, to determine which are primitive, which are generalised and which are derivative. In view of this no attempt has been made to place *Sminthopsis psammophilus* in a phylogenetic position within the genus.

Ride, (1970) however, in a popular account of *Sminthopsis* divides the genus for convenience into two sections; those with fat tails and those with thin tails. Under such an arrangement *Sminthopsis psammophilus* falls naturally into the thin-tailed section, where it may be placed adjacent to *S. murina*, with which it shares a distinct similarity of foot-pad structure, dental development and overall cranial conformation.

REFERENCES

Cockrum, E. L., 1955. *Manual of Mammalogy*, Burgess, Minneapolis.
Ride, W. D. L., 1970. *A Guide to the Native Mammals of Australia*, Oxford University Press, Melbourne.
Ridgway, R., 1912. *Colour Standards and Colour Nomenclature*, author's publication, Washington.

Acknowledgements

I am particularly indebted to Mr. Mervyn Andrews of Mamblyn, Mrs. J. Luscombe of Kyancutta, Mr. Dennis Eichner of Bailey Plains and Pastor Ivan Wittwer of Cleve, without whose co-operation the recent material of *Sminthopsis psammophilus* would never have reached the South Australian Museum. I am also extremely grateful to Mr. John McNally, Director, and Miss Joan Dixon, Curator of Vertebrates at the National Museum Victoria, for allowing me to examine the Holotype.

CORRECTION

"In *Vict. Nat.* **88**: 74 (March 1971), under *Exhibits, Scaevola* sp. aff. *linearis* is mentioned as a new record for Victoria, being found near Annuello. Further, more complete material, of this plant has now been received at the National Herbarium, and full examination shows that the identification given above is incorrect. The Annuello material and also some from Hattah Lakes National Park, should be placed with *S. aemula*. In Victoria this species is more commonly known from southern and eastern localities."

Field Naturalists Club of Victoria

Established 1880

OBJECTS: To stimulate interest in natural history and to preserve and protect Australian fauna and flora.

Patron:
His Excellency Major-General Sir ROHAN DELACOMBE, K.B.E., C.B., D.S.O.

Key Office-Bearers, 1971-1972.

President:
Mr. T. SAULT

Vice-Presidents: Mr. J. H. WILLIS; Mr. P. CURLIS

Hon. Secretary: Mr. D. LEE, 15 Springvale Road, Springvale (546 7724).

Hon. Treasurer & Subscription Secretary: Mr. D. E. McINNES, 129 Waverley Road, East Malvern 3145

Hon. Editor: Mr. G. M. WARD, 54 St. James Road, Heidelberg 3084.

Hon. Librarian: Mr. P. KELLY, c/o National Herbarium, The Domain, South Yarra 3141.

Hon. Excursion Secretary: Miss M. ALLENDER, 19 Hawthorn Avenue, Caulfield 3161. (52 2749).

Magazine Sales Officer: Mr. B. FUHRER, 25 Sunhill Av., North Ringwood, 3134.

Group Secretaries:

Botany: Mr. J. A. BAINES, 45 Eastgate Street, Oakleigh 3166 (57 6206).

Microscopical: Mr. M. H. MEYER, 36 Milroy Street, East Brighton (96 3268).

Mammal Survey: Mr. P. HOMAN, 40 Howard Street, Reservoir 3073.

Entomology and Marine Biology: Mr. J. W. H. STRONG, Flat 11, "Palm Court", 1160 Dandenong Rd., Murrumbeena 3163 (56 2271).

Geology: Mr. T. SAULT.

MEMBERSHIP

Membership of the F.N.C.V. is open to any person interested in natural history. The *Victorian Naturalist* is distributed free to all members, the club's reference and lending library is available, and other activities are indicated in reports set out in the several preceding pages of this magazine.

Rates of Subscriptions for 1971

Ordinary Members	$7.00
Country Members	$5.00
Joint Members	$2.00
Junior Members	$2.00
Junior Members receiving Vict. Nat.	$4.00
Subscribers to Vict. Nat.	$5.00
Affiliated Societies	$7.00
Life Membership (reducing after 20 years)	$140.00

The cost of individual copies of the Vict. Nat. will be 45 cents.

All subscriptions should be made payable to the Field Naturalists Club of Victoria, and posted to the Subscription Secretary.

Published by the
FIELD NATURALISTS CLUB OF VICTORIA
In which is incorporated the Microscopical Society of Victoria

Registered in Australia for transmission by post as a periodical.
Category "A"

F.N.C.V. DIARY OF COMING EVENTS

GENERAL MEETINGS

Monday, 10 May — At National Herbarium, The Domain, South Yarra, commencing at 8 p.m.
1. Minutes, Reports, Announcements.   2. Nature Notes and Exhibits.
3. Subject for the evening — "New Guinea Highlands": Mr. J. H. Willis.
4. New Members.
   Ordinary:
   Mr S E. J. Mayhew, Flat 5, 28 Walsh St., South Yarra, 3141.
   Mr. R. B. Withers, Flat 14, 55 Clow St , Dandenong, 3175.
   Miss Anne Owen, Flat 13, 19 Riversdale Road, Hawthorn, 3122.
   Country:
   Mrs. A. D. Bathard, 63 Dimboola Road, Horsham, 3400.
5. General Business.
6. Correspondence.

Wednesday, 9 June — "The Life of the Badger": Mr. Robert Withers (formerly of British Mammal Society).

Monday, 12 July — "Antarctica in Mid-Summer": Dr. M. Beadnall.

F.N.C.V. EXCURSIONS

Sunday, 9 May — Botany Group excursion to Blackburn Lake. Meet 2 p.m. at Blackburn Station.

Sunday, 16 May — Toolangi district. Leader: Mr. J. H. Willis. The coach will leave Batman Avenue at 9.30 a.m. Fare $1.60, bring one meal and a snack.

Saturday, 21 August to Sunday, 5 September — Flinders Rauges. The coach will leave Melbourne Saturday morning and travel via Bordertown, Adelaide, Quorn, Wilpena (2 nights), Arkaroola (4 nights), Wilpena, Port Augusta (3 nights), Renmark, Swan Hill, Melbourne. Most of the accommodation will be on a dinner, bed and breakfast basis but Adelaide and Wilpena will be bed and breakfast only, dinner à la carte. A certain amount of motel accommodation was available at Arkaroola, the remainder of the party will be in huts with bunk only but provisions can be obtained from the store or meals at the motel. Approximate cost $195.00 with motel at Arkaroola, $150.00 with bunks. Deposit of $50.00 to be paid when booking, balance by 3 July. Please do not send deposit before checking if there is a vacancy. Will any member whose name is on the list but who has not yet paid a deposit, please confirm booking. Cheques should be made out to Excursion Trust.

The Excursion Secretary would welcome offers to lead excursions.

GROUP MEETINGS
(8 p.m. at National Herbarium unless otherwise stated)

Thursday, 13 May—Botany Group. Mr. B. Fuhrer will speak on "Victorian Orchids".
Wednesday, 19 May—Microscopical Group.
Friday, 28 May—Junior meeting at 8 p.m. at Hawthorn Town Hall.
Wednesday, 2 June—Geology Group.
Wednesday, 2 May—Geology Group.
Monday, 7 June—Entomology and Marine Biology Group meeting at 8 p.m. in small room next to Theatrette at National Museum.
Thursday, 10 June—Botany Group. Speaker: Miss H. Aston.

The Victorian Naturalist

Editor: G. M. Ward

Vol. 88, No. 5   5 May, 1971

CONTENTS

Articles:

Some Observations on the Lowan at Wychitella.
    By Bob Johnson   116

F.N.C.V. Excursion to Cann River District.
    By Margery J. Lester   120

Burramys parvus Broom from Falls Creek area of the Bogong High
    Plains, Victoria.   By Joan M. Dixon   133

Aboriginal Paintings at Muline Creek. By Aldo Massola   139

Feature:

Reptiles of Victoria — 1.   By Hans Beste   117

Field Naturalists Club of Victoria:

Group Report .   ..   141

Diary of Coming Events   114

Front Cover:

The Freshwater Crocodile (*Crocodylus johnstoni*), inhabits the freshwater billabongs, lagoons, and rivers of far Northern Australia. Note the narrow snout as distinct from the broad snout of the larger Esturarine Crocodile. The Freshwater species is generally harmless to man.

                                                                                Photo: John Wallis.

Some Observations on the Lowan at Wychitella

by Bob Johnston

Early in June of 1970, I made a survey of the Lowan nests, and found four of these active. The activity was evident due to the birds having opened up a crater in each nest. This is a tapered hole into the mound, about 3 ft. 6 in. to 4 feet deep, and about 6 feet wide at the top.

This work is done from mid-June to July. There then appears to be a rest period of a week or two until about August, when the scratching and sweeping of the leaves and twigs begins. These are not gathered into the nest at once, but scratched or swept into a row which may extend sixty to eighty feet from the nest. This is, I believe, so that a good amount of the filling can be done in a short time. About three-quarters of the compost* is filled in during the one operation, and no further filling takes place until a fair rain has fallen. Some scratching is then carried on into this compost, as though to examine it for dampness. At this stage, more compost, composed of coarser materials than in the first case, is scratched up.

If there is a fall of rain following, the Lowan begins to fill in the loose earth of the crater, and this is generally mixed with old rotted compost which is damp. The filling begins well down the side of the compost heap, as though the birds make sure that this outer perimeter of soil is well packed. This careful filling is done in stages, taking a week or two; and if there is a dry break in the weather, slightly longer. But if the rain comes about this time, the Lowan is able to cover the top of the nest very quickly. In a day or so he can put up to two feet of soil over the crown of the nest.

I have noticed that the Echidna will burrow into the Lowan's nest. This year I found Echidna burrows in two nests. This was when the nests were about filled with compost. These burrows appeared to go well down into the base of the nests, as I put a 3 to 4-foot stick down each.

The Lowan does not appear to bother about this intrusion into its nest, and makes no effort to fill in the burrow; but covers up the entry with the earth used to cover the rest of the compost heap. I noticed these burrows about August, and reasoned that the Echidna would hibernate about this time, because it is about then that the Echidna hatches its young and feeds them in the pouch. Such a pile of loose earth and debris is often used by the Echidna, but not always, for I have found the burrows in hard earth on the gully banks too.

I wondered on noticing these burrows in the nests, if it was at all possible that the Echidna was after the eggs of the Lowan.

On the last Sunday in September there was a good fall of rain, and the bush had a good soaking. On the following Wednesday evening, I went with two friends to a Lowan nest on the west side of the road, and to my surprise I found a large Echidna dead by the side of the nest. This was one of the Lowan nests in which I had noticed the burrow

---

*The writer uses this term to describe the litter of leaves and twigs which is used to fill the nest.

early in August. The Echidna's snout was very badly damaged, as though it had been heavily pecked. In the soft earth beside the nest the soil was torn about as if there had been a great battle between the Lowan and the Echidna.

There were no other marks on the Echidna that I noticed, but it would have been hard to find these, unless there were any large scars.

Some time in April last I examined an Echidna in the bush area, about half a mile from this Lowan nest, and noticed that it had the end of the snout damaged, as though a spike had gone through and torn the end of the bill. I was puzzled at that time as to how this injury could have happened.

All I can conclude is that the dead Echidna we found beside the nest was killed by the Lowan.

I am anxious to hear if any readers can help with more evidence, and whether these two—the Lowan and the Echidna—are natural enemies at nesting time.

When you consider the possibility of a battle between these two, it would only be the Lowan who could be the aggressor, as he is the one with the natural weapons, a strong beak and stout legs with handy talons to match. This bird, too, must be of great strength and stamina, because of the way he works when building the nesting mound.

I have seen large dogs try to kill the Echidna, but they always gave up before doing any damage.

Perhaps the Lowan waits until the Echidna unrolls, and strikes the snout with a strong blow from its beak, thus stunning the anteater.

reptiles of victoria – 1

by HANS BESTE

INTRODUCTION

With the recent publication of P. A. Rawlinson's revised list of the reptiles of this state, this seems an opportune time to make the readers of this magazine more familiar with the species which can be found within the boundaries of Victoria. Australia is sadly lacking in books which deal with the identification of reptiles. The works which are available, are mostly out of date (especially with regard to nomenclature) and usually treat only selected species, or confuse the laymen with descriptions, which become quite useless in the field. Most of us identify animals with our eyes, and the picture which we get by imagining a species after reading its description, usually looks completely different from the actual animal. The most useful guide is therefore still the one that shows a photograph or drawing, supplemented with a description.

There are in the vicinity of one hundred species of reptiles found within the boundaries of this state, the distribution of which is not very well known; and many records are lost due to the inability of the average person to identify them. This is not entirely their fault; but is also because of the lack of field guides.

It is hoped to help fill this gap by publishing photographs and descriptions of our reptiles in serial form, not in a strict order, but as they come to hand, so that additions can be made from time to time, which may eventually cover most of the species on the Victorian checklist. So far, the writer has photographs of about half the species in hand, but with further work in the future this should give a good start in filling a gap which has existed for too long.

One thing, however, must be realized when comparing a live animal with the photographs, and this is the fact that many of our reptiles vary considerably throughout their range of distribution, and that colour changes can be quite drastic, especially when trying to identify dragons. Therefore, whenever in doubt of the identity of a specimen, consult the National Museum in Melbourne, whose staff should be able to answer your queries.

Plate 1

Ctenotus lesueurii — Striped Skink

A slim fast moving skink with a very long tail.
Length: to 12 inches.
Head pointed, distinct from body. Ear-opening as large as eye. Five fingers and five toes. Tail long and tapering to a point. A richly marked species with lines and spots. Central vertebral stripe dark-brown, edged with fine white lines. A lighter brown stripe each side of this, again edged with a fine white line. Sides brown spotted with white. A broader white band from below eye, encircling ear-opening to hindleg. Under white.

Found: in sandy areas, mallee and open country under rocks and among ground vegetation.

Best distinguishing features — stripes and shape. Distinguished from closely related species (*Ctenotus uber*) by absence of spots on upper surface. (Latter species has row of spots along light-brown stripe.)

Plate 2

Tiliqua occipitalis — Western Bluetongue

A large robust skink with a blue tongue.
Length: to 18 inches.
Head pointed, distinct from body. Snout rounded off slightly. Large ear-opening. Five fingers and five toes. Legs thick but short. Lizard drags body along ground when walking casually. Claws short but sharp. Brightly marked lizard with chocolate-brown and honey coloured cross bands. Brown line through eye. Normally slow mover but can move smartly if pursued.

Found: in extreme western desert areas of state.

Best distinguishing features — bold colouration, blue tongue.

Plate 1

Plate 2

F.N.C.V. Excursion to Cann River District
26 Dec 1970 — 3 Jan 1971

by MARGERY J. LESTER

Note: Common names of plants have been used in this report, but the scientific names are listed at the end as indicated by the number following each name. This report and its attached list necessarily omit many plants seen during the trip.

Thirty-seven members boarded the chartered coach at Melbourne on Saturday morning 26 December and set out on the 300-mile journey to Cann River. The hot weather and after-Christmas lethargy caused many nodding heads but, after the long-desired lunch stop at Sale, all were sufficiently alert to appreciate the view as we topped the hill to look over Lakes Entrance.

While passing through the tiny settlement of Cabbagetree, somebody asked about Victoria's only palm but we could see no Cabbagetree Palms[1] anywhere. We have since learned that the palms are several miles away and cannot be seen from the Princes Highway.

We came into the township of Cann River at about 5.30 p.m.

To Tamboon Inlet, 27 December

On this day and all subsequent days we were joined by three members who were camping in the area and by two members from the Bendigo FNC. Our party from Melbourne included a member of the Frankston FNC and a country member from Cope Cope. It was good to have these people from different clubs, and members of other affiliated clubs would also be welcomed at these excursions.

In spite of the heat, or perhaps because of it, we were all aboard the bus soon after 9 a.m. and set out due south for the head of Tamboon Inlet. The road went through forest all the way. The eucalypts consisted mostly of Silvertop[2], White Stringybark[3] and Messmate Stringybark[4]. For those to whom all eucalypts are still simply "gum trees" the Saw Banksia[5] was readily recognisable and occurred in abundance. There is something very Australian about Saw Banksias. The immense number of fruit cones on these trees suggests that they flower in abundance but, probably, many of the fruits were years old. Some had a few large, silvery heads of flowers still at the bud stage.

After a few miles the bus stopped while we all rambled through a more open area. The first excitement was the Large Duck Orchid[6]. There were several patches of them. The quaint resemblance to flying ducks is enchanting and, when a touch on the "head" caused it to snap down into the "body", there was a chorus of surprised delight. The head of the duck is actually the labellum which, in most orchids, is towards the bottom of the flower; and the body of the duck consists of the broadly-flanged column which is usually towards the top. But some orchids have the whole flower turned upside-down, and the Duck Orchids are among them. The lightning movement of the duck's head into the body sweeps an insect that touches the head (i.e. the labellum) down to the column where it is trapped. Unable to escape but able to move about, the insect carries out its role in the pollination system of the flower. Presently the labellum is lifted to release the insect which, we hope, goes off to another Duck Orchid to continue its pollination service. The labellum is then ready to be triggered by the next insect that alights on it. But what it is that attracts an insect to that chocolate-brown head does not seem to be known.

Later we found the Small Duck Orchid. Other orchids included the Horned Orchid³, Austral Leek Orchid⁴, Large Tongue Orchid⁵ and the Furred Tongue Orchid ⁶ which carries its flower vertically instead of more or less horizontally. Like Duck Orchids, the flowers of Leek Orchids and Tongue Orchids are inverted — the labellum is above the column.

We had enjoyed the soft colour of Blue Dampiera⁷ en route for it occurred in scattered patches by the roadside. It was plentiful at this stop too but was outshone by the Large Fan-flower⁸. These flowers are 1″-1½″ across with the five petals spread out on one side like a fan, so the common name could hardly be more apt. Some of them were a comparatively pale blue, some a mauve colour, and many were a wonderfully rich purple-blue.

There were several other flowers in the blue to purple range. There was the richly coloured Tufted Blue-lily⁹, the Blue-spike Milkwort¹⁰ (the pink Heath Milkwort¹¹ was there too and in much greater quantity), a low-growing mauve Fringed Lily¹² and, at a later stop near a tiny stream, we came on several clumps of Fairies' Aprons¹³. If these dainty little flowers were white instead of blue they would surely be called Ballerinas.

Of the many other plants at that spot Golden Spray¹⁴ was probably the most spectacular. Along a swampy area bushes up to six feet high carried long sprays of yellow pea flowers. The pink Swamp Heath¹⁵ also flourished in that damp part. And the tall fruiting spikes of a Grass Tree²¹ dominated a drier area.

There were several small, spiked, jewel spiders here but a larger spider captured the attention of us all. It was about 1″ long — the body striped black and yellow, the head grey. It was far too absorbed in feeding on a fly to be disturbed by cameras only 6″ away or by an occasional thrust against the grass blades to which its web was attached. It was a St. Andrews Cross Spider.

At another stop we found great quantities of Wedding Bush¹⁶. Some bushes were heavily loaded with 1″ spiky balls, but others still had a few flowers. Members were interested to learn that male and female flowers are produced on different bushes. They observed the difference between the two flowers and were then a bit shaken when they found male flowers on a bush bearing the ball-like fruits! But plants have an awkward tendency to produce exceptions that confound the general rule.

At various places along the road were lovely sprays of Prickly Tea-tree or Manuka⁰. This plant is so well-known and so widespread that we tend to overlook it in our eagerness to see less familiar things, but it is surely one of the attractive features of our bush — whether in far Gippsland or near-to-home Dandenongs. Or would this Gippsland species be the old *Leptospermum scoparium*?

Near Tamboon Inlet we were met by a member of the club who was camping there. He joined us again the next day at Tonghi Creek.

May, 1971

121

We lunched under the shade of a dense grove of Swamp Paper-barks" on the estuary of Cann River before it enters Tamboon Inlet. We watched the launching of several pleasure craft and did not envy the occupants as they advanced out to the unshaded water. And later, when some of our party went swimming, their anguished entry (and return) on the muddy bottom of the first few yards consoled us that we were not with them.

The protracted lunch time permitted individual rambles and provided some interesting observations. An Azure Kingfisher flashed in front of us; a 3-foot goanna was disturbed on the ground and promptly raced up a tree; a water-dragon raised itself with projecting elbows better to observe the interlopers; a large skink was lying on an anttrack so that the ants followed along the skink's tail and body to the head where they were smartly snapped up; and one member made a discovery we all went to see — Blue Olive-berry" in full flower. The 15-foot tree could scarcely have crowded on any more flowers.

The flowers of Blue Olive-berry are the most charming imaginable — ¼" white bells with a finely fringed edge —and here the little bells were hanging in hundreds in the cool shadow of the foliage. We were to see these flowers several times again but never in such astonishing abundance. If every one of those fairy-weight bells were fertilised and became a solid berry, we felt that the tree could hardly remain erect under the load. From later observation it would seem that many flowers are not fertilised so we can rest easy about that tree breaking!

To Tonghi Falls, 28 December

Aboard our bus soon after 9 a.m. we went five or six miles along the highway westward to Tonghi Creek Road. One of our members with a car took the bus driver to examine the road to the Falls. Apparently it resulted in an adverse report for, when they returned, we packed our

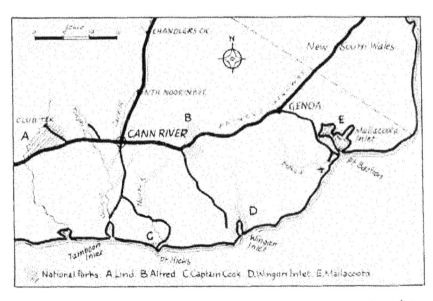

lurches into the car and walked. Later, the bus made it up the hill to our lunch spot, but we would have missed many interesting things had we missed that walk.

Again our route took us through forest, but a very different forest from yesterday. The eucalypts were mostly Mountain Grey Gum*, Swamp Gum*, White Stringybark again, and River Peppermint*.

There were extensive patches of the Derwent Speedwell* with its small white flowers gathered into long cone-shaped heads, and many flowers and plants familiar to us. A very interesting find was the Wombat Berry* in flower. The pendant ½" flowers were a very pale mauve, the three inner perianth members were fringed, and the stamens formed a projecting yellow tube in the centre. During the week we found more of these climbers bearing the familiar ⅜" orange balls, but few of us had seen the flowers before. Although the berries are far more conspicuous than the flowers, the latter would be hard to beat for delicacy of colour and form.

Further on we found the strikingly unusual flowers of Bearded Tylophora* The ¼" plum-coloured flowers were hanging in loose clusters. The five pointed petals, edged with fine hairs, formed a very precise star. In the centre were small raised parts making a darker star; and in the centre of that was a tiny, blunt-pointed white star. This geometrical appearance recalled the flower of Hoya, and we discovered that they both belong to the family Asclepiadaceae. All members of this family have a complicated system of pollination — sometimes as involved as the pollination of orchids — and the inner stats of Tylophora are part of that system. At various places during the week, we saw more of these scrambling plants with their opposite leaves, but never again did we see the flowers.

A large goanna with zebra-striped legs attracted comments, and everyone stopped to look at an unusual moth. The wings were a bright light green and biscuit, the large areas of each colour being separated by a fine white line; the legs were very broad, and the body had a blunt, turned-up end. About 1½" long, it was hanging motionless from a blade of grass.

The car had taken our lunches to a spot where the road met the creek. This S-bend of Tonghi Creek was chock-a-block with plants, large and small, and of wonderful variety. Muttonwood* carried its match-head size cream balls closely packed along the stems. Twiggy Heath-myrtle* crowded against the bridge with masses of 3/8" white flowers, white Lilly-pilly* carried its nondescript flowers in large loose clusters, and another member of the myrtle family puzzled us for a while as it had neither flowers nor fruit. It was Kanooka*. In Victoria, Kanooka occurs only in east Gippsland but it continues into NSW and Queensland. There were many plants more readily recognised and it was interesting to see Sweet Pittosporum * in its natural habitat instead of as a suburban hedge.

Climbers seemed to be everywhere. Wonga Vine* displayed its large, creamy, red-throated bells. The tough, conspicuously-veined leaves of Smilax* were very noticeable and at last we came on clusters of the blue-black berries. Water Vine* was also easily recognised by its leaves—about 4" long in a palmate formation.

Masses of the small Rock Felt-fern* were growing on an old tree trunk — *not* on rocks. Who but the knowledgeable would realise this was a fern? The "fronds" were quite unlike a fern for they had no pinnae and were simply circular plates about 1" across and very thick. The fertile "fronds" were about the same width but up to 1½" long and the undersides were covered with rust-red sori.

After lunch we all set out to walk the remaining one or two miles to the Tonghi Falls, but we departed at different times and at different speeds. The route took us through a farm, and Mrs. Morrison, wife of the farmer, led the first comers. Having no hesitation about the route, they moved away quite quickly and were lost to sight, so that many of the later ones got bushed and some never reached the Falls at all! Also, there was a bit of a hold-up soon after leaving the farm where the creek had to be forded. Many persons promptly removed shoes and socks, while the reluctant ones were finally lured over by the gallantry of two members who carried them across. This gallantry was extended not only to the ladies of the party.

There was a stiff climb over a couple of hills and we continued in private property where most of the timber had been removed. The long grass and weeds made us alert to possible snakes. In one of these clearings along the creek some members saw a pair of black chick. We re-joined the creek at the top of the Falls.

The Tonghi Falls consist of a series of cascades in a granite outcrop. Some of the boulders are up to 20 feet high.

Growing in the cracks of rocks alongside the Falls were several bushes of the Tonghi Bottlebrush". They carried crimson brushes 1½" to 3" long. Tonghi Creek is one of the few places in Victoria where this Callistemon occurs, but it is less rare in NSW.

When clambering down the rocks, one member rested his hand on a piece of bark. The bark moved and from under it slithered a black snake. Another member picked up a piece of greenish-coloured quartzite. And another saw several water dragons in the area. Most of the dragons hurried into the water when he tried to get nearer but one was less nervous. Our naturalist approached cautiously — on his stomach. He took a photo at a distance of three feet; advanced some more and some more and finally got a shot from about twelve inches. At first the biscuit-coloured dragon was lying out flat to get all benefit from the sun, then it raised the front of the body with elbows outwards and watched the photographer crawling forward.

Returning from the Falls, Mrs. Morrison showed us where petrified wood could be found. Actually, a large "log" of petrified wood had been placed on an oil drum to ensure that we saw it, but we went to the Falls by a different route and did not find it until we returned. The log looked just like wood, but lifting one end was enough to convince me it was wood no longer. Mrs. Morrison said that, when first in the area, she sometimes went to pick up a piece of wood and realised it was far too heavy to serve as fuel. She gave members small pieces to take away. The following week when at home, one member cut and polished a piece of this petrified wood and put it under the microscope. It clearly showed the cellular structure of wood.

In the evening a member showed slides of marine specimens and of the progress of the Cardinia Dam. Another member showed slides of coastal plants.

To Pt. Hicks, 29 December

Most of our bird enthusiasts were early risers, and this morning, a group went off at 6 o'cock with a timber worker to see a lyre-bird's nest in Lind National Park. On various mornings they came back with news of their early rambles. One member reported hearing a peculiar noise and found it was coming from a young Kookaburra. Perched alone on a bough, the youngster was trying to laugh. From a tree nearby the parent bird showed the proper way to do it and junior tried again. The result was still very odd—and very laugh-provoking to our observer.

On this pleasantly cool morning we followed the road towards Tamboon Inlet and, after about ten miles, took the left hand turning towards Pt. Hicks. We had an appointment with the lighthouse officers for 10.30 so we crossed the Thurra River without pausing but promised to stop there on our return.

We crossed the Thurra again at its estuary where there were several campers. About midway between the estuary and the lighthouse is the beginning of the Captain Cook National Park — only about 1½ miles in depth. Why more of this natural forest is not included in the Park is a mystery.

At the gate to the lighthouse enclosure we were met by a lighthouse officer in his jeep. We followed him along the narrow road to the group of houses clustered near the lighthouse at Pt. Hicks.

On getting out of the bus at Pt. Hicks, we discovered it was very windy and quite chilly so jackets and coats were hastily donned and securely buttoned. We all trooped down the slope to the extensive flat rocks to see the monuments that commemorated the sighting and naming of Pt. Hicks by Captain Cook on 20 April, 1770.

We returned up the slope to the lighthouse, about 100 yards inland from the monument. In groups of ten, we climbed the 150 steps to the light where another pleasant officer met us, supplied loads of information and answered innumerable questions. One of our members came away with a souvenir — a 1000 watt lamp that had seen only three-quarters of its life. At that stage the lamp is replaced by a new one so as to minimise the chance of light failure and hasty emergency action. The prisms around the light magnify the power 1000 times.

One of the officers very kindly led us to a good beach within the lighthouse enclosure where we lunched and remained a couple of hours. Another returned at 3 p.m. and unlocked the gate. Our president spoke for us all when he expressed appreciation of the friendly, generous attention these officers had given us.

The broad expanse of flat rocks at Pt. Hicks where the monuments stand is well above sea level, and these granitic rocks continue down to the sea in boulders of various sizes. One member followed them out as far as the tide permitted but was disappointed to find so little marine life. There were no pools among the rocks and limpets seemed to be almost the only things that could stand the perpetual pounding.

The stretch of flat rocks carried spreading patches of plants that one might expect to see such as Angular Noonflower", Sea Celery", and a plant we did *not* expect to see — a small Lobelia. These plants formed low-

125

growing mats sprinkled with pale blue flowers. We decided it was probably Angled Lobelia" made prostrate by its windswept position.

At our lunch spot we found almost all the common coastal plants that had been shown on the slides the previous night, and it was interesting to see that Saw Banksia had given way to Coast Banksia". We were surprised to find Lilly pilly only twenty yards from the beach, and Water Parsnip" supplied a real novelty.

At this lunching place there was a strange track in the sand near the rocks. It was about 6" across with four pointed marks slightly dragged. Our marine people decided it was made by a crab. A Sooty Oystercatcher was active near these rocks and there were Crested Terns and Cormorants. In the bush behind the beach were many honeyeaters—the Yellow-winged and the Yellow-faced Honeyeater and the Little Wattlebird.

On the return journey we made several stops including the promised one at the upstream crossing of the Thurra River. This was a delightful spot and would be immensely appreciated on a hot day. There were widely spreading Kanookas, trails of Smilax and several bushes of Blue Olive-berry, one with pink flowers instead of white.

More stops revealed more treasures: Forked Sundew" with a surprisingly tall stem bearing a cluster of surprisingly large white flowers; the lovely, deep blue Tall Lobelia"; flowering bushes of Pink Boronia"; and the intriguing Bushy Club-moss". This was about one foot high and almost every stem ended in a slender, creamy-green, male "cone". It looked like a miniature Christmas tree complete with candles. Conspicuous along several parts of the road were the tall, almost black heads of the Giant Sedge". Many had orange-red seeds generously distributed along the 2-3 ft. dark spikes.

Now and again a wallaby was seen in the bush and a black one leapt across the road in front of the bus —to the combined shout of all who saw it. Rainbow Lorikeets screeched attention to their brilliant colours, while a Wedge-tail Eagle sailed high seemingly aloof to the world below it.

Lind National Park, 30 December

Our bus took us west along Princes Highway and turned off to the old highway into the Lind National Park. Shortly after entering the Park we all got out of the bus and walked up the old highway with the valley of the Euchre Creek on our right. The drop from the road to the creek was steep and densely covered with vegetation in jungle-like profusion; we never saw the water of the Euchre except where other little creeks crossed under the road to join it.

There were many familiar plants here, some less familiar, and the ferns by the roadside were particularly striking. There were delightful stretches of Fan Fern", sometimes Coral Fern", and continually Gristle Fern" and the light green fronds of Rainbow Fern" in association with Bracken".

Again we saw several trees of Blue Olive-berry, some with white flowers, some with pale pink flowers, and one with berries. Attractive as the berries are, the flowers are so charming and unusual that we feel they should be acknowledged in the common name. We called them Fringed Bells.

We had learned to recognise the Water Vine by its palmate leaves and here it was carrying clusters of flower buds. Wonga Vine was there too but we found no flowers.

The Golden Everlastings[27] were attractive roadside features at the less ferny northern end, and just south of Club Terrace we found flowers of the Dusky Coral Pea[39]. The large pea flowers recall Sturt's Desert Pea but, as the common name suggests, the colour of Dusky Coral Pea is a muted red rather than a bright red.

All along the road our bird observers found much to interest them. There were Gang Gangs, Black Cockatoos, Dusky Woodswallows, Leaden Flycatcher, a female King Parrot, Rufus Fantails, and a young Rufus Whistler. One member watched a strangely selective incident. A young Grey Thrush was making a lot of noise which had gathered several female Thrushes and a female Golden Whistler. It is known to bird observers that a crying baby bird often draws the attention of some adult females but here an unusual thing happened. A young Cuckoo arrived and it was chased away. Was it chased away because it was a cuckoo or because it was young? The latter seems the more likely reason but even that seems strange.

We lunched at Olive Branch Creek where the bus had followed up with our food. We boarded the bus again at Club Terrace and arrived back at Cann River comparatively early but only just before heavy rain.

There were slides again this evening, slides of a holiday in New Zealand. Mr. Cliff Bueglehole was a very welcome visitor, but the plant enthusiasts had the poor man identifying specimens right up to midnight. Not that I think he minded, and the botanists were most appreciative of his help.

To Mallacoota, 31 December

We set out eastwards along the Princes Highway to Genoa and then turned off south-east for Mallacoota.

We noted the jungle-like growth as we passed through the Alfred National Park and wished we had somebody who knew the place and could take us investigating there. But, no such person being available, we consoled ourselves with thoughts of the leeches to which we would *not* be acting as blood donors.

A couple of miles before Genoa we were met by two club members who guided us to a good botanical area just north of the highway. Right by the roadside were tall spikes of Golden Spray, and further in were bushes of Crimson Bottlebrush[30] in full bloom, Wiry Baurea[31], and both Burgan[61] and White Kunzea[42]. There were several other flowers here including lovely clumps of the now very familiar Blue Dampiera and Large Fan-flower, an Onion Orchid[10] and some Large Tongue Orchids.

Another stop just before entering Genoa gave us a chance to put our glasses on some extensive ponds. There were Swans with cygnets, Black Duck, Spur-wing Plovers, White-faced Herons, and a wheeling flock of 20 to 30 Ibis, both White and Strawnecked.

The road from Genoa to Mallacoota took us through forest similar to that seen on previous days when we travelled south, but here there was the addition of Bloodwoods[44], recognised by their tesselated bark.

Mallacoota Inlet is certainly a lovely spot.

Mr. A. B. Peisley, a country member of the FNCV, met us at Mallacoota and, after the food shops had received our patronage, he led us to Betka River. The bus was parked among Swamp Paperbarks and we walked across the bridge to lunch on the ocean beach a few hundred yards further on. En route we found several Gum Myrtles", Victoria's only Angophora. They looked just like eucalypts but some of these Gum Myrtles had flower buds and we could see they lacked the little cap to the bud which is the main distinguishing feature of the eucalypts.

On one of the eucalypts a Mistletoe" carried orangy-yellow flowers 1½" to 2" long. But what caused even more interest than the handsome flowers was the presence of another mistletoe growing on it. The hunter hunted! The buds and stems of the second mistletoe were covered with hairs forming a sort of yellow felt. This was the Golden Mistletoe". Although the Golden Mistletoe occurs in NSW and Queensland, in Victoria it is confined to the Mallacoota area. It is always parasitic on another mistletoe, and usually on a member of the genus Dendrophthoe as in this case.

At the rocks near our lunching spot there was a basalt intrusion through the strata. Some of the rocks had fantastic formations and shapes. But again, as at Pt. Hicks, marine life was practically non-existent.

After lunch our party divided. Many members took advantage of the several cars to go to the heathlands near the aerodrome, while others stayed to bird-watch or to walk along the beach.

At the aerodrome heathlands the thickets of Southern Mahogany" drew our attention. Although Mahoganies are not large trees, here they were dwarfed and growing like Mallees, low and many-trunked. Two plants caused excitement for they were new to almost all of us: Blue Howittia" and the procumbent Lily Lily". But these names did not coincide with our interpretation of colour! The 1" flowers of Blue Howittia were a mauve-pink rather than blue, and those of Little Lily were pale pink! Orchids included the Large Tongue Orchid, Horned Orchid and the Hyacinth Orchid". The Hyacinth Orchid has no leaves and must get its food in an unusual way. It is probably saprophytic on humus in the soil, but its roots are often near the roots of a eucalypt to which, perhaps, it is connected by a fungus, thus drawing some of its food by a sort of secondary parasitism.

Three o'clock found us all in the bus again bound for Bastion Point to do some rock-pool hunting while the tide receded. But the tide was still rather high and there was little of interest to the marine enthusiasts. Also, the strata of the rocks were tilted almost vertical and made walking distinctly difficult. From Pt. Bastion one looks inland to the narrow entrance to Mallacoota Inlet.

We arrived back at Cann River soon after 3.30 p.m. Our social committee promptly got busy and remained busy right up to 9 p.m. in preparation for the New Year party.

The highlight of the party was the competition "natural arrangements". Actually they were most unnatural, very amusing and extraordinarily varied. They ranged from a large nestling made from sponge gathered on the beach to a small echidna of a weather-worn but still spiky Banksia cone. An enjoyable evening ended with the greeting of the New Year and singing "Auld Lang Syne".

To Wingan Inlet, 1 January

Our bus took us about ten miles eastward along the highway and turned south along the road leading to Wingan Inlet. It was a rough road and the hills went only as far as advised — to the turn-off at the Gateshill Track three or four miles from the coast. At this turn-off were huge Silvertop trees, seemingly in the process of being harvested.

At the turn-off various people went various ways. Many took their lunches and continued along the very rough "road" to Wingan National Park and Wingan Inlet. Some, also with lunches, took the Gateshill Track towards the Elusive Lakes. Others, reluctant to carry lunches, walked a couple of miles along one or other of those routes and returned to the bus for a meal.

We all saw the Giant Triggerplant" for it was fairly plentiful on both routes and those persons who missed observing it were urged to come and look at this astonishing thing. The young plants resembled introduced pine tree seedlings, but the bigger ones branched several times and at the end of each branch was an inflorescence 8" to 12" long. Most of them carried fruit but there were enough flowers to convince us they really were trigger plants. The majority of these plants were three or four feet high but a few went up to six feet. Writing in *"The Age"* some time ago, Mr. Norman Wakefield says that the Giant Triggerplant rarely lives more than two years. He says that, after the first flowers have died, four more shoots grow out from the top of the stem, each shoot bearing a panicle of flowers. If the plant lives to another season each of those four branches produces another four branches.

Along the Wingan road the rapid changes of vegetation were a bit bewildering. Dropping into a slight gully we would find Blanket-leaf" and other moisture-loving plants while on the hill only a hundred yards away was Saw Banksia! Further on we found Hill Banksia" and more of our beloved Fringed Bells, but several of these warranted the usual common name of Blue Olive-berry. There were lovely spreading bushes of Pink Boronia. The Narrow-leaf Geebung" was in flower, some had green berries, and the inner bark of the slender trunks roused interest— paper-thin layers of a bright rust-red colour, almost scarlet.

Stayers on the Wingan route were delighted to find the Lilac Lily again. This time the flowers were 1" to 1" across and mauve rather than the pale pink of yesterday. Towards the Inlet were huge Yellow Stringybarks" mixed with Bloodwoods. Very dark butterflies fluttered round the walkers but nobody knew what they were.

The track alongside the Inlet was marshy and continued right to the sea. About 200 yards out beyond the entrance to Wingan Inlet are "The Skerries", an irregular row of granite rocks. A colony of seals was at one end of low rocks and a colony of cormorants on higher rocks at the other end. As one member succinctly remarked "the colony of cormorants has obviously been there for very many years"!

Those who followed the Gateshill road towards the Elusive Lakes walked five or six miles but never arrived at any lake. At the crest of each rise they saw the lakes ahead through the trees but, arrived at the next crest, the lakes were still beyond them. Did our walkers follow the wrong road? Are the lakes only of seasonal occurrence? Or has the name been given to the mirages which, perhaps, are characteristic of that locality? Those people would certainly like to know.

The seekers of the Elusive Lakes disturbed the domestic duties of a pair of Spotted Pardolotes. The group was having lunch on a bank by the road when the pardolotes appeared on a nearby Geebung. Each had something in its beak and remained on the bush or fluttered near it and watched the munching Field Nats. Observing the birds observing them, the group decided they ought to move. The birds promptly flew down to the bank to a nesting tunnel that had been hidden by legs!

All parties from all routes were back at the bus by 4 o'clock and we returned over the rough road to Cann River.

That evening we observed an elusive lake of our own — and without walking half a dozen miles. Looking out the back window from the first floor of the hotel, we saw the valley a few hundred yards away covered with mist. The mist was low and flat, not at all furry, and looked just as water looks at twilight. There seemed to be no mist anywhere else.

Cann Valley Highway, 2 January

Our bus took us along the Cann Valley Highway going north. Just out of the township of Cann River we stopped to ramble in the small local reserve. There were many interesting plants in the reserve but it was much overgrown with blackberry, and it was difficult to understand why everything had been cleared to leave a ten-yard bare strip each side of the river. A tumbled-down old shack was interesting because of the thick slabs of stringybark used for the walls.

Bushes of blueish foliage with square stems made us aware of the presence of Blue Gums". There were Blackwoods" and Black Wattles", and hanging from a Lilly-pilly were "three climbers in one tangle" as a member expressed it. One of the climbers was Smilax, another was Wombat Berry with both flowers and green (not yet orange) berries, and the third we thought to be a species of Marsdenia. Clematis" made a fourth climber and it was covered with its feathery fruits.

There were several Rufus Fantails flying nearby and one was found on a nest. There were butterflies galore, though far outnumbered by the flies, The butterflies were Common Browns and they were almost all males. We were told that these were probably the first hatching; the next hatching would include many more females, while a third hatching would consist mostly of males again.

Leaving the reserve, our route followed the Cann River. At first it was some distance away to our left and the river flats were under pasture. These pasture lands with the hills behind made an attractive rural scene. Approaching the hills we crossed the river and followed it closely, the river then being on our right. As we went further north the river banks became steeper and, further up, the water formed tumbling cascades over grey rocks of gneissic nature.

Soon after entering the hills a cutting along the highway drew the attention of the geologists. It was mostly granitic rock with the associated aureole. Gneiss and a vein of yellowish-green rock were discovered. Meantime somebody had observed a White-naped Honeyeater feeding young in the nest.

We went as far north as Chandlers Creek where we lunched—all of us under one huge Blackwood. There it was that Mr. Archie May found us. On the way we had left a letter at his house to which we now returned. Several tall and handsome Blue Gums edged the side road that led to Mr. May's house.

Mr. May has a remarkable collection of moths, butterflies, phasmids and beetles, and outside he has a walled pond and island with tortoises, goldfish and several lizards. A small water-dragon on the roof apparently defied such enclosure.

Mr. May was about to lead us to an area where the Black-stem Maidenhair" is plentiful when the heavens opened and we received our only heavy rain at daytime. It poured and hailed. Fortunately we were all in the bus and moving off. Unfortunately, water got into the engine and brought us to a halt. Our driver managed to remedy the matter while several members braved the downpour and went out to see the orchid collection of Mr. May's nearby neighbour. Fearing further engine trouble and the rain continuing, the majority decided with the bus driver to return to Cann River. A few enthusiasts transferred to Mr. May's car.

The car turned off the highway and, after crossing a small bridge, the occupants got out to slither and weave their way between, under and over dripping vegetation to the Black-stem Maidenhair. There was quite a lot of it, and the property owner picked some fronds to be taken back to the hotel.

Return to Melbourne, 3 January

The grey weather moderated our regret as we departed for home soon after 9 a.m. Unfortunately, we left two members behind with a gastric complaint. Two other sufferers faced the long journey home, a journey made even longer by the witnessing of an accident near Drouin. And next day, two more members went down with the same complaint. All attributed it to the water at Cann River. At least we can be grateful that the trouble did not occur earlier and impair the enjoyment of the daily trips.

Instead of going via Lakes Entrance we took the more northerly route through Bruthen and stopped at Stony Creek to find Ladies Tresses". These are sometimes called Spiral Orchids because of the spiral arrangement of the closely-packed flowers up the stem. We lunched at Bairnsdale and came into Melbourne about 6 p.m.

Everybody who went on this trip had a thoroughly good time and they join me to thank our excursion secretary, Marie Allender. Miss Allender plans such a trip months ahead, contacts local and other naturalists, makes our bookings and sees to many other things. Well before all arrangements are complete for one trip, plans for another are forming in her mind. And the monthly day excursions also keep her busy. We are indeed lucky to have such a competent and enthusiastic person as excursion secretary. Also, we think that Marie has some sort of pull with the clerk of weather. We went to one of the wettest places in Victoria but, while Melbourne endured rain day and night to total 4½ inches, we had sunny days and most of our rain fell at night. Some very heavy showers during our return to Melbourne emphasised our good luck—or Marie's pull with that clerk. Thank you Marie Allender for a happy, carefree, naturalist-rich ten days—and for the good weather!

Botanical names of the plants that are mentioned in the foregoing account:

1. Cabbagetree Palm, *Livistonia australis*.
2. Silvertop, *Eucalyptus sieberi*.
3. White Stringybark, *Eucalyptus globoidea*.
4. Messmate Stringybark, *Eucalyptus obliqua*.
5. Saw Banksia, *Banksia serrata*.
6. Large Duck Orchid, *Caleana major*.
7. Small Duck Orchid, *Caleana minor*.
8. Horned Orchid, *Orthoceras strictum*.
9. Austral Leek Orchid, *Prasophyllum australe*.
10. Large Tongue Orchid, *Cryptostylis subulata*.
11. Furred Tongue Orchid, *Cryptostylis hunteriana*.
12. Blue Dampiera, *Dampiera stricta*.
13. Large Fan-flower, *Scaevola ramosissima*.
14. Tufted Blue-lily, *Stypandra caespitosa*.
15. Blue-spike Milkwort, *Comesperma calymega*.
16. Heath Milkwort, *Comesperma ericinum*.
17. Fringed Lily, *Thysanotus juncifolius*.
18. Fairies' Aprons, *Utricularia dichotoma*.
19. Golden Spray, *Viminaria juncea*.
20. Swamp Heath, *Sprengelia incarnata*.
21. Grass Tree, *Xanthorrhoea resinosa*.
22. Wedding Bush, *Ricinocarpus pinifolius*.
23. Prickly Tea-tree or Manuka, *Leptospermum juniperinum*.
24. Swamp Paper-bark, *Melaleuca ericifolia*.
25. Blue Olive-berry or Fringed Bells, *Elaeocarpus reticulatus*.
26. Mountain Grey Gum, *Eucalyptus cypellocarpa*.
27. Swamp Gum, *Eucalyptus ovata*.
28. River Peppermint, *Eucalyptus andreana*.
29. Derwent Speedwell, *Veronica derwentia*.
30. Wombat Berry, *Eustrephus latifolius*.
31. Bearded Tylophora, *Tylophora barbata*.
32. Muttonwood, *Rapanea howittiana*.
33. Twiggy Heath-myrtle, *Baeckea virgata*.
34. Lilly-pilly, *Eugenia (Acmena?) smithii*.
35. Kanooka, *Tristania laurina*.
36. Sweet Pittosporum, *Pittosporum undulatum*.
37. Wonga Vine, *Pandorea pandorana*.
38. Smilax, *Smilax australis*.
39. Water Vine, *Cissus hypoglauca*.
40. Rock Felt-fern, *Pyrrosia rupestris*.
41. Tonghi Bottlebrush, *Callistemon subulatus*.
42. Angular Noonflower, *Carpobrotus rossi*.
43. Sea Celery, *Apium prostratum*.
44. Angled Lobelia, *Lobelia alata*.
45. Coast Banksia, *Banksia integrifolia*.
46. Water Parsnip, *Sium latifolium*.
47. Forked Sundew, *Drosera binata*.
48. Tall Lobelia, *Lobelia gibbosa*.
49. Pink Boronia, *Boronia muelleri*.
50. Bushy Club-moss, *Lycopodium deuterodensum*.
51. Giant Sedge, *Gahnia clarkei*.
52. Fan Fern, *Sticherus lobatus*.
53. Coral Fern, *Gleichenia microphylla*.
54. Gristle Fern, *Blechnum cartilagineum*.
55. Rainbow Fern, *Calenia dubia*.
56. Bracken, *Pteridium esculentum*.
57. Golden Everlasting, *Helichrysum bracteatum*.
58. Dusky Coral-pea, *Kennedya rubicunda*.
59. Crimson Bottlebrush, *Callistemon citrinus*.
60. Wiry Bauerea, *Bauera rubioides*.
61. Burgan, *Leptospermum ericoides*.
62. White Kunzea, *Kunzea ambigua*.
63. Onion Orchid, *Microtis unifolia*.
64. Bloodwood, *Eucalyptus gummifera*.
65. Gum Myrtle, *Angophora floribunda*.
66. Mistletoe, *Dendrophthe sp.*
67. Golden Mistletoe, *Notothixos subaureus*.
68. Southern Mahogany, *Eucalyptus botryoides*.
69. Blue Howittia, *Howittia trilocularis*.
70. Lilac Lily, *Schelhammera undulata*.
71. Hyacinth Orchid, *Dipodium punctatus*.
72. Giant Trigger-plant, *Stylidium laricifolium*.
73. Blanket-leaf, *Bedfordia salicina*.
74. Hill Banksia, *Banksia spinulosa*.
75. Narrow-leaf Geebung, *Persoonia linearis*.
76. Yellow Stringybark, *Eucalyptus muelleriana*.
77. Blue Gum, *Eucalyptus bicostata*.
78. Blackwood, *Acacia melanoxylon*.
79. Black Wattle, *Acacia mearnsii*.
80. Clematis, *Clematis aristata*.
81. Black-stem Maidenhair, *Adiantum formosum*.
82. Ladies' Tresses, *Spiranthes sinensis*.

Burramys parvus Broom (Marsupialia) from Falls Creek area of the Bogong High Plains, Victoria

by JOAN M. DIXON*

ABSTRACT

A live female specimen of *Burramys parvus* was collected on 23 February 1971 by the author when carrying out a mammal trapping programme close to Falls Creek at approximately 1798 m. (5900 ft.). This is only the second specimen of this species collected in Victoria. A study of the species and its ecology is in progress.

HISTORY OF SPECIES

Robert Broom, a Scottish doctor and naturalist, located bones of a small marsupial in the Wombeyan Caves, New South Wales in 1894. He described these as a new genus and species, *Burramys parvus* Broom, 1896. At that stage the affinities of this species were not known, and despite work on the Wombeyan fossil material which was carried out by Ride (1956, 1964), and on subsequently located Buchan Caves material by Wakefield (1960), the status of *B. parvus* remained conjectural.

In 1966, a small possum was found in a ski hut at Mt. Hotham, Victoria, by Dr. K. Shortman who took it to the Fisheries and Wildlife Department, Melbourne (Ride 1970). It was identified as a living *Burramys* by Mr. N. A. Wakefield. This specimen, a male, was kept in the hope that more specimens would be located, but attempts to do this were unsuccessful, and it died nine months later. Its skin and skeleton are lodged in the National Museum of Victoria, registered number C7290.

Early in 1970, a live *Burramys* was trapped in the Kosciusko National Park, N.S.W., by Dr. I. and Mrs. J. McT. Cowan, from Canada, who were working there with a team from C.S.I.R.O., Canberra. Further field work led to the capture of two more specimens which were taken to Canberra for study.

On 19 and 20 February, 1971, the author attended an Alpine Forum conducted by the Natural Resources Conservation League of Victoria and held at Mt. Beauty in north-east Victoria. Following this, she spent five days in the field trapping mammal species at increasing altitudes from Howmans Track 1128 m. (3700 ft.) to the granite strewn peaks above Falls Creek at 1798 m. (5900 ft.). It was in the latter area, approximately 7.2 km. (4.5 miles) southwest of Falls Creek (Lat. 36° 53' S; Long. 147° 15' E) that a specimen of *Burramys* was trapped. (Plate 1).

During the following week, three more specimens of *Burramys* were trapped at Mt. Hotham by officers of the Fisheries and Wildlife Dept.

MATERIALS AND METHODS

Traps were of the folding aluminium type, and included Sherman's 23 x 8 x 9 cm. and Elliott's 32.5 x 9.5 x 10 cm. Twenty-one traps were set in the area on this occasion. Bait used was walnut, a peanut butter honey and oatmeal mixture, or both. The *Burramys* was captured in a

*Curator of Vertebrates, National Museum of Victoria

Plate 1.
Hill close to study area showing characteristic granite tors.

Photo: Joan M. Dixon.

Sherman trap which contained both baits. No other mammal species was captured on that night. However, on subsequent occasions *Antechinus swainsonii* was collected in large numbers, and skinks *Egernia whitii* and *Sphenomorphus tympanum* were also trapped. There was no evidence from trapping of the presence of the allied rat, *Rattus fuscipes assimilis* in the area. Large scats deposited on rocks and tracks appear to be from the fox *Vulpes vulpes*. Large tracks throughout the area are made by cattle which browse herbs and larger shrubs, and clamber among the

Plate 2. Study area showing alpine shrubs and snow gums. *Burramys* capture site is arrowed

Photo: Joan M. Dixon

1. *Eucalyptus pauciflora* var. *alpina* — White Sallee
2. *Orites lancifolia* — Alpine Orites
3. *Prostanthera nivea* — Snowy Mint-bush
4. *Drimys xerophila* — Alpine Pepper
5. *Euphrasia glacialis* — Glacial Eyebright
6. *Oxylobium alpestre* — Mountain Shaggy-pea
7. *Hovea longifolia* var. *alpina* — Alpine Hovea
8. *Poa australis* — Tussock grass
9. *Stylidium graminifolium* — Grass Trigger Plant
10. *Phebalium squamulosum* var. *alpinum* — Phebalium
11. *Olearia frostii* — Daisy-bush
12. *Brachycome aculeata* — Daisy
13. *Pimelea* sp. probably *P. alpina* — Alpine Rice-flower
14. *Bossiaea foliosa* — Leafy Bossiaea
15. *Pimelea axiflora* var. *alpina* — Bootlace-bush
16. *Olearia phlogopappa* var. *subrepanda* — Daisy-bush
17. *Ranunculus* sp. close to *R. lappaceus* — Common Buttercup
18. *Oreomyrrhis* sp. — Carraway
19. *Asperula* sp. probably *A. gunnii* — Mountain Wood-ruff
20. *Scleranthus biflorus* — Twin-flower Knawel
21. *Cotula alpina* — Alpine Cotula
22. *Celmisia asteliifolia* (broad-leaved form) — Silver Daisy
23. *Acaena anserinifolia* — Bidgee-widgee

BURRAMYS IN CAPTIVITY

When collected, the animal was docile. It was a juvenile, the pouch being scarcely visible. Water offered to it soon after capture was readily accepted. The rat-like tail was fully extended, and at that stage it made no attempt to coil it. Its weight soon after capture was 27 gm.

Diet: A mixed diet was offered to it, and seeds of the following native plants accepted: *Bossiaea foliosa, Hovea longifolia, Orites lancifolia, Oxylobium alpestre*. The tops of pods of these species were neatly decapitated and the seeds extracted. Fruit of *Leucopogon suaveolens* (Mountain

Plate 3.

Vegetation in close proximity to *Burramys* site, which is indicated by paper marker at left of photograph.

Photo: Joan M. Dixon

Beard-heath) which was not recorded from the study area was also eaten.

The area abounded in insect life, especially moths and grasshoppers. Various insect species from the study area and from other locations have been offered to *Burramys*, and the following species taken: *Orthodera ministralis* (praying mantis), *Teleogryllus commodus* (brown cricket), noctuid moths, grasshoppers Fam. Acrididae, *Antheria eucalypti* (emperor gum moth). It appears that insects are a desirable if not essential constituent of the diet of *Burramys*.

Artificial foods such as parrot seed, nuts, grapes are well tolerated by the animal, while honey and Penta-Vite have been given as dietary supplements. Nuts are carried away and stored by the animal. This gives some insight into its feeding habits in winter. After five weeks the animal weighed 36 gm, and showed evidence of maturation. The pouch, tightly closed at first, is now (April) quite marked.

Sound production: Occasionally small guttural noises are uttered, usually in response to some adverse situation which causes the animal some degree of stress.

Movements: Well adapted for climbing and jumping. *Burramys* makes only occasional prehensile use of its long tail. However, the distal end may curl lightly but firmly around a support when the animal is preparing to move from one place to another. When it is awakened during the day, it often has the tail tightly coiled, in such a way that the caudal vertebrae are particularly apparent through the skin. When it runs and jumps, *Burramys* usually has the tail held almost straight out behind it like a rudder.

Grooming activities: These are particularly common when the animal has been repeatedly disturbed as in photographic work. The pattern of activity is similar to that shown by a number of small marsupials and native rodents. It squats on its hind feet and rubs one front paw after the other across the snout. The tail may be brought forward between the hind feet and carefully cleaned with the mouth and front feet.

Nest building: The animal has been housed in a stainless steel cage 42 x 31 x 23 cm, with an attached nest box 13 x 13 x 13 cm. Stainless steel mesh occupies the top and one side of the cage, and glass the other side. The cage appears to offer reasonable space for the animal except in the vertical dimension. Experiments will be conducted with a larger cage. Although it has been provided with the nest box, tussock grass, bark, leaf litter, pieces of shrubs and small rocks as well as cardboard cylinders a few inches in diameter, *Burramys* usually prefers to sleep in an open nest constructed of loosely woven tussock grass. It may cover this partially with other vegetation. No special runways have been made in the cage, which is to be expected with an arboreal or semi-arboreal species. It is apparent that larger observation areas are necessary for this type of information to be obtained.

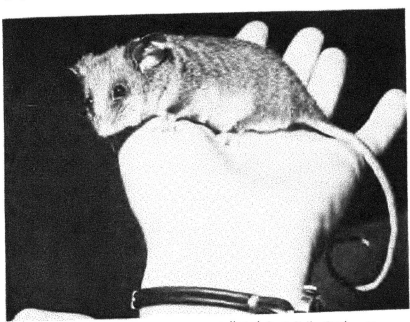

Plate 4. *Burramys parvus* female, collected from Falls Creek area, Victoria, 23 February, 1971.

Photo: By courtesy "The Age".

MORPHOLOGICAL COMMENTS

The colour and form of the animal are very similar to the first specimen from Mt. Hotham, except that it is much smaller. The most obvious feature by which the species may be recognised externally is the long tail. The body fur continues for about 1 cm. along the tail, and beyond this the tail is almost naked (Plate 4). The following are the dimensions for our specimen: Total length 235 mm., tail 142 mm., ear 16.5 mm., hind foot 16.0 mm.

The pouch has been examined and nipples found to be four in number, arranged in two pairs, one pair on each side situated postero-laterally with one nipple of each pair located behind the other.

The teeth have been examined while the animal was anaesthetised with ether, and the premolar teeth seen to exhibit the sectorial form typical of *Burramys parvus* (Plate 5).

Acknowledgments: The author extends thanks to the following persons who assisted in various ways before, during and after collection of the specimen. Dr. I. McT. Cowan, University of British Columbia; the Spargo family, Falls Creek; Mr. T. B. Muir, National Herbarium; Mr N. A. Wakefield, Monash Teachers College; Mr. E. D. Gill, Deputy Director, National Museum of Victoria.

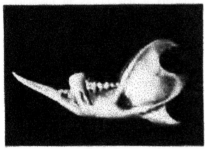

Plate 5.
Left mandible of Mt. Hotham specimen of *B. parvus*, male, C7290, Nat. Mus. Vict., showing sectorial premolar. x 2½.
Photo: F. Guy.

REFERENCES

Broom, R., 1896. On a small fossil marsupial with large grooved premolars. *Proc. Linn. Soc. N.S.W.* 10: 563.

Ride, W. D. L. 1956. The affinities of *Burramys parvus* Broom, a fossil marsupial. *Proc. Zool. Soc. Lond.* 127 (3): 413-429.

Ride, W. D. L., 1970. A Guide to the Native Mammals of Australia, Oxford University Press.

Wakefield, N. A., 1960. Recent mammal bones in the Buchan district. 1. *Vict. Nat.* 77 (6): 164-78.

F.N.C.V. PUBLICATIONS AVAILABLE FOR PURCHASE

FERNS OF VICTORIA AND TASMANIA, by N. A. Wakefield.
The 116 species known and described, and illustrated by line drawings, and 30 photographs. Price 75c.

VICTORIAN TOADSTOOLS AND MUSHROOMS, by J. H. Willis.
This describes 120 toadstool species and many other fungi. There are four coloured plates and 31 other illustrations. New edition. Price 90c.

THE VEGETATION OF WYPERFELD NATIONAL PARK, by J. R. Garnet.
Coloured frontispiece, 23 half-tone, 100 line drawings of plants and a map. Price $1.50.

Address orders and inquiries to Sales Officer, F.N.C.V., National Herbarium, South Yarra, Victoria.

Payments should include postage (11c on single copy).

Aboriginal Paintings at Muline Creek

by ALDO MASSOLA*

Muline Creek is the most northerly of the larger creeks issuing from the western scarp of the Grampian's Victoria Range. For our purpose it is best reached from Hall's Gap by crossing the Victoria Valley to Victoria Gap, and after negotiating the Glenisla Crossing of the Glenelg River, by continuing south along the foot of the range until the creek is met.

The paintings, which will be known as Muline Creek No. 1, are about three-quarters of a mile in a north-easterly direction from the point where the creek crosses the road, and, of course, can only be reached on foot. However, the nature of the country is such that walking through it is not the strenuous exercise usual in the Grampians. It is nevertheless just as panoramic as elsewhere in these mountains, and nature lovers will find the walk a delightful excursion into the realms of botany and zoology, not in any way inferior to any in the more precipitous and rugged parts of the ranges.

The paintings are on the north-east face of a large tor, at least 50 ft. high by 100 ft. long. They are in red ochre on a relatively smooth part of the rock face, low down, close to the sandy floor; and are partly protected from the rain, but not from the wind and the sun, by the overhang of the tor.

The smooth surface on the face of the rock is only about 2 ft. 10 in. by 4 ft. 6 in., and is cracked and flaked here and there, and the section closest to the sandy floor is abraded and sand blown. This weathering process must already have been in progress in Aboriginal times, because part of the design is over the "new" face of the rock, or should I say, was painted, or perhaps repainted, over old flaking scars. However, flaking has continued into more recent times, to the detriment of the painting.

The paintings consist of a small group of abstract symbols, and these are always difficult of interpretation, even without the added problems caused by weathering and fading. Beginning from the left there is an upright long oval design, 12 in. by 3 in. in size, bisected by a perpendicular stripe. To the right of the oval there are three perpendicular stripes, respectively 14 in., 13 in., and 14 in. long. To the right of these there is a design somewhat like a long legged letter P 9 in. in length with the belly of the P protruding 3 in. Across the top of the group of stripes and the letter P there is a 7 in. long stripe.

A little below and to the left of the whole group there is an isolated rendition of a "bird track", almost entirely painted over a flaking scar; and on the abraded and sand blown section at the bottom of the rock face, below the "bird track", traces of an unrecognizable design are visible.

The long oval design is similar to two depicted in the shelter known as the Camp of the Emu's Foot, which

*4, 18 Waverley Str., Mont Albert, Vic., 3127

is not far from the Muline Creek site, but over the crest of the range. This design has a world-wide distribution, and in Australia is believed to represent a ceremonial head gear made from possum fur.

The stripes and the letter P are probably all part of the one design, which could have represented a pubic fringe. This fringe, commonly known as an "apron", formed part of the Aborigine's ceremonial "dress", the components of which, besides the head gear and the apron are: the nose bone, leg and arm ornaments, and a particular style of facial and body painting.

A similar, but much better rendered and preserved apron was described by me from the Black Range

Plate 1

The paintings of Muline Creek.

Photo: Author

Shelter No. 3, to the west of Muline Creek, but probably in the same tribal territory.

Bird tracks are relatively common in painted shelters in the Grampians. They may, or may not, be totem symbols, or represent the tracks of a bird totem ancestor.

If any "Lizards" were painted on this site they have not survived, but the extant paintings are in the Lizard Style and obviously belong to the same art context as all the other known sites in the Victorian Range, the majority of those discovered in the Black Range, and the only site so far found in the Mount Difficult Range.

A notable particular of the present site, other than the similarity of motifs to Black Range No. 3, is the bird track being painted over a scar left by surface flaking. Though in itself insignificant because of our present limitations in dating methods, this point, like the hard green fungoid growth on Black Range No. 2 and Boggy Creek No. 1, may in the future prove to be of assistance in dating these Aboriginal antiquities. In any case, the Muline Creek site is a valuable link in the chain of evidence for the study of the distribution of art motifs in the Grampians.

We are indebted for the discovery of this new site to the indefatigable Ellis Tucker, of Brit Brit, who already has to his credit several other painted sites, including the Camp of the Emu's Foot and Black Range No. 2.

Members of our party, besides Mr. Tucker who guided us, were: Ian McCann, of Stawell; Graeme Kinross Smith; my brother Silvio, and myself, all of Melbourne.

REFERENCES

Massola Aldo. The Shelter at the Camp of the Emu's Foot. *Vict. Nat.* vol. 77, p. 188. 1960.

———— Black Range Shelters. (for B.R. No. 2) *Vict. Nat.* 79: p. 352. 1963.

———— Records of New Shelters in the Black Range. (for B.R. No. 3) *Vict. Nat.* 81: p. 15. 1964.

Field Naturalists Club of Victoria

Botany Group
11 March, 1971

Mr. Alan Morrison gave a most interesting talk on "Uncommon Native Plants", illustrated by a great number of superb photographs taken by the speaker in many out-of-the-way places in remote areas of Victoria. These colour slides brought to reality plants that to many of the members were known only as names in the floras published in book form by Ewart, and more recently by Willis. Areas particularly well represented were the Alps, the Grampians and East Gippsland, and the flowering plants were supplemented by examples of rare ferns, clubmosses and fungi. Mr. Morrison has also photographed rare plants in other parts of Australia, and has had many botterizing trips to the Centre and the far North. The speaker was thanked by the chairman (Mr. K. Kleinecke), and the 21 members present showed their appreciation by acclamation.

Mr. Len Fell spoke on his exhibits of pressed Western Australian wildflowers, including Qualup Bell (*Pimelea physodes*), *Regelia*, *Platytheca*, *Latrobea juncea* and *Acacia hynesiana*, a wattle with huge phyllodes and single flowers sessile on stems.

The secretary showed the latest issue of "Geelong Naturalist", a special Olways number that includes very good botanical surveys as well as other aspects of the natural history of the region. A nature show committee, consisting of Miss M. Allender, Mr. Kleinecke, Mr. J. Baines, and others to be co-opted, was elected.

The excursion proposed for the Baxter to Frankston line had to be cancelled, as the railway authorities had burnt off the vegetation on each side

Hawthorn Junior F.N.C. Annual Report

Key Office-Bearers 1970/1.

President — Mr. D. McInnes.
Secretary — John Hindle, 12 Marden St., Canterbury (83.4011).
Treasurer — Jenny Forse.
Editor — Barry Cooper.
Excursion Secretary — Pam Conder.
Publications Officer — Michael Howes.

Membership at the end of 1970 was comparable to numbers in 1969, being about 150.

Meetings — 1970

30 January	— "Members' Night".
27 February	— "Gum Trees, Tea Trees and Bottlebrushes", by M. Lester.
20 March	— "Hopping Mice and their adaptation to Desert Life", by M. Stanely.
24 April	— "The Coolart Story" by W. Davis.
29 May	— "Earthquakes" by P. Bock.
26 June	— "Wildlife in the Little Desert" by C. Crouch.
31 July	— "Fairy Penguins" by P. Reilly.
28 August	— "Members' Night and Celebration of 27th Birthday".
25 September	— "Moulds" by P. Kelly.
30 October	— "Shells of the Pacific" by M. Harrison.
27 November	— "Bull Ants" by J. Forse.

An innovation at the January meeting was the delivery of short lectures by members.

Excursions.

The following outings were organized by the Club in 1970, three in conjunction with the Geology Group and one with the Montmorency Juniors:—

1 February	— Ricketts Point (Marine Life).
5 April	— Waurn Ponds (Geology).
26 April	— Studley Park (Geology).
2 August	— Arthur's Seat-Kings Gully (General).
6 September	— Eden Park (Fossils).
27 September	— "Coolart", Balnarring (Birds, Aquatic Life).
4 October	— Kangaroo Ground (General).
31 October	— Studley Park (Mammal spotting).
13 December	— Ironbark Ranges and Werribee Gorge (Geology and general)

The Arthur's Seat and Kangaroo Grounds trips were long nature rambles up to 10 miles long, designed for more active Club members.

Easter Camp.

The highlight of the year was a five day camping trip to Wyperfeld National Park and the Little Desert at Easter. A bus was hired and 45 members attended.

The party camped for two nights at Wyperfeld and two nights at

Broughton's Waterhole in the Little Desert.

At Wyperfeld, the Club was ably assisted by the late "Rudd" Campbell, the Ranger, in visiting points of interest. In the Little Desert, the local people at Kaniva spent almost two days showing us the area by day, whilst in the evenings, a Fisheries and Wildlife party camped at the Waterhole, took the party spotlighting.

Overall, the camp was an outstanding success. Another was planned for Easter 1971.

Publications.

The Club's monthly magazine 'The Junior Naturalist" completed another year. Features of the Journal in 1970 were special reports of the Easter Camp and a Members' Trip to Tasmania in January, 1970.

Two booklets were published during the year for sale at the Nature Show. These were:—
 "How and Where to Collect Fossils" by Machael Howes.
 "Introducing Frogs" by Palm Conder.

Nature Show.

The Club organized live exhibits at the F.N.C.V. Nature Show in September, including "Fossil Areas around Melbourne" and "Collection and Study of Pond Life".

In August, the Club elected Barry Cooper and Paul Gahan as Honorary Members of the Club in recognition of past services.

Correction

Victorian Naturalist 88 (4) p. 86. col. 2, line 8—for "typical", read "atypical".

ENTOMOLOGICAL EQUIPMENT
Butterfly nets, pins, store-boxes, etc.
We are direct importers and manufacturers,
and specialise in Mail Orders
(write for free price list)

**AUSTRALIAN
ENTOMOLOGICAL SUPPLIES**
14 Chisholm St., Greenwich
Sydney 2065   Phone: 43 3972

**BIOLOGICAL, GEOLOGICAL
AND ASTRONOMICAL CHARTS**

* * * *

STANDARD MICROSCOPE
7X, 10X, 15X eyepieces
8X, 20X OBJECTIVES $54.00
CONDENSER $22.00 extra

* * * *

40X WATER IMMERSION
OBJECTIVES $18.00

* * * *

A Wide Range of Plastic
Laboratory Apparatus
available at

**GENERY'S SCIENTIFIC
EQUIPMENT SUPPLY**
183 Little Collins Street
Melbourne: 63 2160

May, 1971

Field Naturalists Club of Victoria

Established 1880

OBJECTS: To stimulate interest in natural history and to preserve and protect Australian fauna and flora.

Patron:
His Excellency Major-General Sir ROHAN DELACOMBE, K.B.E., C.B., D.S.O.

Key Office-Bearers, 1971-1972.

President:
Mr. T. SAULT

Vice-Presidents: Mr. J. H. WILLIS; Mr. P. CURLIS

Hon. Secretary: Mr. D. LEE, 15 Springvale Road, Springvale (546 7724).

Subscription Secretary: Mr. D. E. McINNES, 129 Waverley Road, East Malvern, 3145

Hon. Editor: Mr. G. M. WARD, 54 St. James Road, Heidelberg 3084.

Hon. Librarian: Mr. P. KELLY, c/o National Herbarium, The Domain, South Yarra 3141.

Hon. Excursion Secretary: Miss M. ALLENDER, 19 Hawthorn Avenue, Caulfield 3161. (52 2749).

Magazine Sales Officer: Mr. B. FUHRER, 25 Sunhill Av., North Ringwood, 3134.

Group Secretaries:

Botany: Mr. J. A. BAINES, 45 Eastgate Street, Oakleigh 3166 (57 6206).
Microscopical: Mr. M. H. MEYER, 36 Milroy Street, East Brighton (96 3268).
Mammal Survey: Mr. P. HOMAN, 40 Howard Street, Reservoir 3073.
Entomology and Marine Biology: Mr. J. W. H. STRONG, Flat 11, "Palm Court", 1160 Dandenong Rd., Murrumbeena 3163 (56 2271).
Geology: Mr. T. SAULT.

MEMBERSHIP

Membership of the F.N.C.V. is open to any person interested in natural history. The *Victorian Naturalist* is distributed free to all members, the club's reference and lending library is available, and other activities are indicated in reports set out in the several preceding pages of this magazine.

Rates of Subscriptions for 1971

Ordinary Members	$7.00
Country Members	$5.00
Joint Members	$7.00
Junior Members	$2.00
Junior Members receiving Vict. Nat.	$4.00
Subscribers to Vict. Nat.	$5.00
Affiliated Societies	$7.00
Life Membership (reducing after 20 years)	$140.00

The cost of individual copies of the Vict. Nat. will be 45 cents.

All subscriptions should be made payable to the Field Naturalists Club of Victoria, and posted to the Subscription Secretary.

JENKIN BUXTON & CO. PTY. LTD., PRINTERS, WEST MELBOURNE

F.N.C.V. DIARY OF COMING EVENTS
GENERAL MEETINGS

Wednesday, 9 June — At National Herbarium, The Domain, South Yarra, commencing at 8 p.m.

1. Minutes, Reports, Announcements. 2. Nature Notes and Exhibits.
3. Subject for the evening — "The Life of the Badger": Mr. Robert Withers (formerly of British Mammal Society).
4. New Members.

Ordinary:
 Mr. David E. Body, 8 Range View Road, Boronia, 3155
 Miss Janet MacEwan, 9 Stanhope Grove, Camberwell, 3124. (Interest—Botany-Marine Biology.)
 Miss Jill Maplesden, 9 Mountainview Parade, Rosanna, 3084. (Interest—Mammal Microscopy Survey.)
 Miss Ruth Netherway, 1/33 Park Drive, Parkville, 3052
 Mr. Thomas A. Burchell, 41 Empress Ave., West Footscray, 3012. (Interest—Geology.)
 Miss Audrey Oliver, Flat 6, 87 Studley Park Road, Kew, 3101.
 Mrs. Jean I. Hewett, 1259 High St., Malvern, 3144 (Interest—Botany.)

Joint:
 Sir Henry and Lady Somerset, Flat 10—1 Domain Park, Domain Road, South Yarra, 3141.
 Mr. John R. and Mrs. Sue Brownlie, 20 Brentani Avenue, Elsternwick, 3185. (Interest, Flora and Fauna.)
 Mr. F. Western, 141 Rosanna Road, Rosanna, 3084.
 Mrs. J M Agar, 282 Latrobe Terrace, Geelong, 3220

Country:
 Mr Richard D. Purdy, 21 Childe Street, Stawell, 3380

5. General Business. 6. Correspondence.

Monday, 12 July — "Antarctica in Mid-Summer": Dr. M. Beadnall.

F.N.C.V. EXCURSIONS

Sunday, 20 June—An inspection of the old homestead of "Emu Bottom" at Sudbury in the morning and a bush ramble in the afternoon. The bus will leave Batman Avenue at 9.30 a.m. Fare, including admission to Emu Bottom, $2.20. Bring one meal and a snack, barbecue facilities are available at Emu Bottom if desired.

Saturday, 21 August-Sunday, 5 September—Flinders Ranges. The coach will leave Melbourne Saturday morning and travel via Bordertown, Adelaide, Quorn, Wilpena (2 nights), Arkaroola (4 nights), Wilpena, Port Augusta (3 nights), Renmark, Swan Hill, Melbourne. Most of the accommodation will be dinner, bed and breakfast but Adelaide and Wilpena will be bed and breakfast only, dinner à la carte. Extra motel accommodation has been obtained at Arkaroola to accommodate most of the party but a few will be in the huts. The approximate cost will be $195.00 with motel at Arkaroola $150.00 with bunks. Deposit of $50.00 to be paid when booking, the balance by 3 July, cheques to be made out to Excursion Trust. Please do not send deposit without checking if there is a vacancy.

GROUP MEETINGS
(8 p.m. at National Herbarium unless otherwise stated).

Thursday, 10 June—Botany Group. Speaker: Miss H. Aston will continue her series of Explanations of Botanical Terms.
Wednesday, 16 June—Microscopical Group.
Friday, 25 June—Junior meeting at 8 p.m. in Hawthorn Town Hall.
Friday, 2 June—Junior meeting at 8 p.m. in Rechabite Hall, 281 High St., Preston.
Monday, 5 July—Entomology and Marine Biology Group meeting at 8 p.m. in small room next to theatrette at National Museum.
Wednesday, 7 July—Geology Group.
Thursday, 8 July—Botany Group. Miss L. White will speak on "Proteaceae".
Thursday, 1 July—Mammal Survey Group meets at 8 p.m. in Arthur Rylah Institute for Environmental Research, 123 Brown Street, Heidelberg.

The Victorian Naturalist

Editor: G. M. Ward

Vol. 88, No. 6 4 June, 1971

CONTENTS

Articles:

A Note on the Tides in Bass Strait.
By R. A. Pollock 148

Boggy Creek Painted Shelter. By Aldo Massola 152

F.N.C.V. Camp-out in Wyperfeld National Park and the Little Desert.
By J. A. Baines 160

Notes on Fauna at Wyperfeld and Broughton's Waterhole.
By W. R. Gasking 170

Feature:

Victorian Non-marine Molluscs — 4. By Brian J. Smith 155

Reptiles of Victoria — 2. By Hans Beste 156

Obituary:

A. G. Hooke — 1898-1970 158

Letter to the Editor: 154

Field Naturalists Club of Victoria:

Reports 173

Montmorency Junior F.N.C. 174

Front Cover:

This head study of the Mangrove Monitor (*Varanus semiremex*) was taken by John Wallis. Note the quite evident earhole. It lives in hollow branches of mangroves and other trees along the coast of Northern Australia.

147

June, 1971

A Note on the Tides in Bass Strait

by R. A. POLLOCK*

Amongst the members of the Marine Study Group of Victoria there had been considerable discussion on some apparent anomalies of times of low tide at various localities in Bass Strait, and speculation that the wave caused by the tidal forces, swept south around Tasmania, in some way affected the tides in Bass Strait. It was therefore decided to analyse the tidal movements in and around Bass Strait by detailed examination of the tide tables in order to test these speculations.

The basic data for the analysis were contained in the Australian National Tide Tables—in Part I for Predicted Ports and in Part II for Secondary Ports. From these tables a list of 21 localities was compiled for which times of high tides were available. Of these, Sydney was included to provide a zero time: Hobart and Portland and Warrnambool were to provide information on tide wave travel around Tasmania: five Secondary Ports were available along the island chain from Wilson's Promontory to north-east corner of Tasmania, two only available at western entrance to the Strait; the ten remaining available localities were within the area of Bass Strait.

Co-tidal lines for High Tides in Bass Strait

Based on the tabulated information co-tidal lines were drawn for localities whose time of high tide was I, II, III and IV hours after the time of high tide at Sydney. This was done for several monthly periods. Sydney was selected to provide the zero time base because the tide wave crossing the Tasman Sea from the direction of New Zealand approaches Australia's east coast on a nearly north-south line. This has recently been described by F. G. Walton Smith (Sea Frontiers 14, 6) in an article on the tides of the Pacific.

Table 1 shows co-tidal lines for selected days in the lunar month commencing 23 December, 1969. The figure in the circle within the outline of Tasmania represents the number of days after new moon. Where it occurs, the moon's declination is also shown as "Decl. S." "Decl. zero" of "Decl. N". Both the age of the moon and its declination may dominate the tidal forces and will be responsible for any observed changes from normal.

Of the 30 days for which calculations were made only 12 are shown, these being somewhat different from the normal, most frequent configuration of which 11 January is a typical example. The constant repetition of this simple pattern is held to be the most important characteristic of the tides in Bass Strait. This simple, most frequently met with configuration consists of two co-tide lines representing the same time running north-south at approximately each end of Bass Strait. This means that localities on these lines, though miles apart, all have the same tide times. Over the periods studied this was by far the most common configuration.

The breakdown of this pattern was

*M.I.F. Australia
Executive Leader, Marine Study Group of Victoria, c/o National Museum of Victoria

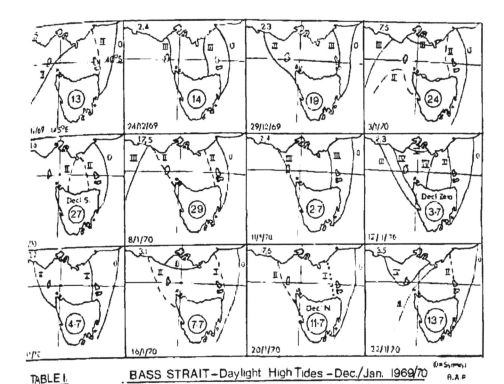

TABLE I.   BASS STRAIT—Daylight High Tides—Dec./Jan. 1969/70

23 December reflects the much stronger tidal forces due to full moon coinciding with earth's closest approach to the sun (perihelion). A similar but reduced effect is seen at the next full moon (22 January) when the sun receded from perihelion. What might have been regarded as a normal tide wave travel only shows up on 20 January when the moon's declination is north.

The figures at the top left hand corner of each map indicate the co-tidal times for Portland and Warrnambool—whose high tides are often "held up" long after the tidal forces have passed.

Table II shows similar data for the lunar month commencing 19 June, 1970. Because the sun has receded during the winter the III hour co-tidal lines become very repetitive and re-inforces the view that this simple pattern is characteristic of Bass Strait.

Speed and direction of Tide Waves around Tasmania.

Given that the oncoming tide wave from the east runs approximately on a north-south line it is evident that Tasmania provides the alternates of a path through Bass Strait and a southerly course around Tasmania. Fig. 1 shows the extent of the abyssal plains and the continental shelf in the vicinity of Tasmania and Bass Strait. It has been shown that the speed of travel of a tidal wave front is proportional to the depth of water, being fast in deep water and slow in shallow water. The abyssal plains between New Zealand and around southern Tasmania are around 15,000 ft. deep which corresponds to a tide wave-front speed of 360 m.p.h. Bass Strait on the other hand is about 200 ft. deep, shelving down very sharply to the east. For this depth of water the tide wave-front speed is reduced to about 50 m.p.h.

The geographical distances along the abyssal plains south of Tasmania are such that the tidal wave-front travelling that way and deflected round by the presence of Tasmania would reach the western end of Bass Strait at approximately the same time as the northerly wave front would reach the eastern entrance of Bass Strait. This latter wave-front has been greatly reduced in speed since leaving the abyssal plain and travelling over the shallow waters of the continental shelf. However because of the deflection of the southerly wave-front by Tasmania, these two waves would be travelling in opposite directions.

The stationary wave theory

Based on the above consideration it can be postulated that the tide wave-front entering Bass Strait from the east and another tide wave-front entering from the western end would interfere with each other and form a stationary wave. This would be directly analogous to two simple harmonic motions of equal amplitude and frequency travelling in opposite directions. This is the most simple way of explaining the observed equal co-tidal lines which occur commonly

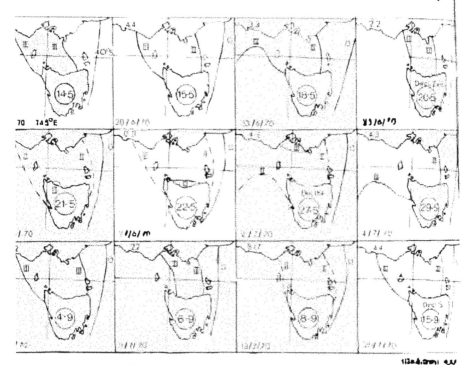

BASS STRAIT—Daylight High Tides — Jun/Jul, 1970

at each end of the Strait. Let us now look and see if other observed phenomena also support it. It should follow from the theory that *at the entrances to Bass Strait* the difference in height between high and low tides should be smaller than localities in Bass Strait remote from the entrances. The Tide Tables do not contain sufficient information on which to base a judgement. To the extent that the stationary wave may be somewhat removed from the "perfect" condition, the differences between high and low tides at the entrances may have only superficial significance.

From the theory the nodes of the stationary wave should occur at the "entrances" to Bass Strait. It must not however be assumed the positions of the nodes lie along the lines Cape Otway-King Is.-Hunter group and Wilson's Promontory-Flinders Is.-Swan Is. Fig. 2 outlines the geography of Bass Strait and these two lines define the geographical entrances. It seems more likely that the nodes of the stationary wave lie further to the seaward of these two lines. It is not the coastlines that define the interference between the two tide waves

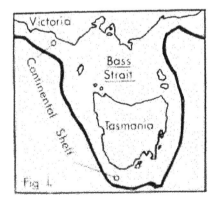

but rather the dimensions of shallow water inside the continental shelf (100 fm. line) and, of course, the distance around the abyssal plains.

Discussion.

If this theory is correct then it suggests that the interchange of water between Bass Strait and the ocean outside may be severely restricted for a large part of the year. This would also mean that the water interchange from Port Phillip and Western Port Bays would also be very severely restricted, a large percentage of the water flowing out would re-enter these two enclosed water bodies on the next tide. Such an action would have serious consequences when considering pollution problems in these two Bays and along the Bass Strait coasts generally.

I would therefore like to urge that some professional body take this study up immediately. It is essential that work be started at once to determine—

> the type, the quality and the quantity of mass flow of water in and out of Port Phillip and Western Port Bays;

a closer mathematical approach to the tidal dynamics in Bass Strait and around Tasmania and to determine quantitatively the type of interference between the tide waves; the type of further work needed to complete a definition of the mass water flows in Bass Strait itself which would be a valuable contribution to exploiting the marine resources of the Strait and a significant aid to the preservation of this marine environment.

Acknowledgement.

The author's thanks are due to Sir Robert Blackwood, Chairman of Trustees of the National Museum of Victoria and to Dr. B. J. Smith, Curator of invertebrates at the same institution, for their encouragement and assistance. To the members of The Marine Study Group of Victoria the author is indebted for their real interest in the tides of Bass Strait without which this analysis might not have eventuated. To Miss R. Plant for help with diagrams.

REFERENCES

1. Australian National Tide Tables 1969, 1970 pub. The Hydrographer Roy. Aust. Navy. Canberra.
2. "Sea Frontiers" Vol. 14, No. 6 Nov./Dec. 1968 pub. International Oceanographic Foundation, Florida
3. "Bass Strait — Australia's Last Frontier" 1969 pub. Australian Broadcasting Comm. Sydney.

Boggy Creek Painted Shelter: A New Locality Record

by ALDO MASSOLA*

I was privileged to be amongst a small group of people** taken by Mr. Ian Smith, Assistant District Forester, to a rock shelter bearing Aboriginal paintings, which he had recently discovered in a new locality for painted shelters, the Mt. Difficult Range of the Grampians.

This is a densely timbered, horseshoe shaped range, the "inside" flanks of which slope and converge funnel-like towards a central depression, Wartook Reservoir.

The name of this range, Mt. Difficult, best explains its natural configuration; and though not as rocky as the Victoria Range it is bisected by deep gullies and numerous unchartered creeks. It is a most difficult country in which to look for Aboriginal antiquities, or for anything, for that matter, and few people penetrate it. Anthropologically, at least, it has remained unexplored, and the discovery of a painted shelter is a wonderful beginning. To be guided to it by Ian Smith, who knows this country so well, was therefore an opportunity not to be lost.

According to our guide the designs we would see were "lizards"; and this was a subject that greatly interested at least two of our party, Ian McCann and myself. On numerous occasions we had discussed liz-

** The party consisted of our guide, and discoverer of the shelter, Mr Ian Smith; his brother, the well known writer Graeme Kinross Smith; the indefatigable Ian McCann, my companion on so many Grampian tramps; my brother Silvio, and myself.

*4/18 Wolseley Street, Mont Albert, Vic. 3127

152

ards and lizard style and pondered about their range of distribution, since this point would have helped to establish the respective "sphere of influence" of the several tribes frequenting the Grampians.

We had found lizards to be a prevalent motif in the Western Grampians, in the Victoria Range and in the outlying Black Range; if these which we were going to see proved to be genuine examples of Aboriginal art and not just imitative work by a white man (as some of the "Aboriginal" paintings are) the range of the lizard motif would be greatly extended. Hence their importance to us.

After a rather incredible and amphibious Jeep ride and a long climb punctuated by having to cross creeks, we reached the shelter. It is on a ridge running N.N.E. on the west side of a long gully on the northern fall of Boggy Creek (Grid. ref. Australia Topographic Survey 7323, series R 652, XD322.934). The creek is unchartered and seems to disappear in boggy country to the south east of Wartook Reservoir, which it must ultimately reach. The shelter, which is on the east base of a tor about 50 ft. long by 30 ft. high, is 16 ft. long (of which only 9 ft. can be used, the rest being encumbered with rock masses) and 12 ft. 7 in. high in the front, sloping down to ground level on an 8 ft. depth.

On the back rock-wall of the shelter, beginning from the south side and moving north, there are a number of indistinct designs in red ochre hidden by a green fungoid growth; then one pencil thin 10 in. long line representing the body of a lizard, crossed top and bottom by two 4 in.

Plate 1. Boggy Creek Shelter. Mt. Difficult Range in The Grampians.

Photo: Author

long lines representing its legs; then a little removed from it there is the rendition of a second lizard, equally thin, measuring 12 in. in length with legs 6 in. across; then another lizard design, its body 10 in. and its legs 4 ins. long. Below it three lines are visible, all that remains of perhaps another lizard.

Although they are the thinnest lizards we have seen and look as if drawn with a red pencil, there is no doubt that they were done by an Aboriginal, and they were accepted as such by all of us. As I have already stated, the first design is covered by a green fungoid growth, which was extremely hard and which we found impossible to remove, and the other designs are pitted by the wearing and flaking of the rock face and discoloured and faded by the elements. There is no question that they are old. The earth floor of the shelter showed no sign of occupancy or traces of old fires; probably it is a fairly recent washed-in floor.

Although the vicinity abounds with shelters, any of which could have been the recipient of drawings, a prolonged and thorough search failed to reveal any further examples. It really looked as if the locality was not suited to lizards, as was jocularly observed by one of our party, as he pointed to the attenuated and starved looking examples before our eyes. Nevertheless Ian Smith's discovery has considerably enlarged the range of the lizard style in these mountains.

LETTER TO THE EDITOR

Dear Sir,

May I assure the Galeshill* track walkers (*Vict. Nat.* 88 (5): 129) that the Elusive Lakes are neither of seasonal occurrence nor mirages. They are very real, the main one, Lake Elusive being 65 feet deep and 80 acres in surface area. It is approachable with some difficulty via a four wheel drive track which leaves the Wingan trail about a mile past the Galeshill track junction. The other lake, unnamed as far as I know, is much smaller and quite difficult of access.

At present I am investigating the limnology of both lakes. One intriguing aspect is the long term fluctuations in water level in Lake Elusive. Drowned trees give evidence of much lower levels in the historic past; also wave-cut benches 6-10 feet above the present level indicate higher previous levels. I would appreciate hearing from any reader who has visited Lake Elusive, so that the timing of recent water level fluctuations can be elucidated.

Yours sincerely,
B. V. TIMMS.

Plate 1

Lake Elusive — looking south from the N.E. arm. The end of the track from Wingan and the author's Christmas 1969 camp is on the left.

*Recorded incorrectly I think as Galeshill in Vict. Nat. 88 (3), Ed.

Victorian Non-Marine Molluscs — 4

by

Brian J. Smith*

Pygmipanda kershawi (Brazier 1872)

This is the largest species of native land snail found in Victoria and is confined to the East Gippsland area of the State. It has a large, fairly thin, conical, spiral shell which can exceed 55 mm. in length. It is light brown in colour and many specimens show some degree of dark longitudinal banding on the shell. This can range from a few thin, dark lines giving a faintly striped effect, to a nearly uniform dark brown shell. In Victoria it is confined to the coast and mountain region of East Gippsland, the westerly limit being just east of Lakes Entrance. Unlike most native snails this species does not seem to be confined to areas of exclusively native vegetation or to one type of enviroment. It has been found in the coastal scrub with many introduced plants around the Lake Tyers area and in the open sclerophyll forest country north of Orbost and Cabbage Tree.

The genus *Pygmipanda* was erected by Iredale in 1933 to separate this group of conical snails from the large, globose *Hedleyella* of northern N.S.W. and southern Queensland. Three supposed species comprise the genus. (1) Our Victorian form (*P. kershawi*), (2) a very similar species with only minor differences in sculpture and a larger maximum size (*P. atomata*) from central and northern N.S.W., and (3) a curious squat, dark form (*P. divulsa*) described from the high country of the Snowy Mountains, and which may occur in the mountain districts of Victoria. Separated from these is the very similar but more fragile snail from northern N.S.W., for which Iredale erected a separate genus *Brazieresta*, which could be called *Brazieresta lareyi*. However, further research may prove that this also should fall into the *Pygmipanda* group.

However the actual status of these "species" must await further research. It may be that these are all forms or subspecies of the one species with the smaller, dark-shelled form being an adaptation to life in the higher, colder mountain areas.

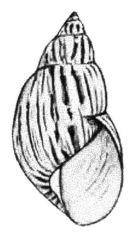

Fig. 1 *Pygmipanda kershawi*
Drawn by Miss Rhyllis Plant

*Curator of Invertebrates, National Museum of Victoria.

reptiles of victoria - 2

by HANS BESTE

PLATE 3

Parasuta brevicauda (Worrell)—Mitchell's Short-Tailed Snake.

A small colourful snake of the western desert areas of this state. Length: to 18 inches.

Head long, oval, slightly distinct from body. Snout rounded. Upper part of head black (like cap), with a black line extending along the spine to the tip of the tail. General colouration, yellowish-brown to orange, very shiny. Each scale outlined with black. Under cream.

Found: rare, in desert areas.

Best distinguishing features—head marking, very short tail (less than 30 scales from anus to tip of tail). Anal scale single.

PLATE 4

Amphibolurus barbatus—Bearded Dragon.

The familiar Jew Lizard of the inland.
Length: to 24 inches.

Head large triangular with enlarged spinous scales forming a collar and encircling the body. Scales on head, body, legs and arms rough and enlarged. Five fingers and five toes. Legs powerful. Toes on hindlegs long. Colour variable—in this state yellowish-ochre to bluish-grey. Under off white with faint grey pattern.

Found: in open as well as forested areas, often seen sunning itself on fence-posts or tree stumps.

Best distinguishing features—general appearance. Uses bluff when cornered; inflates body and erects beard.

Plate 3

Plate 4

A. G. HOOKE — 1898-1970

Arthur Garnsey Hooke was born on 29 May, 1898. His father, Frederick G. Hooke, was well known in Melbourne in the early decades of this century, having conducted his own business as an accountant, auditor and insurance broker from the year 1892.

Garnsey Hooke was educated at Camberwell Grammar School, and then, from about 1913 onward, was employed in his father's firm. He enlisted in about 1917, and went overseas with an artillery unit, arriving in France shortly before the Armistice in 1918.

He had the opportunity to visit Great Britain before returning to Australia, and he continued in his father's employ until February 1923, when the partnership of Hooke & Graham was formed. The original partners were F. G. Hooke, A. J. Graham, and A. G. Hooke, whose salaries, as fixed by the original agreement, were £40, £30 and £30 per month respectively.

F. G. Hooke died in 1942, and J. E. Graham was admitted as a partner in 1946. The insurance broking activities were transferred in 1956 to Hooke Graham & Stevenson Pty. Ltd. (now Galli & Marshall Pty. Ltd.), the accountancy business continuing under the original name. J. Digby was admitted as a partner in 1962, and the name of the firm was changed in 1968 to Hooke Graham & Digby. Garnsey Hooke remained a partner in the firm until his death in 1970.

He followed his father's footsteps in many of his business activities, having been auditor for many years of the Australian Glass Manufacturers Co. (now Australian Consolidated Industries Ltd.), being actively associated with insurance broking until about 1960, and having close association with the Melbourne Y.M.C.A. and with a number of Church of England organizations. He was a long time member (and auditor) of Carry On (Vic.), and was auditor for the Girl Guides Association of Victoria from 1927 to 1961.

In April 1928, Garnsey Hooke was married to Joan Bainbridge, daughter of Joseph P. Bainbridge, Registrar of Melbourne University. There is a son, Simon, and twin daughters, Alison and Barbara.

Garnsey Hooke was elected to membership of the Field Naturalists Club of Victoria on 13 December 1920, and he was Honorary Treasurer of the Club from May, 1922 to June, 1929. After relinquishing the treasurership, he was appointed an Auditor of the Club's accounts, and in that capacity he served from 1930 to 1955. Thereafter he was Honorary Treasurer for a further ten years, until the end of 1966.

On 13 August, 1956, the General Meeting of the F.N.C.V. elected him an Honorary Life Member, "in recognition of his long and valued service to the Club as Auditor and Treasurer".

Garnsey Hooke accompanied Alfred J. Tadgell on three horseback excursions into the Victorian Alps— to Mounts Feathertop and Hotham in the summer of 1921-22, and twice to Mount Bogong in 1923. Tadgell's comprehensive botanical reports, resuiting from these trips, are available in the *Victorian Naturalist* of February 1922 (Vol. 38, pp. 105-118) and August 1924 (Vol. 41, pp. 56-80).

In later years there were a number of family excursions to the Otway Ranges, and interest in that area culminated in a Field Naturalists Club excursion to the region in December 1959. As leader of the excursion, Garnsey Hooke published a report of the activities in the *Victorian Naturalist* of March 1960 (Vol. 76, pp. 277-81), and his description of the eucalypts which were met with appeared a month later (*Vict. Nat.* 76: 319).

Interest in eucalypts, and in ferns too, had developed early, during holiday visits to the Dandenong Ranges. There is a report of an F.N.C.V excursion to Sherbrooke, in the *Victorian Naturalist* of May 1924 (Vol. 41, pp. 4-6), when he led members through one of the local fern gullies and then "to the country house of one of his relatives, a Miss Billing, who had kindly prepared a very acceptable lunch for the party".

Over the years, Garnsey Hooke spent much of his leisure time at "Merrimu", as the house at Sherbrooke was called, and eventually, in 1956, it passed into his ownership. Many members of the Field Naturalists Club became familiar with "Merrimu", for its owner liked nothing better than to have friends along to spend a day or so there with him.

From 1961 onward, he served as a member of the Committee of Management of the Sir Colin Mackenzie Sanctuary, Healesville. He passed away on 12 April 1970, and in a short obituary, which appeared in the *Victorian Naturalist* soon after, a fellow member of the sanctuary committee paid this apt tribute:

"Above all he was characterized by gentlemanliness. He was innately courteous, disciplined, scrupulous, never pretentious—a gentleman in true manner. He will be missed, and he will be remembered".

—N. A. WAKEFIELD.

F.N.C.V. PUBLICATIONS AVAILABLE FOR PURCHASE

FERNS OF VICTORIA AND TASMANIA, by N. A. Wakefield.
   The 116 species known and described, and illustrated by line drawings, and 30 photographs. Price 75c.

VICTORIAN TOADSTOOLS AND MUSHROOMS, by J. H. Willis.
   This describes 120 toadstool species and many other fungi. There are four coloured plates and 31 other illustrations. New edition. Price 90c.

THE VEGETATION OF WYPERFELD NATIONAL PARK, by J. R. Garnet.
   Coloured frontispiece. 23 half-tone. 100 line drawings of plants and a map. Price $1.50.
   Address orders and inquiries to Sales Officer, F.N.C.V., National Herbarium, South Yarra, Victoria.
   Payments should include postage (11c on single copy).

F.N.C.V. Camp-Out in Wyperfeld National Park and The Little Desert
29 August — 6 September, 1970
by
J. A. BAINES

Saturday, 29 August

On the forward journey the bus passed through the old gold towns of Castlemaine, Maldon, Dunolly and Moliagul to St. Arnaud, where lunch was taken, then the route was via Donald, Warracknabeal, Jeparit, Rainbow and Yaapeet, thence northwards to the entrance sign "Wyperfeld National Park". A further 51 miles brought us to the camping ground at Wonga Hut, where the presence of many tents was a reminder that this great Mallee reserve is becoming increasingly known to people who like to "get away from it all". The grassy area near the dry bed of Lake Brimin seemed a perfect setting among the River Red Gums, but some inexperienced campers discovered that the Mallee can be really c-o-l-d at night!

Sunday, 30 August

After a briefing by the ranger, Rudd Campbell (whose death soon after our return to Melbourne was a great shock to us all—he has done a fine job over the years and was most helpful to the F.N.C.V. party), we set out on the Eastern Lookout car trail. An excellent little brochure, descriptive of the natural history of the track, is available, and we also had a commentary from Mr. W. Gasking, Director of Cleland National Park, South Australia, and former Director of the Sir Colin MacKenzie Sanctuary at Healesville, who had driven from Adelaide to join the excursion. He gave the meaning of Wyperfeld as German for "snake field", but this is very doubtful. Viper in German is spelt exactly the same as in English, and snake is Schlange. Wyperfeld is probably an old spelling of the name of a German village, Wipperfeld, in North Rhine-Westphalia, situated south-east of Remscheid, which is itself six miles from Wuppertal, the name of the combined city of Barmen-Elberfeld. Wipperfeld means "field near the Wupper". Wuppertal is "valley of the Wupper", and old documents prove that this Rhine tributary was formerly the Wyper or Wipper. The word is cognate with the Latin "vibrare" (referring to "vibrating" or undulating waves). The park was named after the parish, which may have got its name from the birthplace, or ancestral home, of a Wimmera pioneer. Lowan National Park (suggested early but rejected) would have been a better name.

Eleven emus were seen (the first enthusiasm did not envisage the much larger numbers to be sighted later), and a pair of Major Mitchell Cockatoos. There was a stop on Copi Ridge (from "copai", an aboriginal word for the gypsum found there) to look at a Weeping Pittosporum (*P. phillyraeoides*), with orange-coloured split capsules. The Eastern Lookout tower (built by Ararat High School boys in 1964) gave a splendid view of an "ocean" of mallee stretching far to the north—an appropriate metaphor, as this country was all under the sea in Miocene times. In addition to several mallee eucalypts, plant species noted here (mostly in flower) included Wallowa (*Acacia calamifolia*). *Clematis microphylla*, *Baeckea behrii*, *Lasiopetalum behrii*, *Podolepis capillaris* and (in bud only) *Loudonia behrii*. (Three of these were named after Hermann Behr, a German botanist who made early collections in South Australia when staying at Bethanien in the Barossa Valley.) Mallee Ringneck Parrots were seen: Bill Gasking pointed out gum-nuts chewed up by parrots.

In the afternoon most of the party took a walk over the sandhills near the camping area. Listing species seen would be tedious, as most members of our club own or have access to "The Vegetation of Wyperfeld National Park and a Check-list of its Vascular Flora", by J. Ros Garnet, an excellent guide which lists the species, gives accurate habitat notes and illustrates 100 different plants, while "Birds of Wyperfeld National Park", by H. E. Tarr, lists over 200 species of birds. Nevertheless the keen amateur botanists made lists, as did the bird observers, as all good naturalists should. Many were active in photographing items of interest. At the foot of Flagstaff Hill there were specimens of Mallee Bitterbush (*Adriana hookeri*), which was new to some members: kangaroos are fond of the leaves. Sweet Appleberry (*Billardiera cymosa*) was noticed climbing high into a Yellow Mallee (*Eucalyptus incrassata*), the latter being the commonest of the mallees here. Green Tea-tree (*Leptospermum laevigatum* var. *minus*) is a Mallee form of the species common near the coast. An aboriginal shield scar was seen on a River Red Gum (*Eucalyptus camaldulensis*). Returning to the camp along the northern side of Lake Brimin, it was noticed that here the red gums are higher up from the former inundated area than the Black Box (*Eucalyptus largiflorens*), the reverse of the usual sequence. Some of us joined a couple (who said they had been "rained out of the Gramplans") observing Mallee Ringneck Parrots.

Monday, 31 August

Most members set out on a walk to Lake Brambruk. Fine specimens of Desert Banksia (*B. ornata*), the sole species of the genus in the park, were photographed, some of the stamens being distinctly red above the usual yellowish-green of the flowers. Scrub Pine (*Callitris verrucosa*), with low habit of growth and warty cones, was common, and the taller Slender Cypress Pine (*C. preissii*) less so. A shallow lagoon, Devil's Pools, with much water augmented by the recent rains, was notable for Nardoo (*Marsilea drummondii*) and a grass-like sward formed by a tiny composite, Grass Cushion (*Isoetopsis graminifolia*). A huge depression covered with red gums was the precursor of the flat, treeless area of Lake Brambruk proper, to which small mobs of kangaroos retreated as the human intruders invaded their timbered refuge. A truly remarkable sight (for city dwellers used to excitedly commenting on glimpses of them in ones or twos in other parts of the State), greeted us as we emerged from the forest to see on the grassy lake-bed no fewer than 50 kangaroos and 120 emus, the latter in three detached flocks of 72, 31 and 17 respectively. These areas, of which there are many in the park, show relics of the days when Wyperfeld was part of the old Pine Plains station, with prolific growth here of introduced plants like horehound and mallow, and in other areas sorrel, chickweed, clovers, medicks, nettles, thistles and Tree Tobacco (*Nicotiana glauca*).

At the eastern end of the bed of

Lake Brambruk a flock of about 50 Galahs was observed, until take-off when one photographer tried to approach too near. Other birds observed were White Cockatoos, Mallee Ringnecks, Grey Thrush and Pallid Cuckoo, and a Kookaburra was heard calling. Bill Gasking pointed out, among the red gums, what he called an "emu climb" — their droppings told the story. The excreta of possums and kangaroos, too, reveal much to an experienced observer such as he is, and of course spoor prints also. Dick Morrison photographed a Red and Black Spider (*Nicodamus bicolor*), as it trussed up a grey caterpillar with silk; then, in thick mallee scrub, a Midget Greenhood (*Pterostylis mutica*). Reaching a fire access track, this was followed till it rejoined the road travelled on yesterday.

In the afternoon, two carloads went three miles to Black Flat, a black-soil lake-bed that contained water in 1956 when the Wimmera River flooded and Outlet Creek brought water much further north than usual. Two Spur-wing Plover were seen here. We then followed the Round Lake nature trail, initiated, I understand, like the other one, by Ian Maroske, a prominent member of the Committee of Management. Attention is drawn, by numbered references, to species such as *Eucalyptus largiflorens*, *E. foecunda* (very colourful, and, appropriately, with great masses of terminal fruits), *E. incrassata*, *E. porosa* (Quorn Mallee, normally a South Australian species and rare in the park). Yellow Burr-daisy (*Calotis erinacea*), Oondoroo (*Solanum simile*), Porcupine Grass tussocks (*Triodia irritans*), *Banksia ornata*, *Micromyrtus ciliatus*, *Aotus villosa*, *Clematis microphylla*, Muntries (*Kunzea pomifera*), *Hakea muelleriana*, *Hybanthus floribundus*, *Callitris preissii*, *Pittosporum phillyraeoides*, *Hibbertia virgata* and *H. stricta*, *Leptospermum laevigatum* var. *minus*, *Loudonia behrii*, *Acacia spinescens*, *Billardiera cymosa*, *Ajuga australis*, *Olearia lepidophylla*, *Dianella revoluta*, *Calytrix tetragona*, *Adriana hookeri*, *Vittadinia triloba*, *Dodonaea attenuata*, and *Muehlenbeckia* sp. No-one could find the Erect Rice-flower (*Pimelea stricta*) that was signposted as growing on an eroded sand dune.

Round Lake was beautifully green, with a dead tree standing out from the distance like a prehistoric dinosaur! Before the 1959 bush fire that devastated much of Wyperfeld (its destructiveness has ruined what was a lovely stand of cypress pines not far in from the entrance to the park) there used to be many lowan mounds in the mallee scrub on the far side of Round Lake, but none nest there now.

Tuesday, 1 September

Ranger Rudd Campbell led us from a fire access track along a trail 30 chains through mallee to an active mound of a lowan — no. 58 of about one hundred known in the park. This nest would be ready for laying of pink, thin-shelled eggs in three weeks' time, followed by a gestation period of 60 days. Father's remarkable dedication to the temperature-controlled incubation has been often told (best in H. J. Frith's book "The Mallee Fowl"); his complete lack of paternal solicitude (he has been known to brush aside a newly-hatched chick as an irksome interruption to his working of the mound) is a marked contrast to the devoted mothering by the male emu. Mr. Campbell told of 23 November, 1959, when the big fire, started eleven days before when it escaped from a burning-off at Broughton near Yanac, swept into Wyperfeld. Keith Hately and Graham Pizzey saw one lowan that survived the fire and had returned to its mound to find it too hot to work. No effective measures against the fires then existed, but many access tracks have been put in since. Because of the fox menace, Rudd was given permission to carry a .22 rifle, with which he shot 37 in the first week!

On the walk back from the mound were seen *Stackhousia*, *Ophioglossum*, *Stenopetalum*, *Acacia calamifolia* and *Podolepis capillaris*. Zillah Lee photographed a Mallee Mouse Spider. The bus stopped on the return trip so that observations could be made and photographs taken of over a hundred emus feeding on the grassy lake-bed.

In the afternoon some of us climbed Mount Mattingley, an imposing sand hill commemorating Arthur H. E. Mattingley, a founder member of the R.A.O.U., and President of the Bird Observers' Club and Gould League, whose enthusiastic description of his visit to Pine Plains and Camba-Canya in September, 1907 (see *Vict. Nat.* 26: 64-77, Oct. 1909) led to the first moves for the creation of Wyperfeld National Park. This well-illustrated report is full of interest, and a pertinent quote is as follows:

"The whole place is a perfect paradise for nature lovers, and in view of its probable early opening up for settlement, Wonga Basin, along with Brambrook and the adjoining Jerriwerrup (locally called Cherry-whip), should certainly be reserved . . . Everlasting flowers grow to perfection on the sand-ridges, and were ten or twenty thousand acres of this country set aside as a national park, it would be a most valuable heritage for future generations of nature students".

Mattingley and his companions, J. A. Ross and F. E. Howe, had been inspired to visit the area by reading contributions to Donald Macdonald's "Nature Notes" column in "*The Argus*" by "Mallee Bird", the pen-name of Charles McLennan, who soon after became first ranger of Wilson's Promontory National Park, Mr. McLennan and the station owner, Mr. Poulton, were most helpful to the three visitors from Melbourne, and posterity owes them a debt.

Alan Legg and family (camping, but not of the F.N.C.V. party) guided us to the nest of a Southern Scrub-robin (*Drymodes brunneopygia*), but the bird was not about when we reached it, and the nest was untenanted; they had observed it four times previously. Curiously, this bird, which was obviously common when Mattingley observed it in 1907, is omitted from Tarr's Wyperfeld bird list, or, probably, Southern Scrub-wren has been printed in error. In Roy Wheeler's "A Handlist of the Birds of Victoria" the Southern Scrub-wren's habitat map includes the Wyperfeld area, but neither the Large-billed Scrub-wren nor the White-browed Scrub-wren (*Sericornis* spp.) is found in north-western Victoria, no "Southern Scrub-wren" being included in his book nor in those of Cayley or Leach. The Leggs also took us to a hollow dominated by a veritable "forest" of Desert Banksia, in full flower, with half a dozen species of honeyeaters, including the White-eared, the Yellow-winged, the Spiny-cheeked and the Red Wattle-bird, all gorging themselves with nectar and from time to time flying up to either of two dead Black Box trees that bounded the area on the north and the south, making themselves conspicuous for easy observation through field glasses. Fantails and tree-creepers were also noted.

A botanizing walk through the

nearby scrub of mallee and tea-tree was followed by a climb to the summit of Mt. Mattingley, because George Collis wished to photograph the view towards each point of the compass. From this eminence the sole specimen in the park of Drooping Sheoke (*Casuarina stricta*) stood out atop Flagstaff Hill. It was a pleasure on return to camp to find that Alan and Win Morrison had arrived, in their 4-wheel drive vehicle and caravan, to join us on our visit to the Little Desert. Some members went spotlighting for native mammals in the experienced company of Bill Gasking.

Our visit to Wyperfeld had come to an end without an opportunity (for reasons of time and state of track) of seeing the famous old patriarch River Red Gum known as Be-al (the Wimmera aboriginal word for the species in general, as in the name Warracknabeal, which comes from "warrak", plain; "na", of; "beal", red gum; the Murray River tribes know *Eucalyptus camaldulensis* as "yarrow" or "yarra", as recorded in Major Mitchell's journal). A few members however managed to see it, and were duly impressed. The other disappointment was not seeing a mallee fowl; a camper not in our party was fortunate in seeing one near the track we had walked along.

Wednesday, 2 September

The bus stopped on the way out of the park to enable us to look at a Tawny Frogmouth on its nest in a red gum, perfectly camouflaged as usual. Twice were kangaroos seen in pairs, and many ringnecks were about; a pair of smaller parrots flashed by unrecognized. At Rainbow is the original homestead of the old Hindmarsh station (classified by the National Trust). Unfortunately there was no time to have a look at either Lake Albacutya or Lake Hindmarsh, but the level of both were reported to be low. There was a brief stop at Jeparit (which now has a Pioneers' Museum) for a walk into Sir Robert Menzies Park, with its 60-ft. "thistle" spire at the entrance, the former Prime Minister having attended Jeparit State School from 1899 to 1906. The name Jeparit means (according to Aldo Massola's "*Aboriginal Place Names of South-east Australia*") "shell parrot", and one wonders whether this is a white man's guess as the coincidence seems too unlikely. A local tourist brochure gives it as "home of small birds". Certainty budgerigars, found in the region, are small birds.

On again, and George Wellington stopped the bus to remove a recently killed Galah from the road, to protect its mate, which had remained loyally by its side, in danger of being hit by a car itself. A stop was made at Glenlee Forestry Reserve, an area of original Wimmera scrub, 467 acres on the right side of the road, 1061 acres on the left. Three wattles were here (*Acacia acinacea* and *A. pycnantha* in flower and *A. brachybotrya* in bud), and other plants noticed included *Swainsona procumbens*, *Stackhousia monogyna*, *Olearia* and *Eutaxia*.

Lunch was taken in Nhill's main street gardens, alongside the fine statue by Stanley Hammond of the Clydesdale horse, to which the Wimmera owed so much for its early agricultural development. Erected by the Australian Horse Society, it was unveiled by the Minister for Primary Industry (Mr. J. Anthony) in 1968. A few miles beyond Nhill we passed the memorial to John Shaw Neilson, on the site of the settler's cottage where he wrote many of his lyrical poems.

"The leaves have listened to all the birds so long;
Every blossom has ridden out of a song;
Only low with the young love the olden hates are healed;
Let the tired eyes go to the green field!"

Near twin towers (radio and television) on a rise, there was a roadside botanizing stop, where most of the plants were familiar old friends, but good to see with wheat fields stretching afar. Among them were the orchids *Glossodia major*, *Acianthus reniformis*, *Caladenia carnea* and *C. deformis*, and two sundews, *Drosera whittakeri* and *D. planchonii*, the commonest plant (not in flower) being Malice Broombush (*Melaleuca uncinata*) (the official common name is Broom Honey-myrtle, but Broombush is the only name heard among the locals of the north-west).

Further on a sheoke forest was passed. We were met in Kaniva, at 2 p.m., by Alec Hicks, who lives in the town and has at his home a large herbarium of plants collected in the district over many years, during which he has made new records of species and has sent numerous specimens to the National Herbarium in Melbourne. We were fortunate indeed to have him as our botanical guide throughout our stay in the Little Desert. Alec drove ahead out on the Edenhope road to the junction of a very sandy track leading east, along which we travelled some distance to where P. L. (Perc) Williams was waiting in his 4-wheel drive vehicle. He directed George through the worst sandy stretches and wet patches, and finally the huts and all other vehicles arrived at Broughton's Waterhole. Alf Lewis and family

Plate 1. At Broughton's Waterhole (Little Desert), with Alex Hicks, Avelyn Coutts, P. L. (Perc) Williams, Sept. 1970. Yellow Gums (*Eucalyptus leucoxylon*) in background. Since this photograph was taken, P. L. (Perc.) Williams was killed in a level crossing accident—a sad loss to conservation.

Photo: Ian Morrison.

joined us at Kaniva and Peter Kelly and family at the camp site: Dick and Alan Morrison came in their own vehicles, as did Karma Hastwell (with Chris Walker), whose small car was successfully pulled out of the mire, and Miss May Moon (of the Save the Dandenongs League). P.L. introduced Avelyn Colins, who, like himself, is a Miram farmer very interested in conservation. These two, like Alec Hicks, were most assiduous in helping us, and their good humour was infectious. P. L. Williams is the owner of a 20-acre area that includes Broughton's Waterhole (which he intends to have permanently and unassailably reserved), and his efforts during the 1968 confrontation of conservationists against Sir William McDonald's finally rejected Little Desert farm scheme are too well-known (and appreciated) to need recapitulation.

Tents were pitched, and after tea a sing-song around a roaring campfire (the nights were still cold) was led by Bill Clasking, whose scouting leadership experience made him ideal for this role.

Thursday, 3 September

Most were up early to explore the immediate surroundings of this "oasis" that provides the soil-moist habitat for the tall eucalypts that stand out from all the surrounding low scrub. Most water is in the Middle Dam of the three soaks that were used by pastoralist Broughton in the early days (1879?). There are three wattles (*Acacia pycnantha*, *A. rigens* and *A. mitchellii*). *Hakea flexilis*, *Melaleuca uncinata*, *Hibbertia stricta*, *Ranunculus robertsonii*, a blue Dwarf Daisy (*Brachycome goniocarpa*), Yellow Marsh-flower (*Villarsia exaltata*), and sedges and moss. Bird life was not evident this morning.

Leaving the bus at the camp, the party left in several four-wheel drive vehicles, to be joined later by six senior boys from Kaniva High School, adventurously tackling all tracks in a Falcon sedan. Back along the five-mile sandy track to the Edenhope Road, then along Elliott's Road, with a stop to see Wheel Print (*Gyrostemon australasicus*). Pink Zieria (*Z. veronicea*) and three Guinea-flowers (*Hibbertia stricta*, *H. virgata* and *H. fasciculata*). Further stops were made (enforced) at treacherous parts of track (water, sand and grass-tree hazards) before arrival at Mount Moffatt, a sand-hill with a trig. point on the summit, where Alec showed us *Boronia filifolia* in flower (discovered here in 1949 and rare in the Little Desert). Orchids seen here (after lunch) included *Lyperanthus nigricans*, *Leptoceras fimbriatum*, *Pterostylis nana* and *P. vittata* (unusually rich in colour, so photographed by many). The dominant eucalypt was Brown Stringybark (*E. baxteri*), grass-trees (*Xanthorrhoea australis*) were common, and *Banksia marginata* also. A Bitter Quandong bush (*Santalum murrayanum*) bore several fruits.

Back to the main track, across that and along a similar broombush-lined track south, then stops were made first to look at Wrinkled Hakea (*H. rugosa*) in flower, secondly at a waterhole normally containing much more water than at present. The Kaniva boys, bogged, were able to push their car out, but all hands were called for to extricate the Buick of the Langs, a Brighton family that had joined us by chance "to see the Little Desert": 30-manpower finally got it out! Others crossed all right, with passengers off-loaded. On through a large swampy area with an extensive stand of splendid Yellow Gums (*Eucalyptus leucoxylon*). A brief stop was made to photograph a large mass of *Tetratheca ciliata* in flower around a hakea bush.

Now again in Brown Stringybark country for a while before entering another Yellow Gum area, with *Acacia hakeoides*, at the South Australian border, which several crossed to note the effect of farming, which here comes right up to the boundary. It is good to know that Victoria's side seems reasonably safe for conservation. Not far from here a stop was made at the site of Miles's house, where "old McPhee" used to mind sheeep grazing free in the scrub (up to about 1940), for a search for aboriginal artefacts (a particular interest of Avelyn Coutts). Some small finds were made. A huge Yellow Gum stood near the front of the site, and there were a few surviving garden plants.

Return was made along the northern section of the new track known, inevitably, as "McDonald's Line". A 20,000 acre property, "Bindibu", showed some evidence of experiments conducted to show the effect of fertilizer on the soil, but most of the previously alienated land we saw looked pathetically poor as farming land—a good indication of the correctness of the analysis of agricultural economics experts who condemned the McDonald scheme as potentially unviable.

At night Bill Gasking caught a Mallee Silky Mouse (in a humane, non-killing trap), and kept it temporarily before release so that members of the party could become familiar with this attractive little native rodent. Named by Finlayson in 1932 *Gyomys apodemoides* after its discovery at Coombe near the Coorong, S.A., and given under that name and as the Silky-grey Southern Mouse by Troughton in "Furred Animals of Australia", it is now called *Pseudomys albo-cinereus*, the Asbygrey Mouse, as it is known to be a form of this species first named from Western Australia by Gould in 1845. A specimen from south of Kiata sent to the National Museum, Melbourne in 1957 by Keith Hately was labelled *Mus musculus*, the House Mouse, but its true identity was recognized in 1963 by R. Mark Ryan (see *Vict. Nat.* **79**: 363) and verified the following year by Norman Wakefield when camped at Salt Lake in the same part of the Little Desert. Bill Gasking caught a House Mouse too and the differences could be plainly seen. The native species has light bluish-grey, short, erect, silky fur, whitish underneath the body, with pink feet and tail; like other pseudomice of the inland country, its burrowing habits are of great interest.

Friday, 4 September

After yesterday's rough ride, a few decided not to go on the trip to The Critter, but they missed the best botanical area we saw in the Little Desert. First stop was at Chinaman's Flat, where there were many bushes of *Westringia crassifolia*, with its blue flowers, *Melaleuca wilsonii* (not yet in flower), *Hibbertia sericea* var. *scabrifolia*, *Lasiopetalum baueri* (not in flower), Nealie and Cold-dust Wattle, *Ranunculus sessiliflorus* and *Stackhousia monogyna*. Further on, Alan Morrison stopped to photograph masses of Purple Eyebright (*Euphrasia collina*). A diversion was caused by an emu with seven chicks running straight ahead on the track for about two miles; the vehicles behind all slowed down, but the bird did not seem to realize that leaving the track would be the best tactic! There was a brief stop for Alec to point out *Daviesia pectinata* Passing through treeless heathland a kangaroo made a splendid sight leaping away at top speed.

On arrival at The Crater (first described by St. Eloy D'Alton, *Vict. Nat.* 30: 65-66, August 1913) we saw a large natural depression, non-volcanic, almost certainly caused by subsidence, completely covered in vegetation of a great many species, ranging from trees and shrubs to quite lowly herbs. The dominant tree was Brown Stringybark, which covered the "crater" right across to a sandstone outcrop on the far rim. Plants recorded, in order of coming under notice, included *Pultenaea d'altonii, Pterostylis ?robusta, P. vittata, Persoonia juniperina, Daviesia ulicina* var. *ruscifolia, Phebalium ?stenophyllum, Leptospermum myrsinoides, Baeckea crassifolia, Grevillea ilicifolia, Leucopogon rufus, L. virgatus, Dillwynia sericea, Calytrix alpestris, Acacia calamifolia, Isopogon ceratophyllus, Correa reflexa* (with red flowers), *Prasophyllum nigricans, Banksia marginata, Monotoca scoparia, Acacia mitchellii, Hibbertia fasciculata, Craspedia uniflora, Astroloma conostephioides, Drosera whittakeri, Xanthorrhoea australis, Spyridium vexilliferum, Micromyrtus ciliatus, Chamaescilla corymbosa, Astroloma humifusum, Pultenaea laxiflora* var. *pilosa, Thelymitra antennifera, Pterostylis longifolia, P. barbata, Epacris impressa* (white), *Amphipogon strictus, Leptomeria aphylla, Caladenia carnea, Schoenus brevicalmis, Brachycome ciliata, Hakea rostrata, Dampiera lanceolata,* and *Lomandra glauca*. Bees in a hive in hollow sandstone aroused some interest, and ants' activities were watched.

P.L. told an amusing story of a previous visit by three well known naturalists when it was freezing cold in the Little Desert and they imbibed freely of whisky to get warm! He also told of the visit years ago of Dr. Melville of Kew Gardens, London, whose intense interest in this primitive floral region made him an unwilling victim of P.L.'s "cracking the whip" such as was necessary to get the F.N.C.V. party moving on!

A mallee fowl's mound, not used for five years, was pointed out. A White-lipped Snake (*Drysdalia coronoides*) was promised to Zillah Lee for her junior reptile enthusiasts by Kaniva H.S. boys, Geoff Austin and Bruce Gibson of Broughton (it was duly delivered later through Alec Hicks, together with a specimen of the Legless Lizard (*Delma fraseri*). Seven of these senior science students were with us today; they are genuinely interested in the natural history of the district, and credit for inspiring them must go to their biology teacher, John Parkes, who sought the Kaniva H.S. position after having his interest in the Little Desert aroused during his own student days.

On the return trip a northward route was taken, stopping to see *Pultenaea vestita* (the only Victorian habitat of an otherwise S.A. and W.A. species), *Helichrysum obtusifolium, Cryptandra tomentosa, C. leucophracta, Lomandra leucorrphula* and *Prostanthera aspalathoides*, Scarlet Mint-bush, in flower amid acres of mallee and broombush. Next we climbed Mt Turner, an outcrop of red sandstone named after Thomas Hedington Turner, a surveyor who died in 1918. (The designation of these low sand-hills as "mounts" is of course only comparative.) Noticed here were *Acacia gunnii* (syn. *vomeriformis*), *A. myrtifolia, A. mitchellii, Bertya mitchellii, Hakea ulicina, Spyridium thocephalum, Logania linifolia, Lepidosperma carphoides* and *Cassytha glabella*. Another disused lowan's mound was noted, and sand-heaps made by the Silky Mouse—

from there he burrows 12 feet underground and makes exit holes at several points around.

With the night cold and windy, a large hessian screen enabled the campfire group to enjoy Dick Morrison's tape recordings and whimsical elocutionary efforts. There was heavy constant rain during the sleeping hours of the night.

Saturday, 5 September

After breakfast Bill Gasking's mice were watched and photographed, then there was a demonstration of boomerang throwing by Avelyn Coutts (who makes them in the craftsmanlike traditional way) and Pere Williams, who each then took a group of tyros whose successful throws were not many; one throw left the boomerang high in a tree, defying dislodgement until eventually brought down by P.L.'s good marksmanship! Alec Hicks (who gave a talk in Nhill the night before on his visit to South Africa and Rhodesia) arrived with John and Jill Parkes, then John Teasdale (of A.B.C. television) came with 'Wimpy" Reichelt (of "Little Desert Tours", Nhill) to make a film coverage of the F.N.C.V. camp-out. After filming activities near Broughton's Waterhole, Teasdale came with us to Wild Dog Spring, where shots were taken of naturalists handling *Ruppia maritima*, Sea Tassel, a plant of salt or brackish lagoons, and an orange bracket fungus (*Trametes*) on a banksia branch, then out to the Sundhills on a very rough track, several vehicles (even with four-wheel drive) experiencing difficulties. Filming was completed here.

Back at camp there was further photographing of native animals, and also of the "Three Blind Mice", a contemptuous nickname given to Messrs. Williams, Hicks and Coutts

by some local supporters of the McDonald farm scheme, but which, like the "Rats of Tobruk", has become an honoured name, for they are regarded as "men of vision", among conservationists and other opponents of the alienation of what should become a greatly extended Little Desert National Park.

The first census of Little Desert plants was made by Dr. C. S. Sutton of those noted in the area in 1913 by St. Eloy D'Alton (see the latter's paper, "The Botany of the Little Desert, Wimmera, Victoria", *Vict. Nat.* 30: 65-78), and contained 224 species; the florula compiled by Alec Hicks for the Little Desert totals more than 600 species; he records also the parishes in which each species was collected, and the list indicates the collectors' names for each plant, beginning with D'Alton and including A. J. Swaby, J. H. Willis, A. C. Beauglehole, T. B. Muir and A. J. Hicks as major collectors, with less frequent mention of F. M. Reader, Alison Jordan, Jean Galbraith and A. B. Court. Such floristic richness must be preserved in as near to its primitive state as possible and in a sufficiently large area, as habitats for the native fauna as well as for the flora.

Sunday, 6 September

Camp was struck, our three guides were thanked, then there was quite a task to get the bus up a steepish sandy rise. Miss Moon was also bogged for a while. Dick Morrison drove home via Kiata, calling on Keith Hately of Lowan Sanctuary fame. The bus route home was via Goroke, Mt. Arapiles, Natimuk and Horsham. Lunch was at Sisters Rocks, that monument of human desecrators. Said to be named after three sisters who camped there in the late 1850's on the way to the

Pleasant Creek diggings (Stawell), these granite boulders were saved from destruction in 1862 by Mr. S. J. Davidson, and came under the "protection" of Stawell Council in 1869. The name-carvers and dauber vandals have made a mockery of such a word!

Finally, a word of thanks should be recorded to McKenzie's driver, George Wellington, and to Mark Allender, who, as usual, organized and supervised the trip very well.

* * *

Notes on Fauna at Wyperfeld and Broughton's Waterhole
by W. R. GASKING*

In Wyperfeld National Park Brush-tailed Possums were very plentiful in the vicinity of the camping area. At night every large tree within a half-mile radius seemed to contain at least one pair of shining eyes.

The Black-faced Grey Kangaroo, *Macropus fuliginosus melanops*, also known as Mallee Grey Kangaroo, and now officially named Western Grey, were in evidence at the camp-site and along the various tracks. Few were seen there, but tracks were fairly common. On the walking trail to Lake Brambruk tracks were more frequently encountered, as well as several skeletons. When we reached the Red-gum (*Eucalyptus camaldulensis*) flats bordering the well-grassed dry lake bed, we were rewarded by the sight of a huge mob, which I estimated would contain at least 500, possibly even 1000 'roos. These were thickly dispersed among the trees, and animals were visible for as far as we could see. As we moved slowly towards the lake-bed the grass-and-shrub-covered ground seemed to develop a heaving and undulating movement, so many animals were there, hopping away from us in front and to each side. Many females carried joeys and there were a number of juveniles at foot.

Out on the grassy lake-bed there was a large mob of emus feeding, and here again I would estimate their numbers in excess of 500. This mob seemed to melt away as we appeared, as the alarmed birds commenced to walk away from us towards cover. In contrast to the kangaroos, where young were plentiful, these emus did not have any chicks with them. In fact, although two or three small clutches were reported near the camp, no others were seen despite it being "chick season". In open-range captivity, such as at Healesville (Vic.) and Cleland National Park (S.A.), the nesting pairs tend to seek isolated nesting sites, and keep the newly hatched chicks in the vicinity for up to about a week before leading them out into the open. However, once the chicks appear among the other, non-breeding birds, the latter are liable to attempt to stamp on the chicks. I've never heard a satisfactory explanation of this behaviour, which is also extended to include other small creatures, such as wallabies, ducks, and dogs, and not necessarily in the breeding season. Could it be an inbuilt means of regulating the growth of the population? When there are large

*Cleland National Park, Summertown, S.A. 5141.

numbers of adults, as we saw at Wyperfeld, chicks would have little chance of survival, but when there are few adults, more chicks could survive. Perhaps readers with some experience of wild emu populations could comment?

At Broughton's Waterhole and surrounding country in the Little Desert, there was again evidence of a fair-sized population of kangaroos. Tracks were seen in many places, and several animals were seen from vehicles. No sightings or tracks of emus were reported**.

In order to check on the smaller mammals in the area, a line of 40 traps (borrowed from S.A. Field Naturalists Mammal Club) was set each night. We caught three house mice (*Mus musculus*) and five Silky, or Ashy-Grey Mice (*Pseudomys albo-cinereus*, Ride). The house mice were caught in the more open ground near the camp, while the *Pseudomys* were caught among the clumps of Porcupine Grass (*Triodia* sp.) and also in nearby dense herbage. These interesting little native rodents live

** Emus were sighted on a trip not taken by Mr. Lasking because of indisposition. See report of excursion by J. A. Baines, p. 107

in long deep burrows in sandy soil, with several entrances 10 ft. or 12 ft. apart. They apparently push all the spoil from digging out of one hole, which is then closed, and the resulting mound of soil, as much as 2 ft. or more across becomes a marker to observant seekers, while apparently foiling predators.

It was very disappointing to observe the damage caused by earlier collectors, seeking these mice. Whole clumps of *Triodia*, covering many square yards, had been dug out in the ruthless searches, and I saw three large trees left standing in excavated pits, where the soil had been removed to a depth of about two feet, in a circular pit about 15 ft. radius from the butt, leaving all the roots exposed. One of the locals told me "the University had been up looking for Silky Mice and Hopping Mice". It is bad to leave so much destruction behind, and a great pity that the collectors did not at least replace the soil removed from around the trees. I cannot believe that such methods are justified in collecting specimens, even "in the name of science".

Plate 2.

Ashy-grey Mouse (*Pseudomys albo-cinereus*) taken at Broughton's Waterhole.

Photo:
W. R. Gasking

June, 1971

book reviews

Australian Spiders in Colour
by Ramon Mascord

A. H. & A. W. Reed. Pp. 112, col. pll 192, line drawings. $3.95.

This is one of the most useful of the current Reed books on natural history. There are few sound popular reference books on Australian spiders and the academic ones are rare, costly and specialized in tone. The author has studied Australian spiders for some years partly because of their being little known. His book outlines external anatomy and general biological features of spiders with the aid of well labelled drawings and a handy glossary. The remainder of the book provides excellent coloured photographs of, and basic information on, more than 150 species representing 24 families. Although many of the species are of wide distribution the emphasis, to the detriment of the implication of the title, is upon eastern forms. The notes on species generally include simple descriptions, limited reference to taxonomic history, and notes on habitat, habits, dimorphism, food, egg sacs data for the specimens illustrated and occasional items of particular interest. The plates, which show spiders in natural settings, added to the brief descriptions, provide a means of identification as adequate as can be expected.

Details of the classification of spiders are disappointingly limited and the separation of the Mygalomorphs from other groups is not clearly shown in the family arrangement. The notes however include such interesting points as a description of the observed hunting action of the Bolas Spider. The book is a welcome popular reference to a group known well only by the specialist.

A.R.McF

Annotated Bibliography of Quaternary Shorelines — Supplement 1965-1969
H. G. RICHARDS

(Philadelphia USA: Academy Natural Sciences). 1970. 240 pages.

Study of the ocean is a feature of our time. Its importance is now recognized, the high costs can be met, and the necessary technologies are available. During the Ice Age, sea-levels changed dramatically. Thus Australia was one-third (1,000,000 sq. mi.) larger; New Guinea and Tasmania were extensions of the mainland. The Great Barrier Reef was a landform. So great were such changes, that they are important for geologists, geomorphologists, anthropologists, zoologists, botanists, geneticists, engineers, and many others.

However, this subject is not easy to study because its literature is so disperse, finding a place in the journals of so many disciplines. It is a subject of global interest, so its research is published in every country having a scientific literature. Richards and Fairbridge's *Annotated Bibliography of Quaternary Shorelines* was therefore a real research aid. Now an equally large *Supplement* covering the years 1965 through 1969 has been published. This great increase in the number of articles published reflects the expanding interest in this subject. Over 1800 titles appear, even after omitting lesser papers, abstracts, and the individual articles in encyclopedias. The author is the President of the INQUA Commission on Shorelines.

EDMUND D. GILL

Field Naturalists Club of Victoria

F.N.C.V. 91st Annual Meeting

11 March, 1971

The President, Mr. Sault, opened the meeting and welcomed new members. He announced with regret the deaths of the following members:

Mrs. E. King
Mr. W. C. Woollard
Mr. H. B. Barrett

Mr. Sault spoke of Mrs. King, Mr. Swaby and Mr. Woollard, and Messrs. McInnes and Baines, and Mr. Barrett. Two minutes silence was observed.

Mr McInnes presented the annual balance sheet and explained the salient details. It appears that the financial situation should remain satisfactory this year. Danby, Bland & Co. were re-appointed as auditors.

As no new nominations for Council had been received the following members of the existing Council were re-elected, leaving four positions vacant:

President—Mr. T. Sault.
Vice-Presidents—Mr. J. H. Willis, Mr. P. Curlis.
Secretary—Mr. D. Lee.
Treasurer—Mr. D. McInnes.
Asst. Treasurer—Miss M. Morgan.
Editor—Mr. G. Ward.
Librarian—Mr. P. Kelly.
Asst. Librarian—Miss M. Lester.
Excursion Sec.—Miss M. Allander.

Councillors—
Mr. A. Fairhall
Mr. R. Riordan
Mr. B. Cooper
Miss G. Piper.

Mr. Curlis moved a vote of thanks to Mr. Condron for his valuable assistance over a number of years; seconded by Mr. Morrison; and Mr. Garnet a vote of thanks to all members of Council for their work in 1970, seconded by Mr. Baines.

Mr. Kelly, the speaker for the evening, gave a fascinating introduction to his subject, "Beetles". After describing the typical insect structure, and the distinguishing features of the beetles, he told how to tell the various families from one another, and showed a series of excellent slides of typical Australian members of the main groups. Finally he described his studies into the life history of the Chiropsis beetles and made a plea for more amateurs to help in work of this type.

Mr. Sault thanked Mr. Kelly for an excellent talk, and after a number of questions the audience expressed its thanks in the usual way.

Mr. Sault reported seeing a flock of many thousands of ravens feeding on berries of the Coast Beard Heath.

Mr. Garnet moved that the club support the existing nomination of Mr. Cliff Beauglehole for the 1971 Natural History Medallion. This was seconded by Mr. Fairhall and carried.

* * *

Montmorency Junior F.N.C. Second Annual Meeting

The second annual meeting of the Montmorency Junior Field Naturalists Club, held on Friday, 12 March, 1971, was attended by some 90 members, predominately teenagers and younger children.

The President, Mr. Ray Wilton, in opening the meeting, spoke of the past year's activities, which included: A most successful screening of Photo Flora at Eltham Youth Hall, from which our Club gained $45. This was the first time Photo Flora had been shown in the metropolitan area, north of the Yarra. Several well attended field excursions, and a nature exhibit staged by our Juniors, as part of the FNCV Wild Life Show in the Melbourne Town Hall.

The President thanked the officers, and everyone who had helped during the Year, and paid tribute to the various speakers at monthly meetings.

Attendance figures at these meetings averaged 75, and the junior members maintained a good showing of exhibits.

The Club has a bank balance of $107.00.

Mr. Wilton was re-elected President.

Vice-President—Mr. P. McMahon.
Secretary—Mrs. J. Cookson.
Minute Secretary—Mrs. D. D'Alton.
Treasurer—Mrs. D. Howard.
Committee—Mr. R. DeGrouchy, Mrs. Woodburn, Mrs. McMahon.
Librarian—Mrs. T. North.

The Junior Committee:
President—Laurie Cookson; with Andrew Williams, Robert Callander, Margaret Howard, Julie Cookson.

The speaker for the evening was Mr. W. Davis, who showed two excellent bird films to augment his lecture.

MRS. J. COOKSON, HON. SECRETARY

* * * *

Botany Group
8 April, 1971

Mr. Fred Woodman spoke on his second visit to South Africa and first to Kenya. He was privileged to stay with the warden of Tsavo National Park, and was thus able to photograph elephants, rhinoceros, hippopotamus, lions, cheetahs, zebras, buffaloes, antelopes of many species, and a variety of smaller game at much closer quarters than is possible to the usual tourist. The botanical settings were mostly arid-looking, and one wonders at the ability of such vegetation to sustain a vital part of the food chain that supports such a remarkable fauna. In the Mount Kenya area the rain forest and sub-alpine regions produced a much richer flora, and slides included shots of famous species like giant lobelias, tree heaths and tree groundsel. Species of East African *Helichrysum* and *Acacia* were a reminder of Australia, while in South Africa several Australian acacia species have become noxious weeds and are known there as "green cancers". In South Africa the rich Cape Peninsula flora and the Kirstenbosch Botanic Gardens were featured, with close-ups of many species of magnificent *Protea*. Giant euphorbias, aloes and kniphofias were others that made striking impact on the mind.

The chairman (Mr. K. Kleinecke) thanked the speaker, who was cordially acclaimed by members. Books and pamphlets on the African flora were on exhibit.

The speaker at the next meeting, on 11 May, will be Mr. Bruce Fuhrer, who will take Victorian orchids as his subject.

13 May, 1971

There were 24 members present, including Miss Annie Cooper, a member of the Tasmanian Field Naturalists Club (Hobart). Mrs. Matches reported on the excursion to Blackburn Lake, her species list of native plants still surviving in this area surrounded by suburbia being quite a lengthy one; many members pulled out numerous Boneseed plants (the South African Jungleweed pest, *Chrysanthemoides monilifera*). It is to be hoped that the upper area, at present owned by an independent school, will be preserved as well as the area immediately about the lake.

Mr. Bruce Fuhrer gave an excellent talk on Victorian orchids, showing his fine slides of most well-known species as well as rarer ones and unusual varieties. Of particular interest were shots of *Pterostylis coccinea* (Scarlet Hood) and *P. laxa* (Antelope Orchid), both found by Mr. Cliff Beauglehole during his recent botanical survey of East Gippsland. The genus *Thelymitra* was represented by 13 species. *Calochilus* by 4, *Diuris* 7, *Microtis* 4, *Praxophyllum* 14, *Caleana* 3, *Chiloglottis* 4, *Lyperanthus* 2, *Caladenia* 19, *Corybas* 3, *Cryptostylis* 2, *Glossodia* 2, *Pterostylis* 28 (including one undescribed species with a black and white appearance), and one each of the genera *Orthoceras*, *Dipodium*, *Gastrodia*, *Eriochilus* and *Spiranthes* and *Burnettia*. Epiphytic species of *Dendrobium* (2), *Sarcochilus* (2) and *Thrixspermum tridentatum* (Tangle Orchid) were also shown.

The chairman (Mr. K. Kleinecke) thanked the speaker on behalf of all members.

Arrangements were made for an excursion to Tecoma on Sunday, 6 June. Exhibits shown by Mr. J. H. Willis included some interesting phalloid fungi, *Mutinus* spp., and a Crinoline Fungus collected near Lae, New Guinea, as well as the first record for Australia of a species of *Ijuhya*, a genus of fungi known from Java and Brazil, found in 1967 near Hart's Range, Central Australia, by Mr. Cliff Beauglehole.

GEOLOGY GROUP EXCURSIONS

6 June — The excursion ("In search of sharks' teeth) previously advertised has been postponed and will probably be held in August.

10 July — To the igneous rocks of the Lysterfield-Berwick area. Leader — Mr. Edward Nimmervoll.

Transport is by private car. Spare seats are usually available for those without their own transport. Excursions leave from the western end of Flinders Street Station (opposite the C.T.A. Building) at 9.30 a.m.

BIOLOGICAL, GEOLOGICAL AND ASTRONOMICAL CHARTS

* * * *

STANDARD MICROSCOPE
7X, 10X, 15X eyepieces
8X, 20X OBJECTIVES $54.00
CONDENSER $22.00 extra

* * * *

40X WATER IMMERSION OBJECTIVES $18.00

* * * *

A Wide Range of Plastic Laboratory Apparatus available at

Genery's
Scientific Equipment Supply
183 Little Collins Street, Melbourne, 3000
Phone 63 2160

Vice-Presidents: Mr. J. H. WILLIS; Mr. P. CURLIS

Hon. Secretary: Mr. D. LEE, 15 Springvale Road, Springvale (546 7724).

Subscription Secretary: Mr. D. E. McINNES, 129 Waverley Road, East Malvern, 3145

Hon. Editor: Mr. G. M. WARD, 54 St. James Road, Heidelberg 3084.

Hon. Librarian: Mr. P. KELLY, c/o National Herbarium, The Domain, South Yarra 3141.

Hon. Excursion Secretary: Miss M. ALLENDER, 19 Hawthorn Avenue, Caulfield 3161. (52 2749).

Magazine Sales Officer: Mr. B. FUHRER, 25 Sunhill Av., North Ringwood, 3134

Group Secretaries:

Botany: Mr. J. A. BAINES, 45 Eastgate Street, Oakleigh 3166 (57 6206).

Microscopical: Mr. M. H. MEYER, 36 Milroy Street, East Brighton (96 3268).

Mammal Survey: Mr. D. R. PENTON, 43 Duke Street, Richmond, 3121.

Entomology and Marine Biology: Mr. J. W. H. STRONG, Flat 11, "Palm Court", 1160 Dandenong Rd., Murrumbeena 3163 (56 2271).

Geology: Mr. T. SAULT.

MEMBERSHIP

Membership of the F.N.C.V. is open to any person interested in natural history. The *Victorian Naturalist* is distributed free to all members, the club's reference and lending library is available, and other activities are indicated in reports set out in the several preceding pages of this magazine.

Rates of Subscriptions for 1971

Ordinary Members	$7.00
Country Members	$5.00
Joint Members	$2.00
Junior Members	$2.00
Junior Members receiving Vict. Nat.	$4.00
Subscribers to Vict. Nat.	$5.00
Affiliated Societies	$7.00
Life Membership (reducing after 20 Years)	$140.00

The cost of individual copies of the Vict. Nat. will be 45 cents.

All subscriptions should be made payable to the Field Naturalists Club of Victoria, and posted to the Subscription Secretary.

JENKIN BUXTON & CO. PTY. LTD PRINTERS, WEST MELBOURNE

F.N.C.V. DIARY OF COMING EVENTS
GENERAL MEETINGS

Monday, 12 July — At National Herbarium, The Domain, South Yarra, commencing at 8 p.m.

1. Minutes, Reports, Announcements.   2. Nature Notes and Exhibits.
3. Subject for the evening — "Antarctica in Midsummer"; Dr. M. Beadnall.
4. New Members.

Ordinary:
Mr. Robert W. Beck, 182 Inkerman St., East St Kilda, 3182
Mr. Albert O. Griffiths, 91 Lucerne Crescent, Alphington, 3078
Mr. Arthur F. Land, 111 Blackburn Road, Doncaster East, 3109
Dr. Ross G. Macdonald, 6 Evon Avenue, Ringwood East, 3135.
Mr. John M. Robin, 3 Leawarra Drive, Heathmont, 3135.
Mr. William J. Stockdale, 9 Beamsley Street, Malvern, 1144
Mrs. M. D. Cooper, 56 Rochester Road, Canterbury, 3126.

Junior:
George Smith, 15 Short Street, Kangaroo Flat, 3555.

Country:
Mr. W D Gamble, 8 Myall Street, P.O. Box 441, Renmark, Sth Aust, 5341
Mr. George T. Liddell, 50 Cudgen Road, Kingscliff, N.S.W., 2413.
Mr. Roger D Macaulay, 5 Bayview Parade, Nth. Geelong, 3215.

5. General Business.   6. Correspondence.

Monday, 9 August — "Mammal Survey": A. Howard.

F.N.C.V. Excursions

Sunday, 11 July — Botany Group excursion to Ringwood. Meet 1.50 p.m. at Ringwood Station.

Sunday, 18 July — Sherbrooke Forest. "Lyrebirds and Ferns", led by Miss M. McKenzie and Miss M. Lester. The coach will leave Batman Avenue at 9.30 a.m. Fare $1.40. Bring one meal and a snack.

Saturday, 21 August - Sunday, 5 September — Flinders Ranges. The coach will leave from Flinders Street outside the Gas and Fuel Corporation at 8 a.m. on Saturday, 21 August. Bring picnic meals for Saturday and Sunday. Balance of payment is now due, all cheques to be made out to Excursion Trust. Those who asked for bunks at Arkaroola have been allotted same, the rest of the party will be in the motel section.

Sunday, 26 December-Monday, 3 January — Falls Creek; further details next month.

Group Meetings
(8 p.m. at National Herbarium unless otherwise stated).

Thursday, 8 July — Botany Group. Miss L. White will speak on "Proteaceae".

Wednesday, 21 July — Microscopical Group.

Friday, 30 July — Junior meeting at 8 p.m. in Hawthorn Town Hall.

Monday, 2 August — Entomology and Marine Biology meeting at 8 p.m., in small room next to theatrette at National Museum.

Wednesday, 4 August — Geology Group.

Thursday, 5 August — Mammal Survey Group meets at 8 p.m. in Arthur Rylah Institute for Environmental Research, 123 Brown Street, Heidelberg.

Friday, 6 August — Junior meeting at 8 p.m. in Rechabite Hall, 281 High Street, Preston.

Friday, 13 August — Montmorency and District Junior F.N.C. meets at 8 p.m. in Scout Hall, Petrie Park, Montmorency.

The
Victorian Naturalist

Editor: G. M. Ward

Vol. 88, No. 7   7 July, 1971

CONTENTS

Articles:

The use by Birds of Tools and Playthings. By A. H. Chisholm	180
Mangroves as Land-builders. By E. C. F. Bird	189
The Broad-toothed Rat *Mastacomys fuscus* Thomas from Falls Creek Area, Victoria. By Joan M. Dixon	198
Lyrebirds in Tasmania. By W. R. Gasking	200
Botany Excursion to Wilson's Promontory	201

Personal:

"Merrimu" at Sherbrooke	203

Book Reviews:	202

Field Naturalists Club of Victoria:

General and Group Reports	204
Diary of Coming Events	178

Front Cover:
    This photograph taken by E. R. Rotherham, is of the Broad-toothed Rat, which is the subject of an article on p. 198.

July, 1971

The Use by Birds of Tools and Playthings

by A. H. CHISHOLM*

The eminent American naturalist John Burroughs was somewhat astray when he declared about fifty years ago, in one of his numerous books, "Man is the only inventive and tool-using animal". That sage commentator would probably have been astonished if informed that there are two or three species of wild mammals and at least ten species of birds that are more or less proficient in the use of tools.

This matter of tool using by non-human creatures has been recognized in Australia since about 1840, when John Gilbert, able collaborator of the zoologist John Gould, reported from Western Australia, on the authority of Aborigines, that Black-breasted Buzzards were known to break Emus' eggs with stones in order to eat the contents. Definite reports of the kind were published later from both sides of the continent, together with differing examples of tool using by other species of birds, and in 1954 I summarised these records in an article in the *Ibis* (England).

More recently, the subject has come under notice anew through the medium of Jane and Hugo van Lawick-Goodall, who, after noting tool using by chimpanzees in Africa, turned to watching and photographing Egyptian Vultures breaking Ostrich eggs with stones. Their first report appeared in *Nature* (England) in December 1966. Regrettably, it was marred by the statement that only one other species of bird had been known to use a natural object as a tool, this being the Woodpecker-finch of the Galapagos Islands, which uses a thorn or sharp stick to prise insects from holes. In later writings, however, the African-based recorders "caught up" with some (but not all) of other published material on the general subject.

With this situation in mind, it seems desirable to bring all available records of this nature together. In fact I did so up to a point, in an article in the *Sydney Morning Herald* in July 1967, but other cases in kind have come to notice since then, and thus a review of present-day knowledge of the subject may well be given in a journal of natural history.

Egg-smashing by Buzzards

The Black-breasted Buzzard (*Hamirostra melanosterna*), a large kite-hawk of the Australian inland, is much addicted to foraging on the ground, where it preys on reptiles, small mammals, and young birds. Doubtless ground-haunting made the species, long ago, familiar with the eggs of Emus and Bustards, and, failing to break these, or at least the large and hard-shelled products of the Emu, with its beak, it somehow acquired the habit of using stones or other hard objects as missiles or hammers.

Indeed, so resolute is this predator, and so "well-informed" regarding eggs, that it will drive a brooding Emu off its nest. Gilbert reported action of the kind (in the 1840's) when telling Gould what the latter described as "a most singular story": he said he had been told that a Buzzard would attack a brooding

*History Hoole, 133 Macquarie St., Sydney

Emu "with great ferocity" until the eggs were uncovered, and then it would "take up a stone with its feet and, while hovering above the nest, let it fall upon the eggs and crush them".

Many years later (1912) A. J. North confirmed this report on the authority of K. H. Bennett, an experienced ornithologist living in southwestern New South Wales. Remarking on the extraordinary "cunning and sagacity" of the Buzzard, Bennett said that upon seeing an Emu brooding the predator would approach it with outstretched flapping wings and when the eggs were uncovered a stone would be used to break them. If no stone was available (in one case a missile of the kind had been carried from a long distance) a hard piece of calcined earth would be used.

Bennett did not make clear the manner in which the Buzzard wielded its missile, but another writer, Gordon F. Leitch of northern Queensland, has stated that he actually saw a pair of Buzzards flapping over an Emu's nest and dropping stones, while the owners, which evidently had been driven away, were walking about at a short distance.

Adding that some of the missiles in his case missed their marks, Leitch says that although the predators "definitely dropped the stones from the air", he had heard of a case in which the stones had been used as hammers. This, it is suggested, may be the means adopted with eggs of the Bustard, which, being much smaller and fewer in number than those of the Emu, would present a more difficult target from the air; and moreover they would be easier to break at close range, either with the beak or a small stone held in the beak.

Another interesting sidelight is that fragments of the eggs of both Emus and Bustards have sometimes been found in Buzzards' nests. Perhaps these portions were carried away when there was competition at a raided nest. F. L. Berney reported from northern Queensland in 1905 that he had been astonished to see six Buzzards feeding on a batch of half a dozen Emu eggs, all broken and all fresh. A round stone, the size of a hen's egg, was lying nearby. By what means, Berney asks, was the news of that discovery spread in the case of a species rare in that area? "I would not have thought", he adds, "there were that number of Buzzards within one hundred miles of the spot".

Further to the striking matter of the relatively small predator driving the great Emu off its nest, Mr. John Fitzgerald, of Mount Margaret, W.A., sent me some years ago a picturesque report given him by Jenny, an Aboriginal housemaid employed on a district station. This is Jenny's recital:

"Emu lay plenty feller big eggs. Big feller eagle want 'em. Eagle fly round an' round Emu nothin' get up; keep alla time top of eggs. Eagle fly away old feller blacks' camp, get 'em in ash-heap, make 'imself white. Fly back emu's nest. Walk up emu's nest wings like this" (spreading her arms). "Emu frightened feller. Run away. Eagle pick up yahba (stone), drop 'im on emu egg. Cart 'em off longa nest quick feller. Eatem up".

On the whole, there is ample evidence to indicate that a Buzzard may adopt "shock" tactics to disturb a brooding Emu (and that in turn indicates that the robber realises that eggs are present, without seeing them); but, of course, the "make 'imself white" part of the foregoing quotation need not necessarily be accepted. It can be regarded, perhaps, as an Aboriginal "ghost story".

From the declaration by Jenny the native housemaid we may turn, appropriately enough, to a revealing statement regarding the Buzzard's activities made by Lord Casey, former Governor-General. In his book of 1966, *Australian Father and Son*, Lord Casey comments on what he terms "a bird's surprising ability to use a tool", as reported by his father from Murray Downs station, N.S.W., in the early 1880s. At that time it was frequently found that the eggs of Ostriches, bred on the property for the feather trade, were broken in nests, and ultimately the trouble was traced to a large hawk, undoubtedly the Buzzard.

Here again is evidence of the Buzzard's extraordinary enterprise: for, clearly, it was not misled by the fact that whereas the eggs of the Emu are green, those of the Ostrich are whitish. The determining factor, no doubt, was shape.

Egg-smashing by Vultures

The observations on egg-breaking by the Egyptian Vulture (*Neophron percnopterus*), as published by the Lawick-Goodalls in *Nature*, were made in Tanzania. They indicated that each bird picked up a stone in its beak, and, while standing near an Ostrich nest, projected the missile at an egg. Sometimes the target was missed; but the pattern was repeated until the egg was broken.

Later, the same writers discussed the subject in finely-illustrated articles in the *National Geographic Magazine* (U.S.A., 1968) and *Animals* (England, 1969), and in each instance they gave the results of enlightening experiments, including the fact that the Vultures were not misled when the Ostrich eggs were painted red. As with our Buzzards of the 1880s (which turned from green to white eggs), shape was the governing influence.

There had been an impression that these egg-smashing forays by Vultures were restricted to Tanzania, but reports in kind have since come from both South Africa and Ethiopia; and, in the southern case, Dr. Jane Lawick has advised me that a correspondent has reported stone-dropping from aloft on the part of Vultures. Obviously, egg-smashing predators are distinctly adaptable.

The record of egg-breaking by Egyptian Vultures in Ethiopia has come from two experienced ornithologists who were studying a large colony of Great White Pelicans on an island in Lake Shala. They report in the *Ibis* (1969) that each Vulture would pick up a Pelican's egg and hurl it smartly down on to a rock. Thus, probably because of the relative smallness of the egg, they reversed the process adopted towards Ostrich eggs; and this, the recorders suggest, "would seem to represent quite a different level of mental activity". Four to six Vultures operated at Lake Shala and they broke ten to twelve eggs a day, probably totalling 400 to 500 eggs in a season.

Perhaps it should be added here that the breaking of Ostrich eggs by predators in Africa seems to have been known long ago. No basic record is available but a New South Wales naturalist, A. G. Hamilton, referred to the matter in his book *Bush Rambles* (Sydney, 1937). After mentioning that hawks in Australia had been known to drop stones on the eggs of Emus, Hamilton added, "The same story is told of an African eagle and ostrich eggs". It would be interesting to learn where his "same story" originated.

Darwin's Finches

It has long been known that the woodpecker-finch (*Camarhynchus pallidus*), of the Galapagos Islands, often prises insects from holes by using a thorn or small stick. Its reputation in this respect is in fact widely spread, due to the fact that it has frequently been featured on television. More recently, it has been learned that another of these small birds that so intrigued Charles Darwin, the Mangrove Finch (*Cactospiza heliobates*) has the same ability. A note on the subject is given in the *Ibis* for January 1967 (p. 129) on the basis of German observations made in 1964. "This tool-using", it is added, "distinguishes the two species from all other Darwin's Finches".

Bowerbirds' "Wads"

The remarkable habit of the Satin Bowerbird of "painting" the inside walls of its arbour, using charcoal or a dark berry mixed with saliva—or even bluebags stolen from country laundries—caused much interest when it was first made known in Sydney in 1924; and soon afterwards we learned that the bird usually holds in his beak, while painting, a small piece of macerated bark. Early impressions considered these wads to be in the nature of brushes, but it is now generally accepted that they are "corks", or "stoppers", grasped in the beak-tip in order to prevent the mixture from escaping while the head is held sideways.

Subsequently it was discovered that a similar practice is sometimes followed by the Regent Bowerbird and at least some members of the Spotted Bowerbird (*Chlamydera*) group, in these cases both the paint and the wads being, apparently, macerated dry grass. It is probable, therefore, that six or seven species of bowerbirds are tool-users.

Choughs' Use of "Hammers"

The White-Winged Chough (*Corcorax melanorhamphus*), a crow-sized black bird with white wing-slashes, is the most recent addition

Fig. 1

Illustrating the use of a thorn by the Woodpecker-finch, of the Galapagos Islands. (See text).

to the list of avian tool-users. Distinguished for its social habits and its large bowl-shaped nest of mud placed on a tree-branch, it forages mainly on the ground, taking insects, small reptiles, and mice, and sometimes robbing other birds' nests. In some areas it also attacks freshwater mussels, and in so doing turns itself into a tool-user by gripping a dried half-shell of the same species and bringing it down as a hammer on the sound mussel.

N. H. E. McDonald, who reported this practice to the Sunraysia Naturalists' Research Trust (Vic.) in 1970, says he had often been puzzled by hearing, when no person was near, a light hammering sound beside Lake Hattah (Vic.), and eventually he traced the noise to a Chough. Grasping a broken shell, it repeatedly brought this "hammer" down, almost vertically, on to the highest and softest part of an unopened mussel, and when a breach was made it dragged out the animal living inside. Other Choughs were seen probing the muddy edges of the lake, apparently in attempts to locate live mussels for similar treatment.

The writer adds that he had often seen mussel shells with holes punched in one side, always at the soft "beak", but had previously supposed water-rats to have been responsible.

Support for what McDonald terms the "high intelligence" of Choughs is given in the *Emu* for April 1971 by John Hobbs, a New South Wales ornithologist of wide experience. He relates that in May 1970, near Warren in western N.S.W., he saw thirteen Choughs picking up mussels from creek mud, cleaning each one on dry ground, and then, holding it fast in a foot, attempting to prise it open. Failing in this, each bird would at last seize an empty valve, hold it pointing downward, and repeatedly strike it on the unopened mussel. A side-to-side motion was sometimes used, so that the mussel was struck a double blow. Occasionally an unopened mussel was used as a hammer.

During the thrashing process the "tool" sometimes broke, whereupon the bird continued to strike with the remaining part until it was fragmentary, or picked up and used other pieces. One very resolute Chough rained 42 blows on a mussel before its tool completely disintegrated. Occasionally, one of the birds became a secondary tool-user, for while searching for a hammer it would pause and strike its unopened mussel against a fallen branch, an exposed tree-root, or another unopened mussel. In each case, when a breach was effected the bird removed fragments of the animal through the hole.

As Lake Hattah and Warren are about 650 km apart, it becomes apparent that the breaking of mussels by Choughs is widespread. It may, however, have developed independently in different areas, possibly getting a basis when the birds were gathering mud for nest-building.

The "Fire-hawks"

What is to be said of a report by Aborigines in the Northern Territory that birds which they term "Fire-hawks" (probably Fork-tailed Kites, *Milvus migrans*) deliberately set fire to grass in order to obtain reptiles and small mammals? This claim is made in *I. the Aboriginal* (Sydney, 1966), a book produced by Douglas Lockwood and an intelligent native named Waipuldanya.

Describing these birds as "masters of cunning", the Aboriginal says he has often seen one pick up a smouldering stick in its claws and drop it in dry grass half a mile away, and then wait with its mates for the mad exodus of rodents and reptiles.

By chance, soon after reading this statement a letter from Melbourne gave me experiences in point by a Colombo Plan student newly returned from the Rum Jungle area. The local hawks, he related, sometimes snatched burning sticks from a campfire, when cooking was done at mid-day, and dropped them some distance away. Also, the birds broke off green branches and dropped them on the fire. The visitor thought these actions "very curious".

A significant point in this second report is that the hawks took burning sticks from campfires (as distinct from grassfires), which would seem to indicate a definite use of the sticks as tools. The matter, it is clear, needs investigation.

A Cockatoo's Sagacity

Yet another impressive device adopted by a bird when seeking food (and one which should perhaps have been cited earlier in this discussion) is reported by the notable Alfred R. Wallace in his book *The Malay Archipelago*, first published in 1869. Writing of the large Palm Cockatoo of the Aru Islands — this species, *Probosciger aterrimus*, also occurs in the north-east of Australia—Wallace says that the enormous, complicated beak and remarkable tongue of the bird enable it to deal with, in particular, a certain hard and smooth-shelled nut which no other bird is able to open. Apparently the visiting Englishman gained close views of the "very curious" feeding technique of the big cockatoo, for he describes in detail how the beak, tongue and claws function specifically yet in unison when coping with the hard and smooth nut. Moreover—and here is the most striking feature of the operation—he states that the bird bites off a piece of leaf and uses it as "anti-skid" equipment. This is done by retaining it in a deep notch in the upper mandible, where the nut is "prevented from slipping by the elastic tissue of the leaf".

The suggestion has been made that an object should be classified as an avian tool if a bird uses it as an extension of claws or beak for a specific purpose. On this definition the cockatoo of Wallace's admiration ranks as a tool-user, even though its "mouth-piece" may be only a passive aid. In any event, one wonders how the Palm Cockatoo learned that a certain piece of leaf, when placed in a particular part of the beak, would assist to keep a smooth nut from slipping.

Ants as "Tools"

It will have been noted that with the exception of the bowerbirds all the recognised tool-using birds exercise their peculiar abilities to gain food. What is to be said, however, in regard to anting, the confirmed practice of many birds of rubbing live ants and other acidulous substances on their bodies? The insects are usually discarded (not eaten) after use, and moreover the mediums have been known to extend to berries, aromatic leaves, cigar-butts and the like, thus making it apparent that the practice is aimed at cleansing and/or stimulating the bird's body. Should not, therefore, this be regarded as a form of tool-using? If so, the list of avian tool-users will greatly expand, for scores of species have been known "to ant", or practise "anting", since these words were first introduced into the English language (in Australia) after the practice was discussed in this country in 1934.

Here it would be noted, too, that in a recent issue (Oct. 1970) of the

American *Auk* a writer reports that a young American Robin (*Turdus migratorius*) was observed while anting to pick up a leafy twig on several occasions and sweep the ground with it, apparently in an attempt to locate more ants in the leafy litter. That use of a "broom", however, appears to have been spasmodic, or fortuitous, and not an expression of habit.

Secondary Tool-using

Although some birds (as in the case of the Choughs) may be both primary and secondary tool-users, there appears to be a certain difference in mental activity involved between the two practices. Secondary tool-users are those birds which, instead of holding an object in the beak or claws and using it as a missile or other aid, strike a shell against a rock-face or drop it on to a hard surface from aloft. In Australia and elsewhere gulls freely drop shells on to rocky platforms, and all of our jungle-haunting pittas ("anvil-birds") and some thrushes hold shells in their beaks and thrash them against hard surfaces. Dwellers in south-eastern Queensland were once puzzled by often hearing a tinkling sound emanating from a belt of rainforest; and they remained puzzled until learning that the noise was caused by a Buff-breasted Pitta hammering land-shells on the base of a beer-bottle buried in the soil.

Records of shell-dropping in Britain include one in which Colin MacDonald (*Highland Journey*, Edinburgh, 1943) relates the exploits of a Crow. He says that the bird twice dropped a cockle on to sand without success, and then—whether by chance or design—it changed to gravel and got results; and after that it dropped six more cockles, but not one of them on to sand.

Use of Playthings

Possibly there is affinity between tool-using and the habit of some birds of using sticks or other objects as playthings. There are in Australia numerous records of eagles and smaller hawks taking aloft a stick, or perhaps a dried rabbit-skin, and dropping it for another bird to catch, or, alternatively, swooping and making the catch itself. Country children term this practice "the stick game".

Similar actions have, of course, been observed in other countries. That dedicated English (and Australian) bird-observer Miss Joyce Grenfell advises that she recently saw an exhibition of the kind given by two hawks in Surrey. The sight was, she declares, "absolutely sensational".

The tall and stately cranes which we know as Brolgas also play a "stick game", not in the air but on the earth. It is more or less usual for one bird after another, during a company dance, to pick up a dry branchlet and toss it aloft.

Two other reports of birds using playthings provide an element of novelty. In one statement Galahs are said to have frequently dropped stones on to the iron roof of a country home in New South Wales, apparently for the sole purpose of enjoying the noise created. In the second report Magpies also are said to have dropped stones on a rural rooftop, but in this case the action —performed by about thirty birds— was taken only when, in the early morning, delay had occurred in the regular practice of putting out food for the bird-visitors. If, as the householders believed, the roof-rattling was done purposely to claim attention, the action would appear to lift the pebble-wielding Magpies from players to tool-users—"bell-ringing" variety!

Possibly we should also regard as playthings, or toys, the wide range of objects with which the various kinds of bowerbirds festoon their arbours. These differ considerably with different genera. Incidentally, one report has it that the Spotted Bowerbird sometimes engages in "the stick game"—with bones or shells.

Also in affinity, perhaps, is the practice of certain male birds, in this country chiefly Fairy Wrens (*Malurus*), of carrying about bright-coloured flowers, usually red or yellow, during courtship. As with the Satin Bowerbird, which has a strong fancy for blue objects, these flower-carrying birds appear to be able to appreciate particular colours.

Origin of Tool-using

When reflecting on the degree of "intelligence" possibly associated with birds' use of objects as various kinds of aids (mainly in relation to food), there is a temptation to consider certain aspects of food-stealing. Frigate-birds, for example, consistently attack other seabirds in mid-air in order to rob them of food, and Silver Gulls in Australia have been known to ride on the backs of Pelicans and grab captured fish before the larger birds can swallow them. Perhaps such robbers are, in a sense, using the victimised birds as tools!

Further, to the question of intelligence, an unconfirmed report from the Australian sub-interior credits the Rainbow Bee-eater (*Merops ornatus*) with the occasional use of a sharp stick as an aid to tunnel-digging, and with, as well, the frequent equipping of its burrow with "natural lamps" —white bones and mussel-shells, also in one instance a pearl button.

Other remarkable reports, put forward independently in both Europe and Australia, claim that certain birds—a Snipe in Europe and a Snipe and a Magpielark (Peewee) in Australia—have been known to bind an injured leg with a splint made of feathers and mud. Statements regarding each of these reports have been published.

Without seeking further material bearing on the mental activity of certain birds, it seems desirable now to consider the question of how tool-using originated. There need be little doubt, I think, that Australia's Buzzards and Bowerbirds were using tools even before our Aborigines invented the boomerang; and doubtless the practice by other birds is also deep-seated. But how the ability developed, in each instance, is more conjectural.

John Alcock, of the Department of Psychology in the University of Washington, has recently (1970) presented suggestions on the subject in the *Ibis*—thus giving England an American opinion on the behaviour of birds in Africa! The Egyptian Vultures' enterprise, it is suggested, "may initially have been nothing more than re-directed egg-throwing". That is to say, birds thwarted in their attempts to pick up and throw Ostrich eggs, which are too large, would turn their attention to a smaller object if one were available; and a bird going through the egg-throwing movement with a stone lying nearby might accidentally hit and break the egg, after which the Vulture "might come to associate stone-throwing with food and actively seek out stones on encountering an Ostrich egg". And, as a corollary, observation of the stone-throwing may have led to the transmission of the activity from one bird to another.

"Thus", Alcock concludes, "the origin and spread of tool-using by Egyptian Vultures need not involve special insight or a high level of mental activity on the part of the birds".

Tool-using by the Galapagos finches, it is further suggested, may also have arisen in a conflict situation. A finch reaching for a grub might accidentally stuff a twig in the crevice, so forcing the insect within range of its beak, and consequently the bird might learn to associate placing a twig in the hole with food.

These theories merit respect. They are weakened, however, by the author's supposition, based on the van Lawick-Goodall paper of 1966, that tool-using is confined to the Egyptian Vulture and the Woodpecker-finch. Further thought on the subject is needed in view of the fact that avian tool-using has many facets, and also because egg-smashing by Vultures is not restricted to one area (as was supposed) or even to one species. Consideration is needed, particularly, of the Buzzard's actions in attacking an Emu before seeing its eggs.

As to the origin of the practice by the Woodpecker-finch (and also the Mangrove Finch) of prising grubs from crevices, doubtless Alcock's theory of an "accidental" basis is as valid as any other that can be put forward. Similarly, the bowerbirds that hold wads in their beaks while painting may have begun to do so fortuitously through picking up soft bark while gathering painting material. Even now, it seems, the shredded bark used as "stoppers" sometimes becomes part of the "paint".

Finally, it may be noted that in his book *Dominant Mammal* (Sydney, 1970) Sir Macfarlane Burnet says, "Intelligence is effective in the use of tools to procure a desired result". Having this dictum in mind, we may well reflect anew regarding the level of mental activity in tool-using birds. For, certainly, their actions in each case produce "a desired result".

REFERENCES

Alcock, J., "The origin of tool-using by Egyptian Vultures", *Ibis*, Oct. 1970.
Bennett, K. H., in A. J. North, *Nests and Eggs*, 3, p. 244; 1912.
Berney, F. J., "Black-breasted Buzzard", *Emu*, 5, p. 18; 1905.
Brown, L. H., and F. K. Urban, "Biology of the Great White Pelican", *Ibis*, April 1969.
Casey, Lord, *Australian Father and Son*, p. 60; Sydney, 1966.
Chisholm, A. H., *Bird Wonders of Australia*, p. 263; 1934 and later.
——— "The use by birds of tools or instruments", *Ibis*, pp. 380-83; 1954.
——— "Case of the criminal Buzzard", *Herald*, Sydney, 22 July 1967.
Fleay, D., "Eagles do play and enjoy it too", *Courier-Mail*, Brisbane; 2 April 1958.
Gould, J., *Handbook to the birds of Australia*, 1, p. 47; 1865.
Hobbs, J., "Use of tools by the White-Winged Chough", *Emu*, 71; April 1971.
Lawick-Goodall, J. & H., "Use of tools by the Egyptian Vulture", *Nature*, 24 Dec 1966.
——— "Tool-using birds the Egyptian Vulture", *Nat. Geogr. Mag.*, May, 1968.
——— "Vultures that use tools", *Animals* (England), July 1969.
Leitch, G. F., "Buzzards destroying Emu eggs", *North Queensland Naturalist*, March 1953.
McDonald, N. H. F., "Cases of high intelligence of White-Winged Choughs", *Report Sunraysia Nats. Research Trust*, April 1970.
Wallace, *The Malay Archipelago* (Aru Ids chapter), 1869.

Mangroves as Land-Builders

by E. C. F. BIRD*

Mangroves are shrubs and trees that grow on the tidal shores of estuaries, inlets and embayments, in sectors protected from strong wave or current action. In the humid tropical environment of north-east Queensland there are more than a dozen mangrove species, arranged in zones parallel to the coastline (Macnae 1966), but in Victoria, where mangroves reach their southernmost limit in Corner Inlet (38° 55' S.), there is only one species, the White Mangrove (*Avicennia marina* var. *resinifera*). This is found on the shores of Corner Inlet, Shallow Inlet, Anderson's Inlet and Westernport Bay, and in the estuary at Barwon Heads. It formerly grew at a number of sites around Port Phillip Bay.

In Victoria, mangroves generally form scrub communities up to 8 feet high, at the outer (seaward) edge of salt marshes dominated by *Arthrocnemum* and *Salicornia*. They are usually fronted by mudflats exposed at low tide which bear little vegeta-

*Department of Geography, University of Melbourne.

Plate 1 The rear of the mangrove (*Avicennia*) fringe (left) at Yaringa, with *Arthrocnemum* marsh (right).
Photo: Author

tion apart from a sparse and patchy cover of *Zostera*.

Mangroves can be said to act as land-builders on tidal shores if they trap sediment and build up depositional terrain that would not otherwise have developed. According to one view, mangrove communities bring about the advance of the shoreline by spreading forward over the mudflats in the inter-tidal zone (Davis 1940, Richards 1952). Others have argued that mangroves advance only after the inter-tidal zone has been raised by sedimentation to a level suitable for their colonisation, so that the spread of mangroves is a consequence, rather than a cause, of shoreline advance (Scholl 1968).

Accretion within mangrove communities gradually builds up the land to a level where salt marsh plants can move in and take over (Plate 1), but again there is the question of whether the mangroves actually promote accretion, or whether they simply occupy a zone within which accretion would take place anyway. The extent to which mangroves can advance the shoreline and then promote vertical accretion of sediment may vary with the structure of the mangrove community. *Avicennia*, for example, has networks of vertical

Plate 2

Pneumatophores under *Avicennia marina* at Yaringa

Photo: Author

"breathing-tubes" (pneumatophores) projecting from sub-surface root systems (Plate 2), which will have different effects on sedimentation from the stilt-root structures of *Rhizophora*, which lacks pneumatophores. During the past four years, measurements have been made in and around the mangrove fringe at Yaringa, on the north-western shores of Westernport Bay, in an attempt to assess the land-building effects of *Avicennia marina*.

THE YARINGA SHORE

The geological history of Westernport Bay has been described by Jenkin (1962); it is a large tidal embayment, with a spring tide range of up to 11 feet on its northern shores. The mangrove fringe is still extensive (Fig. 1), despite past fluctuations (Enright 1969) and the recent effects of reclamation for port development in the Hastings district. The *Avicennia* fringe is up to 300 yards wide on sheltered sectors, and it stands in front of a salt marsh zone up to 900 yards wide. Swamp paperbark (*Melaleuca ericifolia*) is often present at the back of the salt marsh.

At Yaringa these features are well developed. A boat channel cut in 1967 is bordered by levees of dredged material dumped alongside, which provide easy access to each of the

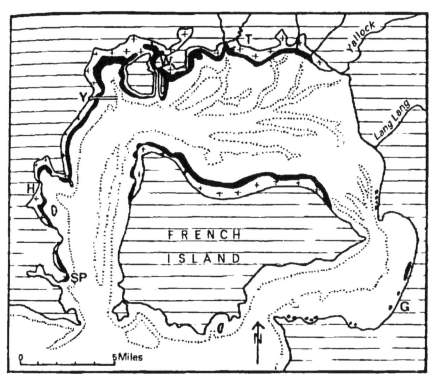

Figure 1 The northern part of Westernport Bay. Mangrove fringe in black, salt marsh shown by crosses, dots indicate low water spring tide line.
H — Hastings.   SP — Stony Point.   Y — Yaringa.
W — Warneet.   T — Tooradin.   G — Grantville.
The extent of mangroves and salt marsh has recently been reduced by land reclamation in the Hastings area.

vegetation zones, and to the mudflats which extend for more than a mile in front of the mangrove fringe, down to low spring tide level. The surface topography here was surveyed in relation to vegetation zones and tidal levels, and sub-surface features were deduced from an examination of the dredged material and from probing the marshland: the results are shown in Fig. 2.

The mudflats have been colonised by *Avicennia marina* down almost to mid-tide level. Mangroves occupy the zone between this level and the line reached by mean high spring tides, where they give place to salt marsh, dominated by *Arthrocnemum* in the outer zone, with *Salicornia* (and some bare or algal-carpeted saline flats) to landward, submerged only during the highest spring tides, which just reach the bordering *Melaleuca ericifolia* zone. The vegetation grows upon a depositional terrace of soft muddy sediment (containing layers of peat and shells), which rests in turn on an undulating floor of quartzose sand, continuous with the low sandy ridges bearing woodland, dominated by *Eucalyptus viminalis*, which form the immediate hinterland.

These low sandy ridges originated as beach ridges, several of which terminate laterally in recurved sandspits now enclosed by marshland.

This sector of the shore of Westernport Bay was evidently sandy about five thousand years ago, when the sea attained its present level after the late Pleistocene low sea level episode (cf. Bird 1968, p. 45). The muddy sediment, derived from material washed into Westernport Bay by rivers, together with organic materials derived from the estuarine shell fauna and plant remains, has accumulated here subsequently.

The Yaringa shore thus provides a good site to examine whether the pattern of sediment accumulation has been determined by colonisation and advance of mangroves, or whether the mangroves have simply spread forward in a pattern determined by the independent accumulation of sediment. It is also a good site for an investigation of rates and patterns of vertical accretion in and around the mangroves.

MANGROVES AND SHORELINE
ADVANCE

Avicennia marina grows in a zone which is at least partly submerged at every high tide, but stands above water level for several hours before and after low tide. The advance of mangroves appears to be limited by the depth and duration of tidal submergence. During the winter there are often large numbers of mangrove seedlings in the mud in front of the mangrove fringe in Westernport Bay.

Figure 2 Section at Yaringa.

where the viviparous seeds, which germinate on the plant, fall on to the adjacent mudflats. Few of these seedlings, survive, however, and the impression of a rapid advance of mangroves on to mudflats is misleading. They persist only where sediment accretion has recently raised the level of the mudflats in front of the mangrove fringe. An example of this is seen alongside the levee banks at Yaringa, where an area of local recent accretion of mud is marked out by mangroves that are now three or four years old (Plate 3).

There are other sectors where a recent advance of mangroves on to mudflats can be demonstrated, but long stretches of the mangrove fringe around Westernport Bay have not advanced at all between 1966 and 1971. The physiological factors which limit seedling survival are not known, but it is clear that mangroves cannot spread forward until the mudflats in front of them have been built up to a suitable level by sedimentation.

There is, nevertheless, one way in which mangroves may contribute to the build-up of adjacent mudflats, and thereby prepare the way for their own subsequent advance. The mangrove fringe provides sheltered conditions in the nearshore zone when winds are blowing off the land during high tides. This was observed while working in boats near the outer edge of the mangroves: onshore winds generate waves which wash sediment into the mangroves (Plate 4), but offshore winds scarcely ripple the water surface within a zone up to 20 yards off the mangrove edge. Where mangroves are not present, offshore winds generate currents which can move sediment away from

Plate 3 Mangrove seedlings on mudflats south of Yaringa.

Photo: Author

the shore, but the mangrove fringe reduces this effect, and therefore creates conditions favourable to more rapid vertical accretion on the adjacent mudflats. In this way, mangroves may be said to contribute to the advance of the shoreline.

ACCRETION ON MUDFLATS

The standard method of measuring vertical accretion of sediment on mudflats or marshlands is to place a layer of identifiable material (such as brick dust) on the surface, and return subsequently to probe for it and measure the thickness of overlying accretion. An attempt was made to use this method on mudflats in front of the mangroves south of Yaringa, but it failed because the brick dust layer was quickly dispersed by waves and currents, and by burrowing crabs, and so did not remain in position long enough to be of much use as a marker horizon.

An alternative method is to insert stakes and measure the level of accretion against these. In January 1968 a set of 12 stakes (bamboo canes) were implanted at various sites up to ten yards seaward of the mangrove fringe south of Yaringa. The mud here is very soft, and can only be traversed on foot with the aid of platcher-boards attached to one's shoes. This results in disturbance of the mudflat surface on a scale which calls in question the validity of measurements made in or near the disturbed area. It proved more convenient to work from a shallow-draught boat, moving towards the mangroves on a rising tide.

The stakes were pushed into the mud until only 20 cm protruded above the surface. Changes of level on the mudflats were then measured at monthly intervals with reference

Plate 4   Waves moving into the mangrove fringe at high neap tide south of Yaringa.
Photo: Author

to the protruding stakes. The method has the disadvantage that the stakes may generate eddies which modify local patterns of accretion, and the results were taken only as a broad indication of the scale of changes that occurred. Monthly measurements during 1968 showed irregular alternations of vertical accretion and surface lowering: over 11 months seven of the sites showed net accretion (maximum 4.4 cm.) and five net lowering (maximum 2.8 cm.). There was no evidence for widespread continuing accretion over the area examined, the changes registered on the stakes being due to movements of mud shoals to and fro across the mudflats by wave action. One site, with a sparse *Zostera* cover, received a capping of mud up to 8 cm. thick during October 1968, but this had largely dispersed a month later. It was noted that when onshore winds accompanied a rising tide there was a definite movement of mud towards and into the zone occupied by mangroves.

ACCRETION UNDER MANGROVES

Vertical accretion was measured within the mangrove fringe with reference to layers of brick dust spread on the surface at ten sites during April 1967. The brick dust was disturbed by burrowing crabs, but persisted sufficiently for measurements to be made over the succeeding three years; its persistance was an indication of the relative stability of the surface under the mangroves, compared with the mudflats already described.

Sites near the inner (landward) margin of the mangroves showed slow mud accretion (up to 0.7 cm over three years), and sites in the centre of the mangrove zone (at high neap tide level) showed more rapid mud accretion (up to 2.4 cm. over three years). Towards the outer (seaward) margin there was much variation, from a minimum of 0.4 cm. to a maximum of 4.6 cm. mud accretion over three years. As all of the sites measured showed a consistent rise in mud level there can be no doubt that vertical accretion is taking place within the mangrove fringe at the present time.

Mud accretion has proceeded more rapidly among the pneumatophores and around the trunks of each mangrove plant (Plate 5), raising the surface up to 15 cm. above the intervening areas of soft wet mud. In this way, mangroves directly influence the *pattern* of mud accretion. A similar effect was observed at Stony Point, where mangroves have recently spread on to a sandy foreshore area, and patches of mud have accumulated around the pneumatophores. When a network of pegs was implanted in the sand here to simulate the effects of pneumatophores, it produced a patch of mud up to 0.3 cm. thick within a month. The pegs were then removed, and the mud soon washed away. Sedimentation among the pneumatophores results from the fact that they produce a calm-water environment conducive to deposition of muddy sediment that would otherwise remain in suspension, or be carried away.

Within the mangrove community the ebb and flow of tides is confined mainly to intricately-branching creek systems, in contrast with the smooth and almost featureless topography of the adjacent mudflats. Accretion of mud around mangrove plants builds up the surface on either side of low corridors, which gradually become well-defined channels. The creek system is thus correlated with patterns of mud accretion that are determined by the presence of mangroves.

Mangroves therefore influence the pattern of sedimentation, but it is difficult to prove that they promote more rapid accretion than would otherwise have occurred. To do this, it would be necessary to make comparisons with rates of accretion measured on similar shore sectors unoccupied by mangroves. Unfortunately, such sectors are obviously different in other respects, and it is not possible to make such a comparison. Where mangroves have died back, or been cut away, the shore is often sandy rather than muddy, and there may be active erosion of the sediment that had been built up beneath the former mangrove cover. This implies that where mangroves have become established they may be responsible for mud accretion on shores that might otherwise have been sandy, or subject to erosion.

Once established, a mangrove fringe tends to stabilise the terrain and impede erosion.

The mangrove ecosystem produces organic materials (leaves, twigs, root remains, shells, crab skeletons) that become incorporated with the underlying sediment, and represent a direct accession of materials that would not have been available in the absence of mangroves. This becomes most obvious towards the landward margin of the mangroves, where very slow mud accretion from infrequent tidal flooding is augmented by a surface accumulation of organic litter. At a higher level, in the *Arthrocnemum* and *Salicornia* zones, the rate of mud accretion is greatly exceeded by the accumulation of organic residues in the form of a superficial fibrous peat.

Plate 5 Tide rising into mangrove fringe at Stony Point.

Photo: Author

CONCLUSIONS

There is evidence at Yaringa for a gradual but continuing accretion of sediment within the mangrove fringe, in contrast with irregular alternations of erosion and accretion on the adjacent mudflats, and with superficial peat formation in the salt marsh. It is deduced that the mangroves are responsible for trapping and fixing sediment that would otherwise have remained mobile, and consequently for the building-up of a depositional terrace to a level where it is taken over by salt marsh. Mangroves are therefore acting as a geomorphological agent in shaping the inter-tidal topography.

Mangroves cannot directly cause shoreline advance, for they spread forward only after the adjacent mudflats have been built up by vertical accretion to a suitable level. By sheltering the nearshore zone at high tide from the effects of offshore winds they may nevertheless facilitate the vertical accretion which will subsequently permit their advance. Once established, mangroves influence patterns of accretion, gradually confining the ebb and flow of tides to residual tidal creek systems. It is likely that they accelerate the accretion of muddy sediment; they also contribute organic materials that augment the accreting sediment. Mangroves have certainly been an agent of land-building on the shores of Westernport Bay.

REFERENCES

Bird, E. C. F., 1968: *Coasts*, A.N.U. Press, Canberra

Davis, J. H., 1940: The ecology and geologic role of the mangroves in Florida. *Publs. Carnegie Instn.*, 524: 303-412.

Enright, J., 1969: Processes and patterns of coastal change in Westernport Bay. *Hons. thesis, Melbourne Univ.*

Jenkin, J. J., 1962: The geology and hydrogeology of the Westernport area. *Dept. of Mines, Victoria.*

Macnae, W., 1966: Mangroves in eastern and southern Australia. *Australian J. Botany* 14: 67-104.

Richards, P. W., 1952: *The tropical rain forest*. Cambridge.

Scholl, D. W., 1968: Mangrove swamps: geology and sedimentology. *Encyclopaedia of Geomorphology* (ed. R. W. Fairbridge): 683-8.

F.N.C.V. PUBLICATIONS AVAILABLE FOR PURCHASE

FERNS OF VICTORIA AND TASMANIA, by N. A. Wakefield.
The 116 species known and described, and illustrated by line drawings, and 30 photographs. Price 75c.

VICTORIAN TOADSTOOLS AND MUSHROOMS, by J. H. Willis.
This describes 120 toadstool species and many other fungi. There are four coloured plates and 31 other illustrations. New edition. Price 90c.

THE VEGETATION OF WYPERFELD NATIONAL PARK, by J. R. Garnet.
Coloured frontispiece, 23 half-tone, 100 line drawings of plants and a map. Price $1.50.

Address orders and inquiries to Sales Officer, F.N.C.V., National Herbarium, South Yarra, Victoria.

Payments should include postage (11c on single copy).

The Broad-toothed Rat *Mastacomys fuscus* Thomas from Falls Creek Area, Bogong High Plains, Victoria

by JOAN M. DIXON*

Summary

A live male specimen of the broad-toothed rat *Mastacomys fuscus* was collected on 23 May, 1971 by the author during a mammal trapping programme in the Falls Creek area at approximately 1798 m (5900 ft.). This is the highest altitude record for this species in Victoria, and also the northernmost record for the State.

THE STUDY AREA

This specimen of *M. fuscus* is recorded from the area where *Burramys parvus* Broom was trapped in Feb. 1971. It is located approximately 7.2 km (4.5 miles) southwest of Falls Creek (Lat. 36° 53' S, Long. 147° 15' E). The hillside faces approximately west, and is studded with granite tors, while the vegetation consists of alpine grasses, shrubs and stunted snow gums, *Eucalyptus pauciflora* var. *alpina*. A more detailed description of the vegetation was given by Dixon (1971). In winter the mammals move about under the snow-covered vegetation. When *M. fuscus* was trapped a small amount of snow was present. This seemed to have little effect on the mammals present, or on their numbers.

The *Mastacomys* specimen was collected in a large aluminium collap-

*Curator of Vertebrates, National Museum of Victoria

Plate 1 Habitat of *Mastacomys fuscus* in the Falls Creek area, Bogong High Plains, Victoria
Photo: Joan M Dixon

Plate 2 *Mastacomys fuscus* C10065 from the Falls Creek area.
Photo: F. R. Rotherham.

sible Elliott trap, 45 x 12 x 15 cm baited with walnut and set on a runway through snowgrass, *Poa australis* and Alpine Orites *Orites lancifolia*. On this same track on previous and subsequent occasions, *Antechinus swainsonii* were also captured. The *Mastacomys* site was within a few metres of where *Burramys* was captured. Although no *Burramys* was caught in May, *Rattus fuscipes assimilis* was trapped less than 1 m from the February trapping site. This is the first record of *R. f. assimilis* in this area despite heavy trapping during February and March. However, fox scats collected from here in February contained *R. f. assimilis* fur.

The study area has many features in common with the part of the Kosciusko State Park, N.S.W., where Calaby and Wimbush (1964) collected *Mastacomys*. The altitude and vegetation of both are similar. The main difference is the less sheltered environment and absence of permanently running creeks in the Falls Creek habitat. Nevertheless, at the latter site many small boggy areas remain wet even in summer. Possibly the *Mastacomys* in this habitat move to higher and drier ground with the approach of winter. This may apply also to *R. f. assimilis*.

This new record of *Mastacomys* extends its distribution in Victoria, and lessens the gap between Victorian and New South Wales populations. The distribution of the species is not as discontinuous as previously thought. Probably these distribution gaps will be filled by further field work.

TAXONOMY

The Falls Creek specimen is identified as *Mastacomys fuscus mordicus*, the subspecies characteristic of mainland Australia, and distinct from the Tasmanian *M. f. fuscus*. This was indicated by Troughton (1941) and has been substantiated by recent work carried out by Wakefield (In press, *Mem. nat. Mus. Vict.*).

The measurements of our Falls Creek specimen (reg. no. C10065) are: weight 125 gm, total length 282 mm, tail 113 mm, ear 20.2 mm, hind foot 32.5 mm. These were made while the animal was anaesthetized with ether.

Acknowledgments

Thanks are extended to Mr. H. Brunner of the Lands Department for his analysis of hair samples, and to Mr. A. J. Coventry for assistance.

REFERENCES

Calaby, J. H., and Wimbush, D. J. 1964. Observations on the Broad-toothed rat *Mastacomys fuscus* Thomas. C.S.I.R.O. Wildl. Res. 9, 123-33.

Dixon, J. M., 1971. *Burramys parvus* Broom (Marsupialia) from Falls Creek area of the Bogong High Plains. *Victorian Nat.* 88 (5): 133-8.

Troughton, E. le G., 1941. *Furred animals of Australia*. Angus and Robertson.

Lyrebirds in Tasmania

Old-timers who recall David Fleay's efforts to transfer some lyrebirds to Tasmania, from 1934 onwards (see *Vict. Nat.* Vol. 69, No. 5, Sep. 1952.) will be pleased to hear that the species now seems to be well established in Mount Field National Park, 50 miles by road west of Hobart.

On a recent holiday (Nov. 1970) to Tasmania I stayed overnight in one of the excellently appointed privately owned cabins just outside the park entrance. Along the sides of the road up to Lake Dobson (about 4,000 ft.) freshly turned soil and leaf-mould was continually noticed, just as it is seen along the road from Healesville to Toolangi. On the drive down again we stopped three times at intervals of about one mile apart, and we could hear a lyrebird calling near the road at each stop, and twice there was another calling some distance away also. If I'd had more time and could have stopped more frequently I'm sure I'd have heard more, as it was real "Lyrebird Country", and the scratchings along the roadside offered the evidence.

The biggest thrill came as we turned a bend and surprised a bird, either a hen or juvenile male, scratching beside the road. I stopped immediately, but the bird fled into the ferns with a leap and flap of wings.

Lower down, in the valley bottom below Russell Falls, I found more evidence of the birds' scratching alongside the walking track and almost to the edge of the picnic area.

For botanists:

I was surprised to find the Adder's tongue (*Ophioglossum coriaceum*) growing in wet moss in the spray from the Falls. Having seen these growing just two months before in Wyperfeld National Park, and previously on the outskirts of Sydney, I couldn't help admiring the little plant's adaptability to such widely varying conditions of soil and moisture.

Note regarding last paragraph about J. H. Willis, in "Handbook to Plants in Victoria", Vol. I, has this to say about the distribution of Austral Adder's-tongue: Throughout Victoria (on moss-covered granitic rocks, loamy pastures, swampy heathland peats, basaltic plains, Mallee sandhills, etc., but often with only very small barren fronds, and consequently overlooked), found in all Australian States, New Zealand and South America.

W. R. Gasking, Cleland National Park in South Australia.

Botany Excursion to Wilson's Promontory
November, 1970

The Botany Group week-end excursion to Wilson's Promontory set off in high hopes on the evening of Friday, 6 November. Wallaby Lodge, with 26 beds was fully occupied and three car loads arrived as well to camp. As usual, the arrangements ran on oiled wheels owing to the efficiency and unselfishness of the F.N.C.V. Excursion Secretary, Miss Marie Allender. The group was lucky to have Mr. Jim Willis in the party to resolve all identification difficulties. Being the springtime, there was a profusion of plants in flower. Perhaps the most spectacular was the beautiful white *Kunzea ambigua* lining the tracks and seeming to cover the flatter places.

On Saturday morning, members walked to the top of Mt. Oberon. This short walk stretched to a matter of hours as so much of botanical interest was encountered on the way. The eucalypts seemed to be of stunted habit, owing partly to the dry hillside and partly to the fact that the bigger trees had been wiped out in the big bushfires. The hillside was not all arid, though, as the King Fern, *Todea barbara*, which likes its feet in water, was growing happily in places. So also was the enormous swamp-loving *Gahnia, G. clarkei*. The *Hibbertias*, especially *aspera*, the Rough Guinea flower, were there in their usual gaiety and the tomato coloured form of the common heath, *Epacris impressa*. There were several patches of two similar greenhoods, *Pterostylis alpina*, with its rather dapper figure, and *P. foliata*, with a small slender flower. Some of the party encountered an Echidna on the track. Early in the morning, two before-breakfast hikers had encountered a koala being frustrated by wire-netting from breakfasting on its favourite manna-gum.

In spite of the blistered feet of soft city-dwellers, everyone set off again in the afternoon, some to Lilli-Pilli Gully, where many ferns and shaded gully-trees and shrubs were encountered and beautiful patches of the Butterfly Flag, *Diplarrhena moraea*. Some walked up and over the track leading back to Tidal River Lookout. This track usually supplies plenty of interest to the orchid lover and a diversity of gay shrubs on the coastal heathlands, including the mauve *Olearia ciliata*. A third group walked round the cliff track past Norman Point to Little Oberon Bay. This way has such beautiful views of the ocean and granite cliffs that it is hard to give enough attention to the smaller botanical specimens, though these were there in plenty. *Thomasia petalocalyx*, the Paper-Flower, being conspicuous on the early part of the track and later the large *Patersonia* and the small one (the purple flags) looking most beautiful.

Next morning a few people walked round the coast track to Squeaky Bay, where they were picked up by the rest of the party in the bus. All the usual coast plants were there, including white correa, *Banksia, Alyxia buxifolia, Rhagodia baccata*, cushion bush (*Calorephalus brownii*), and various *Helichrysums*. The bus stopped again at Darby River and most of the party walked to Tongue Point. This walk is really divided into three or four sections—first up the sheltered hillside with the Ti-trees, Melaleucas and Cassinias, below which grow the smaller shrubs, herbs, and numerous orchids—then over the top and through the rather gloomy she-oak woods (Casuarina) again with many orchids flowering and seeding among the she-oak needles on the ground—and finally out on to the open heathlands of Tongue-Point. These were gay with innumerable bright flowers, many of the pea-family, Pultenaeas, Dillwynias, the primrose Gompholobium huegetii, Running Postman (Kennedya prostrata), the bright purple of the Flag Irises, the purple daisy bush (Olearia ciliata), the Banksias, She-oaks and Acacias dwarfed by exposure to the fierce southerly gales. Finally there is the rugged knob of huge slabs of granite and prickly stunted bushes. A few hardy bodies did this final strenuous climb in record time to find the prized Crimson Berry (Cyathodes juniperina, var. oxycedrus). Later, walking back across the flats and swamps, with their rushes and sedges, various other Thelymitra were found, T. aristata, T. flexuosa T. ixioides, T. media, T. rubra—also Prasophyllum brevilabre, and a plentiful supply of swamp heath, Sprengelia incarnata. One shrub of particular interest to the F.N.C.V. in this area is Olearia allenderae, collected by and named after our Excursion Secretary, Miss Marie Allender. And so back to the city with its noise and smells, but not the aromatic and honeyed scents of a happy weekend.

book reviews

Three further Australian natural history books published in 1970 are useful additions to A. H. & A. W. Reed's now established series.

BENEATH AUSTRALIAN SEAS by Walter Deas and Clarrie Lawler, $3.95, gives interesting information on the history of under-water study and the development of scuba diving, reveals the opportunities to assist marine biological study offered to divers by this modern sport, and illustrates these with excellent photographs of under-water life in colour.

AUSTRALIAN CRUSTACEANS IN COLOUR by Anthony Healy and John Yaldwyn, $3.95, covers both marine and freshwater forms, provides an outline classification, and gives authoritative notes on groups accompanied by excellent colour plates of representative species. References for further reading and photographic information are included. It is a reference book not merely a pictorial one.

AUSTRALIAN NATIVE ORCHIDS IN COLOUR by Leo Cady and E. R. Rotherham, $3.95, is also reference literature as well as a collection of 100 outstanding coloured orchid plates. General notes on the orchid family and the structure of the orchid flower are included and the text covering seventy of the eighty-five genera is very informative. Insect pollination is illustrated and discussed. The insect in plate 19 is a Syrphid but not Eriwalir tenax.

A.R.McF.

"Merrimu," at Sherbrooke

The cottage "Merrimu" is situated off Sherbrooke Lodge Road, close to the edge of the reserved forest. The original building belonged to Robert Graham, who named the locality after his home town, Sherbrooke, in California, and among the trees which he planted is a fine specimen of the Californian Redwood (*Sequoia sempervirens*).

Graham conducted the local post office there, in the early days of the century, and the posting slots are still to be seen in doors of what are now two small tool-sheds. The cottage was subjected to a series of alterations through the years, and the original unit, at the rear, was eventually demolished. Even today though, there is an outside wall consisting of rough-hewn horizontal slabs.

In about 1911, the cottage was acquired by the Misses F. and M. Billing, aunts of the late A. G. Hooke, whose obituary appeared in last month's *Victorian Naturalist*. Probably those ladies named the place "Merrimu". The Misses Billing planted numerous ornamental trees and shrubs on the property, and such natives as the blackwood grew spontaneously. Today the vegetation is mature, with tall conifers, spreading birches, oaks, maples, beeches and the like.

Native birds appreciate the vegetation, and the ground litter too. There are usually crescent honeyeaters and spinebills about, and families of brown-headed honeyeaters come occasionally. Rufous whistlers and olive whistlers are permanent residents, as are yellow robins and white-browed scrubwrens. A pair of ground thrushes live there too, and lyrebirds come occasionally to fossick in the carpet of leaves.

Families of brown and striated thornbills work their way through the trees each day, grey fantails are always about, and the larger fry include kookaburras, crimson rosellas, grey butcherbirds, and, in season, pied currawongs. Now and again something quite uncommon turns up. It might be the large-billed scrubwren, the rose robin, or the rufous fantail. Once a red-capped robin was about the place for a few days!

Native mammals appreciate "Merrimu" too. Long-nosed bandicoots dig in the ground of a night. The trees attract possums — silver-grey, hobuck, and ringtail—especially when two ancient cherry-plums are in flower; and each April little gliderpossums come to the blossoms of an old sugar gum. At certain seasons the hobucks become tame enough to be fed by hand.

Throughout his life, Garnsey Hooke spent many of his leisure days at "Merrimu", first as a visitor to his aunts' place, and latterly as owner of the cottage. He liked to have guests there with him, to take them walking in the forest, or to sit by the wood fire in the loungeroom and watch the avian visitors— rosellas, whistlers, robins, thrushes, wrens—taking turns at the bird tables a few feet away outside the windows.

Also, during recent years, Mr. Hooke made "Merrimu" available on occasion to certain of his friends and their families, who wished to spend a few days, or a week or so, in the quietude of the hills. This was appreciated both by Melbourne residents seeking a respite from the suburban environment, and by persons from much father afield, interested perhaps in the opportunity to see lyrebirds and other denizens of mountain forests.

An endeavour is being made to maintain "Merrimu" as a holiday cottage, for people genuinely interested in the natural environment. Members of the Field Naturalists Club—particularly country members —and of similar organizations are invited to participate; and further information may be had from Norman A. Wakefield, P.O. Box 37, Ferntree Gully, Vic., 3156 (phone 750-1501).

If "Merrimu" can be maintained permanently in this way, it will be an appropriate memorial to Garnsey Hooke, who served the cause of natural history so well throughout his life.

—N.A.W.

Erratum.

In the report on the Wyperfeld Camp-out (*Vict. Nat.* 88: (6) p. 163), in the second column, line 26: read— ". . . Scrub-wren's . . . in place of Scrub-robin's, before the words "habitat map".

Field Naturalists Club of Victoria

General Meetings
10 May, 1971

Summary of talk given by Mr. J. H. Willis on the New Guinea Highlands.

Regions visited by Mr. Willis included the small Chimbu district which has a population of almost a quarter of a million people. Altitudes here rise to fifteen thousand feet. There are range after range of razor-backed ridges, many denuded of trees by the policy of successive cultivation and later desertion by the natives as fertility of shallow soil on the slopes becomes exhausted. Now, throughout the Highlands, steep slopes are seen to be covered with coarse grasses of limited grazing value, instead of the forests that once clothed them — a wasteful economy.

In the highest alps, natural herb fields give way to mosses and lichens. Many plants in the higher parts of New Guinea are closely related to those in our Victorian Alps, e.g., *Astelia papuana* is very like *A. alpina* in Victoria. Ferns are prolific throughout New Guinea alps, and there are several beautiful ranunculi, gentians and potentillas. Pit-pit Grass thrives along the lovely Wahgi Valley where it is used for thatching, the tender shoots being cooked and eaten. Tea, coffee and banana plantations are now replacing pit-pit on the rich, black, peaty soils of this valley, and at least one native farmer here (Don Ments) is now a wealthy man. A slide showed him adorned for an important occasion; on his head was a high magnificent head dress of bird-of-paradise feathers, which is carefully dismantled after each special occasion and the feathers stored between dried banana leaves.

Amongst the limestone knolls and grassy slopes of Wahgi Valley are several kinds of terrestrial orchids, hyacinth-like *Spathoglottis* and a five-foot *Phaius*. There are scores of different tree-ferns in New Guinea and they grow both on mountain slopes and in the swampy areas. Bamboos may grow to sixty feet and their stems have many uses: fencing, water conduits and water buckets.

One charming photo showed a crowd of laughing Wahgi schoolboys who surged down upon the visitor to have their photo taken. However, unemployment is such a problem in the Highlands that there does not seem to be much future for these bright boys when they leave school.

Another slide at Mt. Hagen showed a crowd of natives with elaborate head-plumes grouped around a dais on which Mr. Gorton stood; this was taken a few minutes before the dark sky broke with the suddenness and ferocity of that climate. It took only as many moments for all spectators to vanish, as the deluge of rain would have ruined their precious headgear.

At the Baiyer River sanctuary various mammals are kept in cages, e.g., cuscus, tree kangaroos and possums. Visitors may enter the cages and feed the animals. The beautiful blue goura pigeon has a fanned head-piece of blue and white feathers but, unfortunately, it is regarded as good eating by the natives.

Many interesting rhododendrons occur around Mt Hagen district, including *R. zolleri* which has large blooms of salmon and gold, *R. rarum* with narrow leaves and small scarlet flowers, and yellow-flowered *R. macgregoriae* which is widely dispersed through the Highlands.

The native womenfolk work extremely hard and are quite old at 50; they dig the gardens, carrying heavy produce and care for the children and pigs. Old folk are usually delighted to see visitors and greet them most warmly.

Sweet potatoes ("kau-kau") is the main food of workmen and several pounds per man per day must be taken as part of the baggage on any project involving native labour.

Casuarina oligodon is widely used for building timber and fuel. On its trunks may sometimes be seen the curious anthouse plants, with prickly bottle-shaped stems to 6" wide, hanging on the host tree by surface roots. Myriads of little ants occupy a labyrinth of caverns within these "bottles" and swarm out if their ready-made house is disturbed.

The importance of pigs in the economy cannot be ignored. They are currency, and they can make or break friendships. Pigs roam everywhere and the killing of one by a motorist may bring a very angry reaction from local villagers. On special occasions, at intervals of from three to ten years, pigs are killed and the feasting goes on for days.

Mt. Giluwe, 13,600 ft high, is just across the Papuan border. Up there it is wet, often with an icy wind and swirling mists that obscure the volcanic tops. All rhododendrons growing above ten thousand feet seem to have brilliantly red flowers. *R. saxifragoides* is a curious rosette plant found in subalpine bogs; the single scarlet flowers turn to one side from a stalk up to 6" long. The moss forests of Mt Giluwe support fascinating orchids, including a claret-hued greenhood (*Pterostylis*)

Mr. Willis's lecture was illustrated throughout by a delightful collection of slides showing close-ups of plants, views of magnificent scenery, local markets, interesting native buildings, vine bridges and scanty but often elaborate costumes.

Mr. Willis was thanked warmly for the second talk covering his recent visit to New Guinea.

9 June, 1971

A summary of the talk given by Mr. Robert Withers on the subject "The Life of the Badger".

A newcomer to Australia and to the F.N.C.V. is Mr. R. Withers, and this young man was introduced by Mr. Lee as one who had made a thorough study of the Badger in England.

We heard a most interesting talk, which gave not only a clear and carefully detailed picture of this animal, but also an indication of how much one person can achieve by tenacity and ingenuity, without having prior training or experience.

These nocturnal animals build interconnecting tunnels which may extend one hundred and fifty feet, and openings appear in fields and thickets. Soil preferences are sand pockets in clay-soil or else chalk, but the latter are like fortresses and a pick axe is needed to dig them out. Many farmers have never seen the Badgers; and Mr. Withers spent several months fruitlessly trying to see them, until he eventually learnt that even the metal click of equipment would keep them in their burrows; and that their keen sense of smell would detect if one foot had fallen on their pathways, although perhaps four hours ago. Consequently he set up photographic equipment two hours before the Badgers were due to come out and placed himself downwind.

At sunset the first Badger, usually the boar, comes to the entrance and sniffs the air; then emerges to sit beneath the entrance, frozen in position with one foot raised if suspicious. As successive Badgers come up, they are more relaxed. They scratch themselves, and then play, particularly in snow. Digging follows and the soil is then marked with musk, a strong smelling fluid produced by glands below the tail.

Evidence of Badger presence are the well worn paths, pieces of Badger hair on barbed wire fences, the pug mark (five toes and claws which are parallel), dung areas anything from 15 feet to half a mile from the set, tree trunks ripped open for grubs, the skins of rabbits with all flesh and bones gone, cow dung turned over in the search for grubs, and the marks on the trees where the Badger scratches to remove dirt from its claws.

Food consists of earthworms, snails, honey and grubs from wasps nests, young squirrels and pigeons if found dead. During winter they will dig for the Moles.

Badgers visit other sets, and if arriving after the occupants have gone, they then follow on by smell. Sight is poor, limited to about twelve feet.

The tunnels widen into chambers over five feet in diameter which sometimes are situated so close to the surface in the fields that the cows break through the roof. The "oven" is a terminal chamber in which the young are reared, and it is lined with bedding, which is thrown out when the cubs are old enough to come up and commence using the latrine. In May the cubs are allowed out for an hour early in the evening, and again in the early morning, when they are cleaned and allowed to play for another hour. In June they follow their parents in their foraging. The boar is very friendly and plays with the cubs for hours, even permitting them to toboggan down his abdomen. In October, however, the boar turns the cubs out and the ensuing fights can be heard.

Generally only one family occupies one set, but foxes may share one of the outlying entrances where the vixen rears its cubs. Evidence of this is the food debris and droppings outside the entrance, which Badgers never leave. After the foxes leave, the Badgers clean the old bedding out, leave the chamber to air for about three weeks, then put in fresh bedding. Rabbits may share an entrance with the Badgers but are never touched by them; however, the Badger seeks other warrens to dig up baby rabbits for food.

Many old sets have existed for hundreds of years. One had fifty-seven entrances and may be occupied by one to three families.

As more motorways are being built, it is sometimes necessary to move a number of sets of Badgers. Their usual reaction is to dig deeper if an attempt is made to get them out, and they can dig very fast. Dogs are sent down to drive the Badgers out, but may be buried by the Badgers which block the tunnel in two places. A Badger, if handled, must be grabbed by the tail and held at arm's length, as the animal has a strong bite.

Mr. Withers' talk was illustrated by his excellent colour slides; and he received a hearty vote of thanks for his most interesting and enjoyable lecture.

Botany Group
10 June, 1971

Miss Helen Aston, who was congratulated on her completion of a book on "*Aquatic Plants of Australia*", gave the second of two instructional talks on the meaning of the 200 most commonly used botanical terms, in two main groups: those naming plant parts, and those descriptive of habit of growth, flowers, fruits, leaves, roots and general. Quick clear sketches on a blackboard illustrated the essential differences between terms related in meaning, and the duplicated sheets, containing all the terms but with no definitions, received many annotations by the 19 members present.

Miss Aston recommended a number of most useful books, including *Hunter*

ical Latin, by William T. Stearn (Nelson, 1966), Taxonomy of Vascular Plants, by G. H. M. Lawrence (Macmillan, New York, 1960) and The Language of Botany, by C. N. Debenham (S.G.A.P.). When studying any of the large published flora volumes, recourse should be had to the glossary, as some professional botanists interpret certain terms rather differently. Particularly useful, because of comprehensiveness and accurate line drawings, are the glossaries in Handbook of the Vascular Plants of the Sydney District and Blue Mountains, by Beadle, Evans and Carolin (1962) and A Flora of the Marshes of California, by H. L. Mason (1957). The chairman (Mr. K. Kleinecke) thanked the speaker, and acclamation expressed the appreciation of members.

The regret of the Botany Group on the passing of Mr. Percy (P. L.) Williams had been conveyed to his widow, Mrs. Mary Williams, of "Oaklands", Mirum, from whom a letter of thanks was received, saying that her husband had always enjoyed taking parties of botanists and other naturalists into the Little Desert, for the conservation of which he had worked assiduously for many years.

Plans were made for an excursion to South Warrandyte on 11 July, for an inspection of the fine native garden of Mr. and Mrs. W. King, who are prominent members of Ringwood Field Naturalists' Club and of the Society for Growing Australian Plants.

The subject for the July meeting will be "Proteaceae", by Miss Laura White.

GEOLOGY GROUP EXCURSIONS

Sunday, 10 July — To the igneous rocks of the Lysterfield-Berwick areas. Leader — Mr. Edward Nimmervoll.
Sunday, 8 August — To the Aboriginal Quarries, Mount William, near Lancefield.
Sunday, 5 September — To Waurn Ponds in search of sharks' teeth.
Transport is by private car. Spare seats are usually available for those without their own transport. Excursions leave from the western end of Flinders Street Station (opposite the C.T.A. Building) at 9.30 a.m.

ENTOMOLOGICAL EQUIPMENT

Butterfly nets, pins, store-boxes, etc.

We are direct importers and manufacturers,

and specialise in Mail Orders

(write for free price list)

**AUSTRALIAN
ENTOMOLOGICAL SUPPLIES**

14 Chisholm St., Greenwich
Sydney 2065   Phone: 43 3972

BIOLOGICAL, GEOLOGICAL AND ASTRONOMICAL CHARTS

♦ ♦ ✝ ♦

STANDARD MICROSCOPE
7X, 10X, 15X eyepieces
8X, 20X OBJECTIVES $54.00
CONDENSER $22.00 extra

♦ ♦ ♦ ♦

40X WATER IMMERSION
OBJECTIVES $18.00

♦ ♦ ■ ♦

A Wide Range of Plastic
Laboratory Apparatus
available at

GENERY'S SCIENTIFIC EQUIPMENT SUPPLY

183 Little Collins Street
Melbourne: 63 2160

Field Naturalists Club of Victoria

Established 1880

OBJECTS: To stimulate interest in natural history and to preserve and protect Australian fauna and flora.

Patron:
His Excellency Major-General Sir ROHAN DELACOMBE, K.B.E., C.B., D.S.O.

Key Office-Bearers, 1971-1972.

President:
Mr. T. SAULT

Vice-Presidents: Mr. J. H. WILLIS; Mr. P. CURLIS

Hon. Secretary: Mr. D. LEE, 15 Springvale Road, Springvale (546 7724).

Subscription Secretary: Mr. D. E. McINNES, 129 Waverley Road, East Malvern, 3145

Hon. Editor: Mr. G. M. WARD, 54 St. James Road, Heidelberg 3084.

Hon. Librarian: Mr. P. KELLY, c/o National Herbarium, The Domain, South Yarra 3141.

Hon. Excursion Secretary: Miss M. ALLENDER, 19 Hawthorn Avenue, Caulfield 3161. (52 2749).

Magazine Sales Officer: Mr. B. FUHRER, 25 Sunhill Av., North Ringwood, 3134.

Group Secretaries:

Botany: Mr. J. A. BAINES, 45 Eastgate Street, Oakleigh 3166 (57 6206).
Microscopical: Mr. M. H. MEYER, 36 Milroy Street, East Brighton (96 3268).
Mammal Survey: Mr. D. R. PENTON, 43 Duke Street, Richmond, 3121.
Entomology and Marine Biology: Mr. J. W. H. STRONG, Flat 11, "Palm Court", 1160 Dandenong Rd., Murrumbeena 3163 (56 2271).
Geology: Mr. T. SAULT.

MEMBERSHIP

Membership of the F.N.C.V. is open to any person interested in natural history. The *Victorian Naturalist* is distributed free to all members, the club's reference and lending library is available, and other activities are indicated in reports set out in the several preceding pages of this magazine.

Rates of Subscriptions for 1971

Ordinary Members	$7.00
Country Members	$5.00
Joint Members	$2.00
Junior Members	$2.00
Junior Members receiving Vict. Nat.	$4.00
Subscribers to Vict. Nat.	$5.00
Affiliated Societies	$7.00
Life Membership (reducing after 20 years)	$140.00

The cost of individual copies of the Vict. Nat. will be 45 cents.

All subscriptions should be made payable to the Field Naturalists Club of Victoria, and posted to the Subscription Secretary.

JENKIN BUXTON & CO. PTY. LTD PRINTERS, WEST MELBOURNE

AUGUST, 1971

Published by the

FIELD NATURALISTS CLUB OF VICTORIA

in which is incorporated the Microscopical Society of Victoria

Registered in Australia for transmission by post as a periodical
Category "A"

45c

F.N.C.V. DIARY OF COMING EVENTS
GENERAL MEETINGS

Monday, 9 August — At National Herbarium, The Domain, South Yarra, commencing at 8 p.m.

1. Minutes, Reports, Announcements.   2. Nature Notes and Exhibits.
3. Subject for the evening — "Mammal Survey": Arthur Howard.
4. New Members.

   Ordinary:
   Mr Joseph R Henderson, 213 Civic Parade, Altona, 3018.
   Miss Erica Paddle, 7/633 Malvern Road, Toorak, 3142
   Mr Philip M Rain, 5A Fairmount Road, East Hawthorn, 3123.
   Dr A. J Stapley, Flat 2, 49 Brougham St., North Melbourne, 3051.

   Joint:
   Dr. and Mrs. W. M. McKenzie, 53 Middlesex Road, Surrey Hills, 3127.
   Mrs Adriane Grant, 24 Golden Ave., Werribee, 3030.

   Country:
   Mrs. Winifred Jacobs, 50 Labilliere St., Bacchus Marsh, 3340.

5. General Business.   6. Correspondence.

Monday, 13 September — "Birds of Marsh and Lake": Joan Forster.

F.N.C.V. Excursions

Sunday, 8 August — Botany Group excursion to Mt. Slide area. This trip will be by private cars meeting at Box Hill Station at 10 a.m.

Sunday, 15 August — Cranbourne. An inspection of the site chosen for the Botanic Gardens Annex. The coach will leave Batman Avenue at 9.30 a.m. Fare $1.40. Bring one meal.

Saturday, 21 August - Sunday, 5 September — Flinders Ranges. The coach will leave from Flinders Street outside the Gas and Fuel Corporation building at 8 a.m. sharp. Bring picnic lunches for Saturday and Sunday.

Sunday, 26 December - Monday, 3 January — Falls Creek. Accommodation has been booked at Spargo Lodge which is run by members of our club. Tariff per week ranges from $55.00 to $60.00 full board. Fare for coach, including day trips, $22.00. Full fare should be paid to Excursion Secretary by the September general meeting, cheques to be made out to Excursion Trust. Accommodation to be paid individually at the Lodge.

Group Meetings
(8 p.m. at National Herbarium unless otherwise stated).

Friday, 6 August — A get together evening for those going to the Flinders Ranges at 8 p.m. in the small room next to National Museum theatrette. Miss J. Blackburn will show slides and members are requested to bring books, maps, specimens, etc. for display.

Thursday, 12 August — Botany Group. Mr. N. Walters will speak on "The Role of Fungi."

Friday, 13 August — Montmorency and District Junior F.N.C. meets at 8 p.m. in Scout Hall, Petrie Park, Montmorency.

Wednesday, 18 August — Microscopical Group.

Wednesday, 1 September — Geology Group.

Thursday, 2 September — Mammal Survey Group meets at 8 p.m. in Arthur Rylah Institute for Environmental Research, 123 Brown St., Heidelberg.

Monday, 6 September — Marine Biology and Entomology Group meets at 8 p.m. in small room next to theatrette at National Museum.

Thursday, 9 September — Botany Group. Mr. I. Morrison will show slides of Australian Proteaceae.

The Victorian Naturalist

Editor: G. M. Ward

Vol. 88, No. 8   4 August, 1971

CONTENTS

Articles:

Mt. Howitt and the Macalister. By Ellen Lyndon 212

Study of the Self-pollination of *Thelymitra venosa*.
By David L. Jones 217

Botanical Exploration in the Southern Flinders Ranges, South
Australia (1800-1970). By D. N. Krachenbuehl ..., ... 220

On a Botanical Collection from the Northern Flinders Ranges,
South Australia. By D. N. Krachenbuehl 225

A Botanical Bibliography of the Flinders Ranges, South Australia
(1800-1970). By D. N. Krachenbuehl 231

Feature:

Readers' Nature Notes and Queries 216

Field Naturalists Club of Victoria:

Information on Nature Show 238

Diary of Coming Events 210

Front Cover:

John Wallis took this photograph of the head of the Carpet Snake.

Mt. Howitt and the Macalister

by ELLEN LYNDON

It was the Bairnsdale Club that invited us to spend the long (wet) weekend in January among the high mountains north of Heyfield. As things turned out, the heaviest rains fell mainly on the plains of Gippsland, whereas we, the Latrobe Valley Naturalists, literally lived up in the clouds, or at most, in a light drizzle.

By the time this set in on the Friday evening there were eight separate camps round the hut at Howitt Springs, including that of Mr. Alan Morrison, our official leader, and of the sole representative of our host club, Mr. Keith Rogers, who had managed to slip through before the Avon River rose in its wrath. We were to hear later that a party from the city was marooned in Heyfield when the Thomson closed the roads behind them and the Macalister ripped the roadbeds from round its bridges above Glenmaggie.

Those of us who could manage it, went up on Friday; for the weather forecast was reasonable and the morning's deluge easing off. The foothills were bathed in bright sunshine. Past Glenfalloch a stop was made to look at the log stockade built long ago by one of the Burgoyne brothers as a sheepfold to keep the dingoes off his flocks at night. Great tree trunks were sawn into lengths and upended to stand side by side on their own feet. Others not so heavy were sunk into the earth like posts. The six foot walls enclose perhaps half an acre on the point of a spur, reinforced by living tree boles that happened to be in line. On the two long sides the ground falls sharply away into gullies. This interesting old structure is in the process of decay, both by reason of its age and by the depredations of passing vandals, so that it resembles a sort of wooden Stonehenge.

Along the valley of the Wellington River the usual clouds of purple dust arose from that dry harsh country, a dust that proved useful later in warning us of the approach of the laden timber trucks hurtling down the range. Up on Tamboritha the track is lined by dense walls of Alpine Ash saplings that mark the path of the destructive fire that ravaged that mountain country a few years ago. These young trees are a beautiful soft blue-grey in the juvenile stage, with splashes of bright colour where some of the older leaves have turned crimson. The Howitt road was new to us, a ridge route with some splendid vistas of stone-faced crags and deep gorges where cotton-wool clouds lurked. Every now and then the way crossed flower-bedecked meadows, a constant temptation to stop and browse among trigger-plants and daisies.

The Howitt Plains are long open stretches of snow grass and low shrubbery, sloping from tree-girt rises to central waterways that have been eroded and widened by cattle playing and sliding down the banks. It was a considerable surprise to find these plains mostly fenced and cattle-gridded, some of them privately owned. No doubt the grassland is much altered by nearly a century of trampling by heavy cloven hoofs and white clover is widespread through it. Among other introductions noted were odd plants of St. John's Wort and two ancient herbs of the Roseaceae, Lady's Mantle and the Wood Avens, all three of which feature in the myths of the old world and are probably still in use by pharmacists and herbalists.

Alchemilla or Lady's Mantle likes to trail down over the water. Its shallow-palmate pleated leaves often autumn tinted and the small greenish-yellow flowers borne in packed inflorescences. The alchemists of old are said to have carefully collected the dew from its leaves as this was one of the ingredients of the elixir that bestowed perpetual youth. *Geum urbanum*, the Wood Avens or Herb Bennet, grows in the shelter of the snow gums. At a glance the seed-heads resemble those of the native Bidgee-widgee Burr but the plant is much taller. Each achene is equipped with a long hooked awn just like an old-fashioned buttonhook, a most effective stickfast for man and beast. Along with various other useful properties it was supposed to possess the power of warding off evil spirits.

In spite of the rain and fog plastic-wrapped figures roamed the countryside, delighted to find it well furnished with many of the alpine mat plants in flower and berry. Daisies of many kinds were legion. In its descent down the mountain the little creek below the hut, a head-water of the Caledonia River, had carved for itself a spectacular gorge with remarkable deep and rugged cliffs corrugated in something like organ pipe formation. Where it enters the gorge track the creek is diverted sharply round the point of a narrow spur, a place of black tumbled rocks and foaming cascades, terminating in a long vertical fall. The damp rocks are spotted with crustaceous lichens in big pale patches and Mother Shield-fern trails long green plumes down the cliffs. Here Mr. Morrison had located the tiny Finger-fern, (*Grammitis armstrongii*), but although most of us knew of its existence no one recognised it then, it was so unexpectedly minute and un-fernlike. It grows in pads of moss and root fibre with little hold on the rock and its most robust fronds were barely an inch in length. In the same area the dainty Brittle Bladder-fern (*Cystopteris fragilis*) is not uncommon in crevices, often with the better known Necklace Fern (*Asplenium flabellifolium*), Alpine Water-fern (*Blechnum penna-marina*), and the Shield Fern (*Polystichum proliferum*) are widespread throughout the snow fields.

The Moonwort (*Botrichium lunaria*), was reported from a rocky point across the creek but few of the party were lucky enough to see it. On the opposite side or Wonnangatta fall of the plains Brittle Bladder-fern was found growing prolifically in company with the Common Spleenwort (*Asplenium trichomanes*). A forest road gave access to the foot of a cliff facing the morning sun and these ferns were a truly lovely sight trailing down through cracks and shallow cavities in the basalt, with an occasional water drop splashing down from above. Here again we have, in the Moonwort and the Spleenwort, plants credited by the earliest writers with fabled powers of healing.

Except for a few nondescript toadstools growing on manures, fungi was disappointingly scarce. The small woolly bird's nest type, *Crucibulum vulgare*, was brought in on a dry cowpat and a soft crimson polypore, undetermined, was seen erupting from cracks in the butts of live snow gums. Lichens decorated every rock and tree, revelling in the damp conditions. Birds were few and subdued—Magpies carolled in the early mornings, Bronze-winged Pigeons fed along the tracks, Pipits, Flame Robins in sober plumage, an occasional Currawong or Red Wattlebird—the list was not long.

The sound 'weatherproof' hut with its big ashy fireplace proved a blessing, indeed it was a necessity for a wet weather camp such as this one. It provided a warm haven for the evening get-togethers, dried out clothing and boots and ensured a constant supply of hot water and toast! Bedraggled walking parties squelched in from time to time, made our acquaintance and went their ways dried out and refreshed. The most exciting and adventurous part of this weekend, however, was still to come.

Monday morning found everyone packing up, full of forebodings through listening to too many radio reports of isolated towns and washed out roads. The weather immediately came to its senses and we saw for the first time the beauty of the long plains, the hills and gorges that surrounded us on every hand. At the foot of the range the sleepy Wellington River was a raging torrent, forcing a littered course through the timber of its valley. Beyond Licola, undermined bridges and oozing landslides were the order of the day for the Gippsland bound, but thanks to the road patrol officer and to our dauntless drivers, many of them ladies, all were piloted safely through. Something in excess of 700 points of rain in the Glenmaggie Creek-Macalister catchment certainly left its mark on the landscape that memorable weekend.

There is a charm about the high country that lures the nature lover back again and again whether the weather be good or bad, and two subsequent weekends revealed some of the glories that were missed the first time. On Labour weekend, the 6-8 of March, when three families went back to Howitt hut, we were blessed with lots of hot sunshine during the day and brilliant moonlight by night. The floral scene had faded until the Howitt road traversed a big patch of a most attractive daisy-bush. *Olearia ramulosa* var. *stricta*, which brought all cars to a standstill. Those low shrubs, smothered in native daisies, are the floral highlight of the hills at this particular time. The Gentians, too, were at their beautiful best in every bog. Birds were more plentiful, especially the Red Wattlebirds that were noisy in the trees around the hut. There were signs of small animal life around and under the shrub-sheltered rocks and a mouse-like creature was sighted momentarily. Offerings of the magic peanut butter-honey formula were licked up as soon as dusk fell. Owls and Bats frequently flew through the torch beam.

Thanks to a fatherly Forests Commission, really good motor roads have enabled the family sedan to penetrate the innermost recesses of our wonderful mountains, not always to their advantage, and traffic is heavy at holiday periods and long weekends. Family parties, often lugging young children, think nothing of trudging the last three or four miles up to the final plateau of Mt Howitt, the glories of which let it be said, are, in suitable weather, well worth the effort.

Such a forest road, looping from side to side like a giant snake track falls straight off the end of the Howitt plains to the head of the Macalister. It crosses and recrosses great rock rivers like those of Wombargo, fantastic billows and troughs a hundred yards across, bare of any verdure except the hardy lichens and a tough woolly version of the Common Hemp-bush, *Gynatrix pulchellus*. Alpine Ash and a white trunked gum, either Manna or Candlebark or something in between, grow tall and slender on the slopes above the modest stream, a forest that is fairly free of undergrowth.

A tributary creek coming in beneath a dripping stone wall supported a pretty fern gully shaded by shapely Sassafras and Lomatia trees and a few Soft Tree-ferns. Two Beechnuts grew luxuriantly, the Hard Water-fern and the Ruty Water-fern. Two Spleenworts, the Mother Spleenwort and the Necklace Fern. Austral Filmy-fern (*Mecodium australe*) adorned the cliff face along with different mosses and a small member of the lily family, Slender Wire-lily (*Libertia pulchella*). This was a spreading plant with tufted leaves and long wiry roots, locally plentiful. The dry flower stems bore tiny long-stemmed seedheads, arising in twos and threes from a cluster of bracts. Unfortunately an avalanche of tops and splintered trunks from logging operations on the spur above have almost obliterated this gully.

The weekend of the 26-28 of March began with a dense fog in the high country that deepened to rain at night, but the days were keen and sunny. Cattle mustering had begun and parties of horsemen and dogs disturbed the tranquillity of the hill mountain cattle. They lent a certain picturesqueness to the landscape so that we found ourselves reciting passages from Paterson and Kendall as we regretfully passed by Howitt hut to look for accommodation further up.

The morning was perfect as our small party set out for the final assault on the summit of Mt. Howitt. Pipits perched on every rock to watch us pass and plain-clothes robins and grasshoppers, invisible on the brown-green earth, rose in flocks ahead of every footstep. At Macalister Springs the palatial Vallejo Gantner Memorial Hut is nearing completion, a striking arrangement of copper and glass in the heart of the wilderness. Down at the springs the myriad white "wogs" in the water barrel were duly inspected and near here a specimen of the native slug, *Cystopelta petterdi*, was picked up. Its semi-detached mantle seems to stick out like a tiny wing as it progresses.

On the narrow ridge above Terrible Hollow the rock ledges were flushed with the rose pink stars of *Crowea exaltata* and we wondered what treasures, animal and vegetable, might be hidden among the spectacular steps of the Devil's Staircase just across the way. As we climbed on up the track hundreds of Flame Robins, flight after flight of them, came flipping down the mountain past us. To rest at last on summit, on a day of perfect sunshine; to gaze out across that high plateau patched with golden or white everlastings at the saw-toothed peaks and purple depths that surround it, is a never-to-be-forgotten pleasure and satisfaction.

As we sat enjoying our lunches there was constant passage overhead of big white butterflies with patterned undersides, heading steadily and surely north, without alighting. For all their heavy appearance they proved too agile to be caught for identification. Kestrels hung on the updraught at the cliff edge and occasionally a Wedgetailed Eagle or a Falcon sailed by. Far below on the headwaters of one of the many rivers that rise from the foot of Howitt timber harvesting was in progress. The white trunks of the gums lay criss-crossed on the torn red earth like spilled matches. This was not a long weekend and we had the Mountains to ourselves, the Viking and the Razor, the Crosscut Saw, Cobbler and Speculation, the tailend at least of the famous Barry Mountains.

I am indebted to Mr. J. H. Willis for the following information: *Geum urbanum* (Common Avens) has always been accepted as a truly indigenous plant in S.E. Australia, as it also is in New Zealand where the Maoris had a special name for it, "Kopata". This is one of the few species of typical boreal plants which turn up on mountains of the antipodes (*Botrichium lunaria* and *Carex echinata* an others), but usually these southern populations differ enough to be worthy of varietal rank; thus, our only Geum is more robust and larger-flowered than its European counterpart, and is to be distinguished as the variety *Strictum* Hook. f.

Readers' Nature Notes and Queries

These two interesting notes come from Miss J. Galbraith, of Tyers, in Victoria.

An interesting discovery during a visit to the Chiltern State Forest was the tallest *Calochilus* I have ever seen, with more than twice as many flowers as I had previously seen on a Beard-orchid. It was *C. imberbis*, the Naked Beard-orchid, a single plant amongst many specimens of *C. robertsoni*, which supports Dr. Rogers' theory that it might be a beardless form of that species.

Tree-creeper Notes

A young White-throated Tree-creeper which visited the bird-table for some weeks last autumn, taught me for the first time how very bright brown the back of a young tree-creeper is. It was hardly darker than (and quite as bright as) the beautiful tail of a Rufous Fantail.

It was with another bird, doubtless a parent, in mature plumage, and for some time I thought they were two species. The young bird was evidently old enough to be independent, for it was later driven away by the older bird.

Both always approached the bird-table (usually for coconut) from below, each spiralling round the post in typical tree-creeper fashion, then hopping up over the edge like a Jack-in-a-box. This has been noted on other occasions. All other birds I have watched approach the bird-table from above.

Tree-creepers normally cling parallel to a tree trunk when creeping or feeding, and their body and legs are adapted to this. I doubt whether one could stand in the normal perching position — horizontally. Those that come to my bird-table always perch vertically, sitting erect like tiny penguins (complete even to white "shirt-front"). Probably woodpeckers and other birds with similar feeding habits do the same, but no other perching bird that I know does this so noticeably. One of the tree-creepers often sat erect on the edge of the bird-table for quarter of an hour or more after feeding, as if it enjoyed my company a yard away. I certainly enjoyed its company.

A Study of the self-pollination of *Thelymitra venosa* R.br. and Some Notes on its Implications

by DAVID L. JONES

Thelymitra venosa R.Br. has been previously reported as a self-pollinating species by Rogers' and Nicholls[2], and its mechanism has been admirably dealt with by Rogers[1]. Both authors stated that the species is self-pollinating in bud and by the time the flower opens the process is well advanced. It is difficult to see how this species can be involved in natural hybridization, and yet this has been stated by authors[1,2]. As well I have seen specimens of a possible hybrid involving *Thelymitra venosa* and *T. ixioides* from Rocky Plain in North-eastern Victoria. Pickled material was collected by T. B. Muir of the Melbourne Herbarium and the collection has been reported on by Peterson[3]. Subsequent collections of the same hybrid have been made by A. C. Beaugiehole at Rocky Plain, and other places in North-eastern Victoria.

In an attempt to determine how a species that is self-pollinated early in the bud can be involved in hybridization, I made a comprehensive study of *T. venosa* in the field at Lake Mountain on 25 January 1970. The results are very interesting and will be reported in this paper.

The temperature on 25 January was about 75°F: one of the warm balmy days that make the sub-alpine areas so inviting during early Summer. The Sun-orchids were at their peak with the majority being wide open while a few of the earlier ones had just finished and some of the later ones were still in bud. The survey involved a study of the pollination stage of buds, open flowers and withered flowers. Two hundred open flowers were examined and fifty buds and withered flowers were collected for later dissection.

The examination of open flowers revealed that 35-40% of them had no pollen on the stigma. A closer examination of these flowers showed that in about 20% the anther sacs were completely empty; in 10% the anther itself was abortive while the stigma was still functional and the remaining 10% had anthers which had not yet shed their pollen.

These results show that nearly half of the *Thelymitra venosa* flowers examined at Lake Mountain have not

Thelymitra venosa
Photo: R. Rotherham

been pollinated by the time the flowers have opened. This is in direct contrast to the supposition by Nicholls' and it leaves the way open for natural hybridization to occur in *Thelymitra venosa*. The following observations however will show that this can only occur with *T. venosa* participating as the female parent of the cross.

The absence of pollen from the anthers of 20% of the open flowers can be accounted for in two ways:

a. The species may occasionally be cross-pollinated and insects have removed the pollen.

b. These flowers are male sterile.

My conclusion is that the flowers are male sterile and is based on the following observations.

1. The pollinary mechanism is completely geared to self-pollination and is in no way adapted for insects because:

a. The pollinia are friable, mealy and quickly lose any coherence.

b. The pollinia are free from each other and there is no viscidium uniting them for insect removal. Thus if there was an insect agency it would remove one or two of the pollinia and leave the rest. My observations show that the anthers were completely empty and the stigmas were free of pollen. It would be a poor cross pollinating system that robbed the anthers but did not deposit pollen on the stigmas.

c. The rostellum is poorly developed and does not effectively shield the stigma from the anther.

2. The anther sacs of some unopened buds that I dissected were quite empty eliminating the possibility of insect removal.

These observations prove conclusively that *Thelymitra venosa* is predominately a self-pollinating species. However approximately 30% of the population at Lake Mountain is male sterile and since the stigmas of these are still functional they can be cross pollinated by other species. I am sure that none are cross-pollinated on Lake Mountain since I have not seen any other *Thelymitra* species in the area and certainly none flowering in late January. However where *T. venosa* is growing with a normally cross pollinating species e.g. as at Rocky Plain with *Thelymitra ixioides*, then it is feasible that a pollinating agent bearing pollinia of *T. ixioides* could visit *T. venosa* especially as they grow intermingled together. The chances are low that it will strike a flower with no pollen on the stigma, however this infrequency will be offset by the large number of ovules produced from each cross.

Comments made by T. B. Muir and A. C. Beauglehole indicate that the possible hybrid is not as common as either parent. *T. venosa* however, is especially abundant in the area and from the results of the foregoing study, this seems an essential prerequisite if it is to participate in natural hybridization. The frequency of *T. ixioides* will not be as important but naturally any increase in its abundance will increase the chance of the cross occurring.

I hope to study this natural association further at Rocky Plain in the coming seasons.

The Mechanism of Self-pollination

Dissection of buds and finished flowers provided the complete picture of the mechanism adopted by this orchid. It has been previously published by Rogers' but as this paper may be unobtainable for some I will redescribe the mechanism here.

In the very early bud the column appendages and anther are erect. The anther appears enormous and makes up the bulk of the column. The anther-sacs are completely closed and dissection shows the pollinia to be very pale and coherent. The column appendages are only slightly coiled and also pale at this very early stage. The stigma is small, green, round and moist.

In the bud about five days before opening a number of changes are obvious. The tip of the anther has moved down and is now almost at right angles to the rest of the column. (In some buds at this stage the tip of the column is still at about 60°). The pollinia are pale but are beginning to appear friable. In odd flowers a pollinium will be found already on the stigma but in the majority the anther sacs have not opened sufficiently. The stigma has enlarged and the column no longer appears to be composed entirely of the anther as in the early buds.

Two to three days before opening the anther sacs are quite wide and the majority of flowers have pollen on their stigmas. At this stage the stigmas are fully developed and obviously very sticky. The pollinia have lost much of their coherence and individual grains are easily seen. (In approximately 10% of the buds at this stage, the anther sacs were still closed).

Self-pollination has taken place in the majority of flowers by the time they have opened.

In the expanded flower the anther sacs are wide open and almost horizontal. The majority of pollinia have fallen from them and are adhering to the vestigial rostellum and the stigmatic surface. Under the lens the stigma appears glistening and gluey with the surface becoming distorted. The pollinia appear to be dissolving where they are touching the glue.

The stigma of the withered flower is very swollen and distorted and the remains of pollinia are barely discernible. The anther sacs are browning and curling up at the edges. Occasional pollinia will be found adhering to a sac but these have lost all coherence and are merely a fluffy mass of pollen grains.

Summarizing the procedure shows how simple is the overall process. The downward movement of the column brings the pollinia into a direct line above the stigma. The opening of the anther sacs (possibly also brought about by the downward movement of the column) allows the pollinia to simply drop onto the stigma—a process probably facilitated by movements of the flower stem in the wind.

LITERATURE Cited
1. Rogers, R. S., Mechanism of Pollination in certain Australian Orchids. Trans. roy. Soc. S.A. 37: 48 (1913).
2. Nicholls, W. H., Orchids of Australia 1° Plate 49 (1955).
3. Peterson, J. M., The fascination of orchids. Latrobe Valley Naturalist No 75 (March 1970).
4. Nash, R. C., Some interesting correspondence. The Orchadian 3: 37 (1969).

Botanical Exploration in the Southern Flinders Ranges South Australia (1800-1970)

by D. N. KRAEHENBUEHL

INTRODUCTION

The low undulating hills near the small town of Crystal Brook form the approximate division between the Flinders Range and Mount Lofty Range systems. The Southern Flinders Range extends a further eighty miles north to the hills about Quorn, where many wet country plants are replaced by those of more arid communities.

Botanical exploration in the Southern Flinders Range is dominated by two well known figures—Baron Ferdinand von Mueller and Professor Sir John B. Cleland, to whom we owe much of our present knowledge of the region's flora.

Although no actual census has ever been made of the native flora, the total number of higher plants must be in the vicinity of 500. For details of the endemic plants found in these ranges, the attention of readers is drawn to the June 1970 issue of *Australian Plants*.

TOPOGRAPHY

The many attractive gorges on the western scarp of the range, notably Napperby, Nelshaby, Telowie, Port Germein, Mambray Creek, Alligator and the Buckaringa and Warren Gorges near Quorn are well vegetated and mainly free of despoliation by man. The picturesque Horrocks and Pichi Richi Passes have been sadly invaded by a large number of alien plants.

There are no mountains over 4,000 ft. and the highest and best known peaks are Mount Brown 3,166 ft., Mount Remarkable 3,146 ft., Devil's Peak 2,288 ft. and The Bluff 2,410 ft.

Rocky River the most important stream for the region, rises in the Wirrabara Forest and flows westwards until it joins the Broughton River just north of Merriton. Another large stream the Willochra Creek has its origins in the Mount Remarkable Range, flows past the town of Melrose, then continues on a northerly course, before entering the southern end of Lake Torrens near Kallioota. Other smaller watercourses worthy of mention are Baroota Creek, Waterfall Creek, Mambray Creek, Alligator Creek, and Spring Creek. The extensive Willochra Plain and smaller Beetaloo Valley are two important intermontane plains of the Southern Flinders Range.

RESERVES AND RESERVOIRS

The Wirrabara Forest Reserve and the Mount Brown and Mount Remarkable Forest Reserves are important forested areas under the control of the South Australian Woods and Forest Department. Regional Parks are located at Alligator Gorge, Mambray Creek and Mount Remarkable. Beetaloo Reservoir situated near the head of Beetaloo Valley and the Baroota Reservoir are the two largest catchment areas in the Southern Flinders Range.

OUTLINE OF BOTANICAL INVESTIGATION

1: *Collectors of the Nineteenth Century*

Robert Brown was the first botanist to sample the flora of any part of the Flinders Ranges. On 10 March, 1802, he and two companions, the

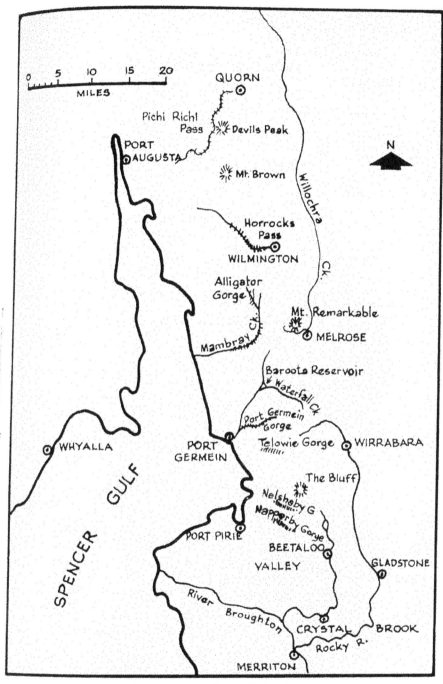

Locality Map — Southern Flinders Ranges.

famous artists Ferdinand Bauer and William Westall, trekked across the plains from their ship the "Investigator", and accomplished the exhaustive ascent of Mount Brown. They returned to ship on 11 March. Among the plants collected at this historic locality "Head of Spencers Gulf" were *Trichodesma zeylanicum, Eremophila scoparia* and *Eremophila oppositifolia*. Robert Brown's South Australian collections are described by him in *Prodomus Florae Hollandiae et Insulae van Diemen* (1810).

Thirty eight years were to pass before Edward John Eyre on his journey north to explore interior districts of South Australia, discovered and named Rocky River, Mount Remarkable and several other historic landmarks. It is fairly certain that he made botanical collections but these were all lost after being sent to Adelaide.

Dr. Herman Hans Behr who botanised extensively in the Barossa Range and north to the River Light had intended to visit Mount Remarkable in 1849 but he left South Australia in the first months of 1850. As Dr. Ferdinand von Mueller and Behr were firm friends it is most likely that it was he who urged Mueller to undertake his momentous journey to the Flinders Ranges in the Spring of 1851. I believe it is significant too that a bilingual advertisement in the English and German languages appeared in an 1850 issue of the Colonial Newspaper *Die Deutsche Post* (Ed. C. Kornhardt) advising that 20,000 acres of land at Mount Remarkable would be soon open for purchase.

Although Dr. Mueller does not lend us an account of his botanical activities in the Southern Flinders Range, he wrote a general version of the Flinders Range Flora "Die Vegetation der Gegenden um den Torrens — See", in the German periodical *Hamburger Garten—und Blumenzeitung* (1853). But it is known from the labels on his botanical collections that he was at Crystal Brook on 7 October, 1851 and in early October had extended his travels to Mount Remarkable, Mount Brown and Mount Arden. He must have botanised at the Elder Range during the middle of October for he had returned to Rocky River and Rocky Creek by 6 November, 1851.

In 1852, Dr. Mueller described the Flinders Range collections in *Linnaea* XXV. In the same journal William Sonder also described a number of Mueller's *Compositae* obtained in this northern expedition. To further illustrate the scope of his work and his collecting localities in the Lower Flinders I have made a selection of some of the novelties he mentions in *Linnaea*.

Phyllanthus sarosus and *Calotis cymbacantha* — Crystal Brook. *Heliotropium elachanthum* and *Solanum eremophilum* — Rocky River. †*Teucrium petrophilum* — Baratta (Baroota) and Crystal Brook; *Glycine tabacina* — Crystal Brook and Rocky River; *Craspedia pleiocephala* — Mount Brown *Veronica decorosa* — Mount Remarkable and Mount Brown. *Ovaria pannosa* and *Daviesia reticulata* — Mount Remarkable; *Isotoma petraea* — Mount Remarkable and Beautiful Valley (Wilmington). He also botanised at Beetaloo Valley.

Before closing my remarks on Dr Mueller's journeys, I deem that the following letter sent by Mueller to J. E. Brown, and quoted by the latter in his *Forest Flora of South Australia* is again worth repeating here:

*Phyllanthus elachanthus.
†Teucrium corymbosum.

"When in 1851, whilst a lonely rider to the Elders Range, I found *Acacia notabilis* near Mount Remarkable, the thought occurred to me that the Latinized name of that mountain then ascended by me might be applied to this new Acacia, because it also was 'remarkable' in some respects particularly in the position of the seeds. Thus the adjective 'notabilis' became first introduced for naming in Botany, since then several writers have utilised the word." Today, we can only marvel at Dr. Mueller's pluck in traversing unknown country and surviving the perils of injury, thirst or surprise attack by hostile aborigines.

John Ednie Brown, Conservator of Forests in South Australia (1879-1890), collected near Beetaloo, Wirrabara Forest, Mt. Brown and Mt. Remarkable. A few plants occurring in the Flinders Ranges were used to illustrate the handsome colour plates for "The Forest Flora of South Australia". *Hakea ednieana*, an endemic *Hakea* from near Parachilna Gorge commemorates his name.

Samuel Dixon, a Foundation Member of the Field Naturalists' Section of the Royal Society, and one of the early pioneers of conservation, was interested in the fodder plants of arid regions. He collected near Gladstone and Rocky River (*Trymalium wayi* was one of these plants) and other places in the Lower Flinders in the early 1880s. Some of his collections are preserved at the State Herbarium, Adelaide. The plant *Newcastlia dixoni* from Crystal Brook, recalls him.

Professor Ralph Tate collected *Eremophila longifolia* at Mount Brown in 1881, but I am unaware of his other movements throughout the Southern Flinders Range in that year. Most of his work was concentrated further north at Arden Vale, the Arcoona Range, Wilpena Pound and towards the eastern shores of Lake Torrens. The results of this journey are embodied in the paper "List of some Plants inhabiting the Northeastern Part of the Lake Torrens Basin".

Walter Gill who succeeded J. F. Brown as Conservator of Forests (1890-1923) discovered *Acacia gracilifolia* at an un-named locality in the Southern Flinders Range, in 1900. This species was later described in 1927 by J. H. Maiden. *Eucalyptus gillii* is named after him.

2: *The Last Seventy Years*

Max Koch collected 200 species of plants near Port Pirie in 1901. As was his custom, duplicate sets of these were sent to many Botanical Institutions in Europe.

In 1902, Charles Johncock published two papers in the *Trans. R. Soc. S. Aust.*, "Notes on the Loranthaceae of the Willochra Valley" and "Further Notes on the Botany of the Willochra Valley". He made a further contribution to the same journal in 1903 entitled "Notes on *Loranthus exocarpi*".

John McConnell Black, author of *Flora of South Australia*, was an indefatigable collector of plants from all parts of South Australia. There are specimens in the State Herbarium, Adelaide, from Hundred of Howe near Hughes Gap 23/9/06; between Gladstone and Beetaloo 9/10/08; Quorn 1/10/16; Gladstone 23/8/21; Telowie Gorge October 1923; Mount Remarkable October 1925 and 16/8/27 and Nelshaby 12/9/32.

Before he resided in Adelaide, Mr. Black's family farmed at Baroota for five years from 1877-1882. He has described a number of new plant species from the region.

William Austin Cannon, an American botanist, explored many areas in the arid north and western districts of South Australia during 1920. He published the results of his surveys in *Plant Habits and Habitats in the Arid Portions of South Australia* (1921). His chapter on the vegetation of the Quorn district is one of the more important ecological papers on the mallee plant associations near Saltia in Pichi Richi Pass.

Professor Sir John B. Cleland has travelled widely throughout the whole of the Flinders Ranges; his copious collections being, possibly the most representative held at the State Herbarium, Adelaide. It is therefore not surprising that Sir John has undertaken more serious botanical collecting in the Melrose district than any other single person since Baron Ferdinand von Mueller.

He visited Mount Remarkable in August 1921, October 1926, August 1927, October 1928 and December 1938. Two years later in the May 1940 issue of *The South Australian Naturalist*, he and Ernest Ising published a list enumerating 189 native and 50 introduced plants for the locality. J. S. Rogers described a new Leek Orchid (*Prasophyllum validum*) which Cleland discovered at Mt. Remarkable on 27 October, 1926. Mention should also be made of his valuable Fungi collections from Quorn, Port Germein Gorge, Mt. Remarkable, including a new species *Peniophora montanum* from the last locality.

Harold M. Cooper was a regular visitor to Mount Remarkable in search of ancient aboriginal artifacts, but it is not well known that he was an enthusiastic plant collector. His first collections date back to about 1941 but even in his advanced years he was still securing plants there at least until 1964. This grand old scientific worker passed away on 14 May, 1970, aged 84 years.

Cliff Boomsma's vegetation study of the Southern Flinders Range, which was published in the *Trans. R. Soc. S. Aust.*, (1946), is the best piece of ecological research carried out in the region. He continues to write occasional notes on the *Eucalyptus* species found there.

Since his first trip to the Flinders Ranges in the early 1950s, T. R. Noel Lothian, the present Director of the Adelaide Botanic Gardens, has taken an acute interest in the South Australian eremian flora and has fostered the introduction of dry area shrub species to suburban gardens. He has made collections at Mt. Remarkable and Alligator Gorge in 1953.

The late Ralph Higginson of Port Augusta championed the original campaign in the early 1950s for the preservation of Alligator Gorge and Mambray Creek. He was a diligent collector of plants and lists 63 species in *The South Australian Naturalist* (1953) which he found at Mambray Creek. The fine native flora plantations in the main streets of Port Augusta are a tribute to his foresight.

Since 1958, members of the staff of the Adelaide Botanic Garden and the State Herbarium have paid occasional visits to the Flinders Range. Ron Hill has collections from Mt. Remarkable in 1960 and Rex Kuchel made investigations in the Lower Flinders during the early 1960s.

Eric Jackson has botanised in Telowie Gorge, Alligator Gorge and Mt. Brown in 1960 and 1961. Paul Wilson accompanied David Whibley on several exploratory trips in 1958 and 1959 to Mt. Brown, Alligator Gorge, Port Germein Gorge, Mt. Remarkable, Wilmington and Pichi Richi Pass. On 2 October, 1958, Mr. Whibley collected *Hovea lanceolata* var. *lanceolata* at Mt. Brown, thereby extending the range of this very localised species. Nicholas Donner has also obtained extensive plant and cryptogram collections from most of the above localities in 1961 and 1962 and at Horrocks Pass in 1967.

Rex Filson, now Curator of Lichens at the Melbourne Herbarium, has interesting higher plants and cryptograms gathered at Telowie Gorge, Mt. Remarkable, Broad Creek Gorge and Alligator Gorge in 1959.

My own botanical activities in the Lower Flinders have been centred at Mt. Remarkable October 1955, September 1960 and October 1969; Alligator Gorge October 1955 and October 1969; Port Germein Gorge 18/9/60; Waterfall Creek near Baroota 18/9/60; Telowie Gorge 10/9/61; Wirrabara Forest 10/9/61 and Willowie 8/3/59. In the November 1969 issue of the *South Australian Society for Growing Australian Plants Journal* 53 native species are recorded for the eastern slopes of Mt. Remarkable and a further 55 occurring at Alligator Gorge.

CONCLUSION

Some of the greatest names in Australian Botanical History have explored the Flinders Ranges in search of plant specimens. Their efforts are praiseworthy and surely an inspiration to future generations. But despite all this work, the flora is still imperfectly known and it is hoped that as the National Parks Commission gradually acquire more parks they will encourage surveys by trained ecologists.

On a Botanical Collection from the Northern Flinders Ranges, South Australia

by DARRELL N. KRAEHENBUEHL

In the Spring of 1962 the writer and two friends, Mr. Jim Burdett of Basket Range and Mr. Ken Preiss of Erindale, spent four days from 13-16 October in the Northern Flinders Ranges with the intention of photographing several of the known aboriginal rock carving sites in the Copley district.

At each of the localities we visited, I obtained a series of botanical collections of a varied number of flowering shrub species regardless of the very harsh drought conditions then being suffered in the interior regions of South Australia.

Very little has ever been published on the flora of this somewhat remote region; the explorer Edward John Eyre climbed Mount Serle on 27 August, 1840 and presumably collected plants there, but these and other specimens from his northern expeditions were tragically lost in Adelaide. He mentions little about the native vegetation in his *Journals*. The Melbourne Herbarium has plant collections from Mount Serle which were obtained by Major Peter Warburton on one of his pastoral-exploration expeditions into the interior in the late 1850s.

Max Koch, a competent German botanist, made valuable collections from the Mount Lyndhurst region between the years 1896 to 1900 and wrote several papers in the Transactions of the Royal Society of South Australia. Two decades later, William Cannon contributed a fourteen page chapter on the "Vegetation of the Copley Region" in his fine publication *Plant Habits and Habitats in the Arid Portions of South Australia* (1921).

The 1956 South Australian State Herbarium expedition which secured many interesting plants in the Northern Flinders Ranges including the virtually unexplored Gammon Range, did not publish an account of their finds. But Dr. Hansjoerg Eichler cites a number of the Gammon Range plants in the *Supplement to Black's Flora* (1965). More recently in 1969, Mrs. Margaret Kenny and Mr. Rex Kuchel in *The Natural History of the Flinders Ranges*, devote two pages to the flora of Mount Serle.

My specimens which have all been lodged at the State Herbarium, Adelaide, were obtained at Red Gorge on Deception Creek (13 Oct.), hills near Mount Serle (13 Oct.), Italowie Gorge (13 Oct.), a small un-named Pound between Angepena Homestead and Nepabunna Mission (14 Oct.) and several other localities near the eastern scarp of the range including Mount McKinlay Creek, Weetootla Gorge, plains near Parabarana Hill and Hamilton Creek Gorge on 15 October.

Red Gorge on Deception Creek is situated about twelve miles east of Leigh Creek. Dr. Herbert Basedow in his paper *Aboriginal Rock Carvings of Great Antiquity in South Australia*, describes in minute detail the countless variety of designs carved on the red rock cliff faces in splendid profusion. We stayed half a day here discovering the fine emu, wallaby and blackfellow tracks, lizard designs, circles and other sundry carvings.

Near a jumbled mass of rock scree at the base of the cliffs I was attracted to the sweet perfumed flowers of *Cynanchum floribundum* 732 AD. This trailing shrub has astringent fruits which are eaten by the natives. Other associated plants were blue flowered *Trichodesma zeylanicum* 759 AD (a member of the Family Boraginaceae) and two native grasses *Digitaria brownii* 758 AD and *Bothriochloa ambigua* 757 AD, the latter tightly wedged in rock crevices. Across the creek on the flats a handsome scarlet flowered Mistletoe *Amyema miraculosa* 731 AD was heavily parasitising trees of False Sandalwood *Myoporum platycarpum*.

At a small group of rocky hills on Mount Serle Station a native Tobacco Bush *Nicotiana velutina* 745 AD with lush green leaves seemed strangely out of place here. Among other plants seen were *Ixodia achilleoides* 747 AD (a composite very widespread in wetter parts of the Mount Lofty Range), *Halorrhagis aspera* 748 AD, an attractive *Hydrocotyle H. trachycarpa* 744 AD and a Greenhood Orchid *Pterostylis* aff *rufa* 743 AD. *Gnephosis foliata* 746 AD and *Blennodia trisecta* 749 AD grew on the open plains near the homestead.

After following a small creek line with Red Gum *Eucalyptus camaldulensis* and Tea Tree *Melaleuca glomerata*, south of the Italowie Gorge road for about a mile, we were delighted to find that it wound its way into a small Pound a mile or so in diameter. To those unacquainted with the Flinders Ranges, Wilpena Pound is not unique, for there are others of these unusual montane formations: Illawortina Pound in the southern Gammon Range is a particularly healthful place and a number of much smaller pounds are located throughout the Northern Flinders Ranges.

In this small pound the dry season did not favour the flowering of some of the larger trees, Bullock Bush *Heterodendrum oleaefolium*, Mulga

Acacia aneura, Royal Acacia *victoriae*, Dead Finish Acacia *Acacia tetragonophylla* and Black Oak *Casuarina cristata*. However on the lower slopes an excellent assemblage of shrubs were in bloom including *Eremophila longifolia* 778 AD, *Eremophila duttonii* 771 AD, *Myoporum montanum* 777 AD, Sandalwood *santalum spicatum* 776 AD, and two native Hop Bushes *Dodonaea lobulata* 765 AD and *Dodonaea attenuata* var *linearis* 767 AD.

Along the higher rock slopes where Porcupine *Triodia irritans* formed almost pure stands, a variety of shrubs, undershrubs and perennials flourished including rounded bushes of *Scleranthus pungens* 762 AD, *Calytrix tetragona* 769 AD, Mint Bush *Prostanthera striatiflora* 772 AD, *Pomax umbellata* 764 AD, *Helichrysum ambiguum* 768 AD, *Helichrysum semipapposum* 763 AD, three Cassia species *C. sturtii* 773 AD, *C. nemophila* 774 AD, *C. artemisioides* 775 AD and *Brachscome ciliaris* 766 AD.

Amyema maidenii 761 AD was parasitic on Dead Finish Acacia and the vigorous trailing creeper *Leichardtia australis* 760 AD with large green gourd shaped pods, later spilling open and revealing the fluffy white seeds, trailed over shrubs and dead branches. The endemic *Goodenia varnicosa* 760 AD, a yellow flowered species with sticky viscid stems and leaves was not uncommon among the rocks.

I first observed the striking Curly Mallee *Eucalyptus gillii* 751 AD in 1955 at the western entrance of Edeowie Gorge, but along the main road between Nepabunna Mission and Italowie Gorge, many trees were resplendent with creamy yellow blossoms. *Acacia ligulata* 750 AD and a small herb *Euphorbia australis* 752 AD were flowering at Italowie Gorge.

After the pleasant surroundings of Italowie Gorge with its huge Red Gums, Native Pines and abundant shrubbery, our sudden arrival at the eastern plains was a stark reminder we were still in the dry outback.

It is hard to imagine two more handsome flowering shrubs than Sturt's Desert Rose *Gossypium sturtianum* 754 AD and *Petalostylis labicheoides* 753 AD, yet here in the most desolate spot imaginable—the dry bed of Mount McKinlay Creek, they presented a massed display. Another attractive herb, the Native Carrot *Trachymene glaucifolia* 733 AD, thrived in this arid habitat.

We did not have time for detailed exploration at Weetootla Gorge, one of the "gateways" to the Gammon Ranges. Near the creek environs the three following shrubs were observed. Flax Lily *Dianella revoluta* 740 AD, *Alyogyne hakeifolia* 742 AD and *Plectranthus intraterraneus* 741 AD, an aromatic Labiate which is apparently quite rare in the Flinders Ranges although it is known to occur at Mount Chambers Gorge and is locally abundant at Mount Illbillie in the Everard Range and at Mount Olga in Central Australia.

Three *Bassia* species *B. lanicuspis* 738 AD, *B. ventricosa* 739 AD, *B. parallelicuspis* 739A AD and *Frankenia crispa* 735 AD, grew on a small knoll near Parubarana Hill and *Senecio odoratus* 736 AD, was quite frequent along Parabarana Creek. The small herb *Velleia paradoxa* 756 AD was one of the few plants then flowering at Mount Fitton Station and *Acacia aneura* 755 AD and *Pluchea dentex* 734 AD occurred at Hamilton Creek Gorge.

Since our visit in 1962, a magnificent Wilderness National Park has been proclaimed in the Gammon Ranges. Providing that Feral Goat populations can be satisfactorily controlled, the survival of unique plant associations and species will be assurred. It is also pleasing that the authorities at the Arkaroola Reserve are taking measures to prevent the destruction of native vegetation by tourists.

A great deal more remains to be accomplished with botanical exploration throughout the whole of the Flinders Range system. It has been shown that even in the driest of years valuable botanical collections can be secured in this fascinating region. Following the record rains of February and March 1971 botanists should confidently look forward to one of the best collecting seasons for many years.

* * * *

ILLUSTRATIONS BY AUTHOR

Native Hibiscus (*Hibiscus huegelii*) Alligator Gorge.

Close up (natural size) of *Eremophila maculata*.

Plate 3

Close up of *Solanum simile*, Mambray Creek.

Plate 4

Leichardtia australis, small pound near Italowie Gorge road.

Plate 5

Porcupine (*Triodia irritans*) association, Northern Flinders Range; small Pound near Italowie Gorge.

Plate 6

Cliffs at Alligator Gorge.

Plate 7

Scene in small Pound south of Italowie Gorge road.

Plate 8

Waterfall on Waterfall Creek near Baroota, Southern Flinders Ranges.

A Botanical Bibliography of the Flinders Ranges South Australia (1800-1970)

by D. N. KRAEHENBUEHL

INTRODUCTION

This Bibliography consists of a comprehensive collection of references to botanical writings on the Flinders Ranges and environs for the period 1800-1970. A broad approach has been intentionally pursued to include a variety of handbooks and articles which will be of value to students of Botany.

It is regrettable that so few taxonomic or ecology papers have been published about the Flinders Range Flora for the last thirty years. Plant lists, small notes on the distribution of species and articles with a horticultural bias far outweigh the scientific material.

New plant species described for South Australia from the Flinders Range region are cited in the "Periodical References" and other footnotes of special botanical interest are included.

SCOPE OF THE BIBLIOGRAPHY

The bulk of the periodical references are from the *Transactions of the Royal Society of South Australia (1877/78-1970)*, and the *South Australian Naturalist (1919-1970)*. Other important Periodicals which have been searched include the following: *Records of the South Australian Museum*, *Proceedings of the Royal Geographical Society (South Australian Branch)*, *Proceedings of the Royal Society of New South Wales*, *Hooker's Journal of Botany and Kew Garden Miscellany*, *Southern Science Record*, *Wildlife Magazine* (Crosbie Morrison), *Victorian Naturalist*, *Australian Plants*, *Emu* and *South Australian* and *Victorian Parliamentary Papers*.

The State Library of South Australia Research Service Bibliography No. 136, "The Geographical Distribution of Native Plants in South Australia and the Northern Territory" has been an important source of information and considerably lightened my task. I am indebted to Dr. Hansjoerg Eichler who kindly drew my attention to papers in the German Periodicals *Linnaea* and *Hamburger Garten — und Blumenzeitung*, and to the staff of the State Library of South Australia, South Australian Archives and Parliamentary Library for their helpful assistance.

1 Audas, J. W., 1929: The Botanical Activities of Max Koch.—J. Proc. R. Soc. West. Aust. XV: 83-86.
2 Beni, G., 1967: The Genus *Ptilotus* R. Br.—Austral. Plants 4: 109-124. *Ptilotus nobilis* var. *angustifolius*, fig. 3., h.
3 Bentham, G., 1863 (30.V): *Flora Australiensis*. Vol. I: (Reeve & Co.: London
4 ——— 1864 (5.X): *Flora Australiensis*. Vol. II.
5 ——— 1867 (5.II): *Flora Australiensis*. Vol. III.
6 ——— 1868 (16.IV); *Flora Australiensis*, Vol. IV.
7 ——— 1870 (VIII.X): *Flora Australiensis*, Vol. V.
8 ——— 1873 (23.IX): *Flora Australiensis*, Vol. VI.
9 ——— 1878 (23-30.III): *Flora Australiensis*, Vol. VII.
10 Black. J. M., 1917: Additions to the Flora of South Australia, No. 11— Trans R. Soc. S. Aust. 41: 31-52.
Kochia coronata, 43-44; 1. IX; *Atriplex campestris australiana* 44-45, 1. X.

11 ———— 1918: Additions to the Flora of South Australia, No. 14.—Trans. R. Soc. S. Aust. **42**: 168-184.
Acacia rivalis, 173-174, t. 18.

12 ———— 1919: Additions to the Flora of South Australia, No. 15.—Trans. R. Soc. S. Aust. **43**: 23-44.
Kochia cannonii, 29-30; *Goodenia vernicosa*, 41.

13 ———— 1922a (VIII): *Flora of South Australia*, Pt. I, 1st Ed. (Gov. Printer: Adelaide).

14 ———— 1922b: Additions to the Flora of South Australia, No. 20.—Trans. R. Soc. S. Aust. **46**: 565-571.
Stipa setacea var. *latiglumis*, 565; *Stipa pubescens* var. *comosa*, 565; *Bassia ventricosa*, 566-567; *Bassia limbata*, 567; *Bassia decurrens*, 567.

15 ———— 1923: Additions to the Flora of South Australia, No. 21.—Trans. R. Soc. S. Aust. **47**: 388-399.
Kochia excavata, 368; *Amaranthus mitchellii* var. *grandiflora*, 368. 369; *Calandrinia remota*, 369.

16 ———— 1924a (VI): *Flora of South Australia*, Pt. II, 1st Ed. (Gov. Printer: Adelaide).

17 ———— 1924b: Additions to the Flora of South Australia, No. 22.—Trans. R. Soc. S. Aust. **48**: 253-257.
Bassia uniflora var. *incongruens*, 254; *Ranunculus parviflorus* var. *glabrescens*, 254; *Swainsona viridis*, 255; *Zygophyllum compressum*, 256; *Zygophyllum tesquorum*, 256.

18 ———— 1925: Additions to the Flora of South Australia, No. 23.—Trans. R. Soc. S. Aust. **49**: 270-275.
Blennodia pterosperma, 272; *Hymenocapsa longipes*, 273-274.

19 ———— 1926 (XII): *Flora of South Australia*, Pt. III, 1st Ed. (Gov. Printer: Adelaide.)

20 ———— 1927: Additions to the Flora of South Australia, No. 25.— Trans. R. Soc. S. Aust. **51**: 378-385.
Swainsona behriana, 379-380; *Goodenia subintegra*, 383-384.

21 ———— 1928: Additions to the Flora of South Australia, No. 26.— Trans. R. Soc. S. Aust. **52**: 225-230.
Plagiobothrys elachanthus, 226-227; *Vittadinia pterochaeta*, 229; *Cassinia complanata*, 229.

22 ———— 1929 (VI): *Flora of South Australia*, Pt. IV, 1st Ed. (Gov. Printer: Adelaide.)

23 ———— 1932: Additions to the Flora of South Australia, No. 30. Trans. R. Soc. S. Aust. **56**: 39-47.
Euphorbia murrayana, 42-43, t.I.; *Frankenia crispa*, 43, t.2; *Clelandia* New Genus, 46; *Clelandia convallis*, 47, t.2.

24 ———— 1933: Additions to the Flora of South Australia, No. 31.— Trans. R. Soc. S. Aust. **57**: 143-159.
Frankenia granulata, 154, t.IX.

25 ———— 1936: Additions to the Flora of South Australia, No. 34.— Trans. R. Soc. S. Aust. **60**: 162-172.
Bothriochloa inundata, 163-164; *Amaranthus grandiflorus*, 166.

26 ———— 1937: Additions to the Flora of South Australia, No. 35.— Trans. R. Soc. S. Aust. **61**: 241-249.
Brachycome lyrifolia, 249, t.XIV.

27 ———— 1939: Additions to the Flora of South Australia, No. 38.— Trans. R. Soc. S. Aust. **63**: 240-247.
Helichrysum adnatum var. *scabrum*, 246.

28 ———— 1941: Additions to the Flora of South Australia, No. 40. Trans. R. Soc. S. Aust. **65**: 333-334.
Stipa breviglumis, 333-334; *Stipa plagiopogon*, 334.

29 ———— 1942: Additions to the Flora of South Australia, No. 41.— Trans. R. Soc. S. Aust. **66**: 248-249.
Lomandra fibrata, 248-249.

30 ———— 1943 (1.V.): *Flora of South Australia*, Pt. I, 2nd Ed. (Gov. Printer: Adelaide).

31 ——— 1945: Additions to the Flora of South Australia, No. 43.—
Trans. R. Soc. S. Aust. 69: 309-310.
Acacia euthycarpa, 310.

32 ——— 1948 (I): *Flora of South Australia*, Pt. II, 2nd Ed. (Gov. Printer: Adelaide).

33 ——— 1949: Additions to the Flora of South Australia, No. 45.—
Trans. R. Soc. S. Aust. 73: 6.
Acacia quornensis, 6.

34 ——— 1952 (IX): *Flora of South Australia*, Pt. III, 2nd Ed. (Gov. Printer: Adelaide).

35 ——— 1957 (III): *Flora of South Australia*, Pt. IV, 2nd Ed. (Gov. Printer: Adelaide).

36 Blake, S. T., 1952: The Identification and Distribution of Some *Cyperaceae* and *Gramineae* chiefly from Australia.
Proc. R. Soc. Qd. 62: 83-100.
Stipa nodosa, 89-90, t.V.; *Stipa brachystephana*, 90-91, t.VI.

37 ——— 1969: A Revision of *Carpobrotus* and *Sarcozona* in Australia. Genera Allied to *Mesembryanthemum* (*Aizoaceae*).
Contr. Qd. Herb. No. 7.
Sarcozona praecox, 34-37, fig. 1, 9, 10. Dist. Map fig. 15.

38 Bonython, C. W., and Daily, B., 1963-64: National Parks and Reserves in South Australia. Proc. S. Aust. Brch. R. geogr. Soc. Aust. 65: 5-15.
Wilpena Pound, 7.

39 Boomsma, C. D., 1946: The Vegetation of the Southern Flinders Ranges.—
Trans. R. Soc. S. Aust. 70: 259-276, t. XLIV, XLV, Two folding maps.

40 ——— 1969: Contributions to the Records of *Eucalyptus*, L'Heritier in South Australia.—Trans. R. Soc. S. Aust. 93: 157-164.
[Distribution of *Eucalyptus viridis* in the Flinders Ranges.]

41 Brown, J. E., 1880-1883: *Woods and Forests Annual Reports of the Forest Board, with Conservator's Progress Report and appendices*. (Gov. Printer: Adelaide).
[Three reports 1880-81, 1881-82 and 1882-83, contain information relevant to the native vegetation of the Wirrabara Forest.]

42 ——— 1883a (before 24.III): *The Forest Flora of South Australia*, Pt. 1. (Gov. Printer: Adelaide.)

43 ——— 1883b (between IV and IX): *The Forest Flora of South Australia*. P. 2.

44 ——— *1884a (II): *The Forest Flora of South Australia*, Pt. 3.

45 ——— 1884b (dated VI, but probably issued not before VII): *The Forest Flora of South Australia*, Pt. 4.

46 ——— *1885a (IV): *The Forest Flora of South Australia*, Pt. 5.

47 ——— 1885b (probably VIII, before 29. VIII): *The Forest Flora of South Australia*, Pt. 6.

48 ——— °1886 (IV): *The Forest Flora of South Australia*, Pt. 7.

49 ——— *1888 (IV): *The Forest Flora of South Australia*, Pt. 8.

50 ——— 1890 (II, not later than 26. III): *The Forest Flora of South Australia*, Pt. 9.

51 Brown, R., 1810: *Prodromus Florae Novae Hollandiae et Insulae Van-Diemen* Vol. I: London.

52 Burbidge, N. T., 1947: Key to the South Australian Species of *Eucalyptus*, L'Herit.—Trans. R. Soc. S. Aust. 71: 137-167.

53 Cannon, W. A., 1921: *Plant Habits and Habitats in the Arid Portions of South Australia*.—Carnegie Institution, Pub. No. 308. Washington, U.S.A.
(a) Vegetation of the Copley Region, 67-81.
(b) Vegetation and environment at Port Augusta, 93-96.
(c) Vegetation and environment at Quorn, 96-108, t. 8-19, 23, 25-28, 30-31.

*Amended dates from Eichler and Eichler (1962) appearing in "Review Notices", *Garden and Field* (1883-1890).

54 Chivers, E. F., 1967: A West Coast and Flinders Ranges journey.—J. Soc. Grow. Aust. Pls. S. Aust. **3:** 3-9.
55 Cleland, J. B., 1934: *Toadstools and Mushrooms and other Larger Fungi of South Australia*, Pt. **1**. (Gov. Printer: Adelaide).
56 —— 1935 (VII): *Toadstools and Mushrooms and other Larger Fungi of South Australia*. Pt. **2**.
57 Cleland, J. B., and Johnston, T. H., 1939: Aboriginal names and uses of Plants in the Northern Flinders Ranges.—Trans. R. Soc. S. Aust. **63:** 172-179.
58 Cleland, J. B., and Ising, E. H., 1940: The Flora of Mount Remarkable.— S. Aust. Nat. **20:** 25-26.
59 Clint, M., 1967: *Crinum luteolum* in South Texas.—Pl. Life. **23:** 114-116. One illustration.
60 Crocker, R. L., and Wood, J. G., 1947: Some Historical Influences on the Development of the South Australian Vegetation Communities and their Bearings on Concepts and Classification in Ecology.—Trans. R. Soc. S. Aust. **71:** 91-136.
    [Distribution of the Sugar Gum *Eucalyptus cladocalyx*, 113-114.]
61 Donohue, F. F., and Ioannou, N., 1969: Rostrevor College Expedition to the Flinders Ranges, 9-14 Sept. Flora Slutty, 16-24.—
    [Processed report (Adelaide) of floristic studies at Bunyeroo Gorge.]
62 Eardley, C. M., and Cleland, J. B., 1953: John McConnell Black: Obituary, List of New Genera, Species and varieties published, and Bibliography —Trans. R. Soc. S. Aust. **76:** i-xii.
63 Eichler, Hj., 1958: The Ranunculus sessiliflorus Group in South Australia.— Trans. R. Soc. S. Aust. **81:** 175-183.
    Ranunculus hamatosetosus, 180-183. fig. [82, 1.1., fig. 2: 1.2.
64 —— 1965: *Supplement to J. M. Black's Flora of South Australia*. (Gov. Printer: Adelaide.)
65 Eichler, Hj., and Eichler, M. L., 1962: Publication dates of J. E. Brown. *The Forest Flora of South Australia*.—S. Aust. Nat **36:** 39-42.
66 Ewart, A. J., White, J., and Tovey, J. R., 1909: Contributions to the Flora of Australia.—J. Proc. R. Soc. N.S.W. **42:** 184-200. *Podocoma*, nota. 192. t.XXXII.
67 Eyre, E. J., 1845: *Journals of Explorations of discovery into Central Australia, and overland from Adelaide to King George's Sound, in 1840-1, including an account of the manners and customs of the Aborigines, and the state of their relations with Europeans*.
    Two Vols. (T. & W. Boone: London).
68 French, R. J., Matheson, W. E., and Clarke, A. L., 19687: Soils and Agriculture of the Northern and Yorke Peninsula Regions of South Australia. Vegetation, II.— Dept. of Agric., S. Aust. (Gov. Printer: Adelaide.)
69 Goyder, W. S. 1876-1880: *Forest Board Reports*. (Gov. Printer: Adelaide) The first four reports of the S. Aust. Woods and Forests Dept., 1876-77, 1877-78, 1878-79 and 1879-1880, contain information on nine tree species in the Wirrabara Forest.
70 Hale, H. M., and Tindale, N. B., 1925-28: Observations on Aborigines of the Flinders Ranges and records of rock carvings and paintings.— Rec. S. Aust. Mus. **3:** 45-60.
    [Notes on Aborigine names for plants.]
71 Hannibal, L. S., 1963a: *Crinum flaccidum*.—JI. R. hort Soc. **88:** 32-34.
    [Notes on Pichi Richi Pass form.]
72 —— 1963b: Variation in *Crinum flaccidum*—Pl. Life **19:** 46-48.
    [Notes on Pichi Richi Pass form.] One illustration. 46.
73 Higginson, A. R. R., 1953: Plants found in Mambray Creek.—S. Aust. Nat **28:** inside back cover. [List of plants collected.]
74 —— 1955a: A new location for *Plectranthus parviflorus*.—S. Aust Nat **29:** 79.
75 —— 1955b: A new *Dodonaea* for the Flinders Ranges, — S. Aust. Nat **30:** 24.

76 Hyde, A., 1964: October Weekend.—J. Soc. Grow. Aust. Pls. S. Aust. 2: 14-16. [List of plants seen in Flinders Ranges.]

77 Ising, E. H., 1969: Six new species of *Bassia* All: (*Chenopodiaceae*).—Trans. R. Soc. S. Aust. 93: 119-125. *Bassia nitida*, 122, fig. 5.

78 Ising, E. H., and Cleland, J. B., 1940: List of Plants near Melrose and Mount Remarkable.—S. Aust. Nat. 20: 30-32.

79 Johncock, C. F., 1902a: Notes on the *Loranthaceae* of the Willochra Valley.—Trans. R. Soc. S. Aust. 26: 7-9.

80 ——— 1902b: Further Notes on the Botany of the Willochra Valley.—Trans. R. Soc. S. Aust. 26: 31-37.
[Plant species found near Wilmington, Spring Creek, Mt. Brown and Willowie.]

81 ——— 1903: Notes on *Loranthus exocarpi*.—Trans. R. Soc. S. Aust. 27: 253-255.

82 Kenny, M., 1969: In Corbett 1969., *The Natural History of the Flinders Ranges*. (Libraries Board of South Australia: Adelaide.)—Key to the Common Wildflowers of the Flinders Ranges. Chapter III: 164-208.

83 Koch, M., 1897-98: A List of Plants collected on Mt. Lyndhurst run, South Australia.—Trans. R. Soc. S. Aust. 22: 101-118.

84 ——— 1900: Supplementary List of Plants from Mt. Lyndhurst run.—Trans. R. Soc. S. Aust. 24: 81-85.

85 Krachenbuchl, D. N., 1954: Wilpena Pound Trip.—S. Aust. Nat. 29: 28-30. [List of plants observed.]

86 ——— 1955: Mt. Remarkable revisited.—S. Aust. Nat. 29: 44.

87 ——— 1969: S.G.A.P., October Day Weekend Excursion to Melrose.—J. Soc. Grow. Aust. Pls. S. Aust 4: 6-14.
[Floristics of Mt. Remarkable, Alligator Gorge and Willowie region, incl. list of terrestrial orchid spp. occurring at Mt. Remarkable, Alligator Gorge and Mambray Creek.]

88 ——— 1970: Plants of the Mount Lofty and Southern Flinders Ranges.—Austral. Pls. 5: No. 43, 295-298, 318-319, 326-333.
[*Callistemon teretifolius*, colour t. 333; *Goodenia albiflora*, t. 328.]

89 Kuchel, R. H., and Kenny, M., 1969: In Corbett 1969, *The Natural History of the Flinders Ranges*. (Libraries Board of South Australia: Adelaide.)—The Vegetation of the Flinders Ranges. Chapter III: 139-163. Four illustrations opp. 151.

90 Lothian, T. R. N., 1953a: Two Rare Trees.—S. Aust. Nat. 28: 23.
[*Codonocarpus pyramidalis*, near Leigh Creek and Quorn.]

91 ——— 1953b: White Quandong.—S. Aust. Nat. 28: 26.
[Tree growing in school grounds at Blinman.]

92 ——— 1953c: Report on Excursion to Mt. Remarkable and Alligator Gorge.—S. Aust. Nat. 28: 28.

93 ——— 1955: Mosses in South Australia.—S. Aust. Nat. 30: 25.—
[Including 8 species collected at Sliding Rock.]

94 ——— 1957a: Regeneration of Mulga (*Acacia aneura*).—S. Aust. Nat. 31: 59-60.
[Notes from Yankaninna Station, Gammon Ranges in 1955.]

95 ——— 1957b: White Hibiscus.—S. Aust. Nat. 31: 60.
[White form of *Hibiscus huegelii* from near Mambray Creek and Cowell (Eyre Pen.)]

96 ——— 1960: In, *Touring the Flinders Ranges*. (Royal Automobile Association: Adelaide.)—Vegetation, 22-24.

97 Maiden, J. H., 1896-1897: On a New *Atriplex* from South Australia.—Trans. R. Soc. S. Aust. 21: 87. *Atriplex kochiana*, 87.

98 ——— 1912: *Critical Revision of the Genus Eucalyptus*. Pt. XV; *Eucalyptus gillii*, 177, t. 67, fig. 6-9.

99 ——— 1919: Notes on *Eucalyptus*.—J. Proc. R. Soc. N.S.W. 53: 57-73. *Eucalyptus gillii*, 68; *Eucalyptus gillii* var. *petiolaris*, 68-69.

100 Maiden, J. H., and Betche, E., 1897-98: On a New *Myoporum* from South Australia.—Trans. R. Soc. S. Aust. 22: 76. *Myoporum refractum*, 76.

101 Maiden, J. H., and Blakely, W. F., 1927: Descriptions of Fifteen New *Acacia*. Proc. R. Soc. N.S.W. 60: 171-196.
Acacia gracilifolia, 191-192. t. XVIII.
102 Meisner, C. F., 1853: Plantae Muellerianae, *Thymeleae*. Linnaea XXVI 345-352.
103 Mincham, V. H., 1941: *Codonocarpus pyramidalis*, a Rare and Lovely Native Tree.—Wild Life, Melb. 3: 408. One illustration.
104 Mueller von, F., 1852: Diagnoses et descriptiones plantarum novarum, quas in Nova Hollandia australi praecipue in regionibus interioribus detexit investigavit auctor.— Linnaea XXV; 367-445.
105 ——— 1853a: Die Vegetation der gegenden um den Torrens.—see.—Hamburg. Gart. Blumenzeit, IX: 340-343.
106 ——— 1853b: The vegetation of the districts surrounding Lake Torrens (Translation from the German by R. Kippist.)—Hook. J. Bot. London. V: 105-109.
107 ——— 1854-55a: Description of fifty New Australian Plants, chiefly from the Colony of Victoria.—Trans. Proc. Vict. Inst. 1: 28-48.
108 ——— 1854-55b: Description of New Australian Plants, chiefly from the Colony of Victoria.—Trans. Proc. Vict. Inst. 1: 114-135.
109 ——— 1858(III)-1859(XII): *Fragmenta Phytographiae Australiae*. Vol. I.
110 ——— 1859: Report on the Plants collected during Mr. Babbage's Expedition into the North Western Interior of South Australia in 1858.—Vict. P.P. No. 1: 21.
111 ——— 1860(III)-1861(V): *Fragmenta Phytographiae Australiae*. Vol. II.
112 ——— 1862(IV)-1863(IV): *Fragmenta Phytographiae Australiae*. Vol. III.
113 ——— 1863(IX)-1864(XI): *Fragmenta Phytographiae Australiae*. Vol. IV.
114 ——— 1865(IV)-1866(XII): *Fragmenta Phytographiae Australiae*. Vol. V.
115 ——— 1867(VII)-1868(XII): *Fragmenta Phytographiae Australiae*. Vol. VI.
116 ——— 1869(VI)-1871(XII): *Fragmenta Phytographiae Australiae*. Vol. VII.
117 ——— 1872(III)-1874(XI): *Fragmenta Phytographiae Australiae*. Vol. VIII.
118 ——— 1875(II-XII): *Fragmenta Phytographiae Australiae*. Vol. IX.
119 ——— 1876(I)-1877(X): *Fragmenta Phytographiae Australiae*. Vol. X.
120 ——— 1878(III)-1881(VIII): *Fragmenta Phytographiae Australiae*. Vol. XI.
121 ——— 1882(XII): *Fragmenta Phytographiae Australiae*. Fascicle No. 94 1st. Pt. of Vol. XII
122 ——— 1886-87: Definitions of Two New Australian Plants.—Trans. R. Soc. S. Aust. 10: 80-81. *Newcastlia dixoni*, 81.
123 Mueller von, F., and Tate, R., 1882-83: Diagnoses of Some New Plants for South Australia.—Trans. R. Soc. S. Aust. 6: 107-109. *Dimorphocoma minutula*, 107-108; *Babbagia pentaptera*, 108; *Babbagia acroptera*, 108-109; *Loranthus murrayi*, 109.
124 Osborne, T. G. B., and Wood, J. G., 1923: On Some Halophytic and non Halophytic Plant Communities in arid South Australia.—Trans. R. Soc. S. Aust, 47: 388-399.
[Investigations in the Curnamona area.]
125 Ratcliffe, F. N., 1936: *Soil Drift in the Arid Pastoral areas of South Australia*.—Aust. Coun. Sci. ind. Res. Pamph. 64. (Gov. Printer: Melbourne.) [Some notes on the Flinders Ranges Flora.]
126 Rogers, G., 1951: The Flowering Ranges.—Wild Life. Melb. 13: 314-320. Eight illustrations.
127 Rogers, R. S., 1927: Contributions to the Orchidology of South Australia. Trans. R. Soc. S. Aust. 51: 1-13. *Prasophyllum validum*, 7.
128 Shaw, E. A., 1965: Taxonomic revision of some Australian Endemic Genera of *Cruciferae*.—Trans. R. Soc. S. Aust. 89: 145-254.
Arabidella filifolia, 188-191; *Arabidella nasturtium*, 191-196, *Arabidella procumbens*, 200-203. Fig. 2, 3, 4, 5, 7.
129 Sonder, W., 1852: Plantae Muellerianae—Beitrag zur Flora Südwestaustraliens aus den Sammlungen des Dr. Ferd. Müller *Compositae*.—Linnaea XXV: 449-530.
130 Specht, R. L., 1961: Flora Conservation in South Australia. I: The Preservation of Plant Formations and Associations recorded in South Australia.—Trans. R. Soc. S. Aust. 85: 177-196.

[31] Specht, R. L., and Cleland, J. B., 1963: Flora Conservation in South Australia. 2: Preservation of Species recorded in South Australia.—Trans. R. Soc. S. Aust. **87**: 63-92.
[32] Tate, R., 1882-83: List of some Plants inhabiting the North Eastern part of the Lake Torrens Basin.—Trans. R. Soc. S. Aust. **6**: 100-106. [Includes plant species collected from the southern portion of the Aroona Range.]
[33] ——— 1883-84: Description of New Species of South Australian Plants.— Trans. R. Soc. S. Aust. **7**: 67-71.
Sisymbrium procumbens, 67; *Kochia pentatropis*, 67-68; *Hakea ednieana*, 70.
[34] ——— 1890: *A Handbook of the Flora of Extratropical South Australia, containing the Flowering Plants and Ferns.*—(Educ. Dept.: Adelaide.)
[35] ——— 1897-98: On some New or Little-Known South Australian Plants.— Trans. R. Soc. S. Aust. **22**: 119-121.
Corchorus longipes, 119; *Acacia papyrocarpa*, 119-120; *Helipterum microglossum*, 121.
[Max Koch collections.]
[36] ——— 1898-99: Diagnoses of Four New Species of Plants from South Australia. —Trans. R. Soc. S. Aust. **23**: 288-292.
Minuriella annua, 288-289; *Zygophyllum hybridum*, 291; *Zygophyllum kochii*, 291; *Eriocaulon submersum*, 291-292.
[Max Koch collections at Mt. Lyndhurst.]
[37] Traub, H. P., 1966: The light primrose yellow *Crinum luteolum*.—Pl. Life **22**: 46-47. [Full description of new species.]
[38] ——— 1967: *Crinum luteolum*.—Pl. Life **23**: 71.
[39] Traub, H. P., and Hannibal, L. S., 1965: *Crinum luteolum*.—Pl. Life **21**: 96.
[40] Ward, G., 1965: Wilpena Pound. Vegetation.—Tarndanya 1: 6. (New Series).
[41] White, S.A., 1912: Field Ornithology in South Australia. Port Augusta District.—Emu XII: 122-130. Two illustrations.
[Short notes on vegetation of Flinders Range gorges.]
[42] Wood, J. G., 1924. Relations between Distribution, Structure and Transpiration of arid South Australian Plants.—Trans. R. Soc. S. Aust. **48**: 226-235. [Experiments at Curnamona.]
[43] ——— 1937: *The Vegetation of South Australia.* (Gov. Printer: Adelaide.)

F.N.C.V. PUBLICATIONS AVAILABLE FOR PURCHASE

FERNS OF VICTORIA AND TASMANIA, by N. A. Wakefield.

The 116 species known and described, and illustrated by line drawings, and 30 photographs. Price 75c.

VICTORIAN TOADSTOOLS AND MUSHROOMS, by J. H. Willis.

This describes 120 toadstool species and many other fungi. There are four coloured plates and 31 other illustrations. New edition. Price 90c.

THE VEGETATION OF WYPERFELD NATIONAL PARK, by J. R. Garnet.

Coloured frontispiece, 23 half-tone, 100 line drawings of plants and a map. Price $1.50.

Address orders and inquiries to Sales Officer, F.N.C.V., National Herbarium, South Yarra, Victoria.

Payments should include postage (11c on single copy).

New Government Botanist

Following our happy relationship with Mr. Pescott during his term as Government Botanist, members of the F.N.C.V. will welcome the appointment of Dr. David Maughan Churchill of the Botany School, Monash University, as our new Government Botanist.

He was born and educated in Western Australia and from 1957 to 1960 was on the staff of that university. In 1960 he accepted an appointment with the Department of Botany of Cambridge University, U.K., but returned to Australia to take up his Monash University post in 1965.

He has been especially interested in changes affecting vegetation in recent geological times, as deduced from fossil pollen, etc., in peat beds, while a contribution on three important W.A. eucalypts (*E. diversicolor*, *E. marginata* and *E. calophylla* in the *Australian Journal of Botany*, Vol. 16, is evidence of his scholarship and thorough field work.

Some of us have met him with pleasure and feel that his term as Government Botanist will be a fruitful one. We hope that under him the pleasant association between the Herbarium and F.N.C.V. will continue, and the new position will be one of interest and satisfaction for him.

He takes up his appointment late this year.

* * *

Field Naturalists Club of Victoria

Botany Group
8 July, 1971

Miss L. M. White gave a talk on Proteaceae, the family which is to be the subject of the Botany Group's exhibit at the September nature show. She issued sheets giving keys devised by Ewart, J. M. Black and Thistle Harris respectively, and characteristics of habitat, habit of growth, leaves, flowers and fruits. There are 55 genera and 1200 species, the type genus, *Protea*, having 130 African species, and Australia sharing the genus *Lomatia* with South America, but the family is strongest in Australia, especially in the West.

The speaker used several charts to show the differences of structure between the genera *Grevillea*, *Hakea*, *Banksia*, *Persoonia*, *Dryandra*, *Isopogon*, *Petrophila*, *Telopea*, *Lambertia*, *Adenanthos*, *Xylomelum*, *Conospermum*, *Lomatia*, *Orites* and *Stenocarpus*. Actual specimens of many species were on exhibit, and were used to illustrate certain points. Among the many fruits shown were those of the Gippsland Waratah. Books and magazines dealing with the subject added to the interest of the talk, and a series of fine colour slides completed a most instructive lecture. After words of praise from the chairman (Mr. K. Kleinecke) the twenty members present showed their appreciation.

A letter from Mr. Alex Hicks of Kaniva revealed that Dr. Melville of Kew Gardens, London, is to make another visit to Australia this year, and will again go to study the flora of the Little Desert. The August excursion will be to Dixon's Creek and Mount Slide.

Flowers and Plants of Victoria in Colour

Copies of this excellent book are still available, and of course would make a wonderful gift. They are obtainable from the F.N.C.V. Treasurer, Mr. D. McInnes

The 1971 Wild Flower and Nature Show

WHERE?
The Lower Melbourne Town Hall.

WHEN?
Monday, Tuesday and Wednesday, 21, 22 and 23 September, from 10 a.m. to 10 p.m.

OBJECT?
The Society for Growing Australian Plants (S.G.A.P.) and the Field Naturalists' Club of Victoria (F.N.C.V.).
The stimulation of a lively interest in all aspects of our natural wild-life and in its conservation.

The main displays will be a large variety of Australian plants grown by members of the S.G.A.P., and a selection of Australian plants from the Maranoa Gardens. The central exhibit, arranged by the National Parks Association, will feature the Alpine areas they hope will be incorporated in an extensive National Park.

The Hawthorn Junior Field Naturalists will arrange exhibits showing economic insects: spiders: and nickel deposits in Australia. They will also show reptiles — confined in glass cases!

The Montmorency Junior F.N.C. will exhibit aquatic plants.

The Preston Junior Club will stage a general exhibit.

A series of microscopes set up by the Microscopical Group will cover many aspects of "the world of very small things". Butterflies and live aquatic insects will be shown by the Entomological Group.

The Botany Group exhibit will cover the large Proteaceae family (grevilleas, etc.). The Geology Group will show a model of the terraces on the Maribyrnong River, and will explain their origins.

At a Publications stand natural history books and magazines may be purchased. If you want to know anything about the F.N.C.V. enquire at that stand.

The State Film Centre has kindly loaned Wild-life films and these will be shown free.

All available help is needed to "run" the Show and to set it up on Sunday, 19 September, from 10 a.m. onwards. Rosters for helpers will be distributed at the August general meeting, and we look forward to seeing many new members' names on these.

Pamphlets featuring Possum Gliders, and posters for distribution will also be available then. Will all members please do their best to publicise our Show?

The Erratum on page 204, July 1971 issue, was itself incorrect, as it made no change from the words as printed on page 163 of June issue. The correct wording should be:
"In Roy Wheeler's 'A Handlist of the Birds of Victoria' the Southern Scrub-robin's habitat map includes the Wyperfeld area".

Field Naturalists Club of Victoria

Established 1880

OBJECTS: To stimulate interest in natural history and to preserve and protect Australian fauna and flora.

Patron:
His Excellency Major-General Sir ROHAN DELACOMBE, K.B.E., C.B., D.S.O.

Key Office-Bearers, 1971-1972.

President:
Mr. T. SAULT

Vice-Presidents: Mr. J. H. WILLIS; Mr. P. CURLIS

Hon. *Secretary:* Mr. D. LEE, 15 Springvale Road, Springvale (546 7724).

Subscription Secretary: Mr. D. E. McINNES, 129 Waverley Road, East Malvern, 3145

Hon. *Editor:* Mr. G. M. WARD, 54 St. James Road, Heidelberg 3084.

Hon. *Librarian:* Mr. P. KELLY, c/o National Herbarium, The Domain, South Yarra 3141.

Hon. *Excursion Secretary:* Miss M. ALLENDER, 19 Hawthorn Avenue, Caulfield 3161. (52 2749).

Magazine Sales Officer: Mr. B. FUHRER, 25 Sunhill Av, North Ringwood, 3134.

Group Secretaries:

Botany: Mr. J. A. BAINES, 45 Eastgate Street, Oakleigh 3166 (57 6206).
Microscopical: Mr. M. H. MEYER, 36 Milroy Street, East Brighton (96 3268)
Mammal Survey: Mr. D. R. PENTON, 43 Duke Street, Richmond, 3121.

Geology: Mr. T. SAULT.

MEMBERSHIP

Membership of the F.N.C.V. is open to any person interested in natural history. The *Victorian Naturalist* is distributed free to all members, the club's reference and lending library is available, and other activities are indicated in reports set out in the several preceding pages of this magazine.

Rates of Subscriptions for 1971

Ordinary Members
Country Members
Joint Members
Junior Members
Junior Members receiving Vict. Nat.
Subscribers to Vict. Nat.
Affiliated Societies
Life Membership (reducing after 20 years)

The cost of individual copies of the Vict. Nat. will be 45 cents.

All subscriptions should be made payable to the Field Naturalists Club of Victoria, and sent to the Subscription Secretary

JENKIN BUXTON & CO. PTY. LTD., PRINTERS, WEST MELBOURNE

victorian naturalist

Vol. 88, No. 9   SEPTEMBER, 1971

Published by the

FIELD NATURALISTS CLUB OF VICTORIA

In which is incorporated the Microscopical Society of Victoria

Registered in Australia for transmission by post as a periodical
Category "A"

45

F.N.C.V. DIARY OF COMING EVENTS
GENERAL MEETINGS

Monday, 13 September — At National Herbarium, The Domain, South Yarra, commencing at 8 p.m.

1. Minutes, Reports, Announcements. 2. Nature Notes and Exhibits.
3. Subject for the evening — "Birds of Marsh and Lake": Joan Forster.
4. New Members.

Ordinary:
Mr. Walter J. R. Barber, 17 James St., Ringwood, 3134
Mr. Ian Beveridge, 2/13 Park St., East Brunswick, 3056.
Miss J. A. Vourne, 36 Westgate St., Pascoe Vale, 3044
Mrs. Phyllis M. Forbes, 46 Hawson Ave., Glenhuntly, 3163
Mr. R. G. Hall, c/o Post Office, Selby, 3159
Miss N. D. Lucas, 423 Middleborough Road, Box Hill, 3128.

Joint:
Mrs R. J. Lawson, Lot 157, Norwood St., Heatherton, 3202
Mr. Douglas R. Myers, Mrs. Nancy R. Myers, Rosemary N. Myers, Bronwyn A. Myers, 7 Clovelly Ave., Glenroy, 3046
Mr. and Mrs. Alan Dove, 510 High St. Road, Mount Waverley, 3149

Country:
Mr. Allan J. Marshand, Bogong, Vic., 3699.

Junior:
Philip Weigall, Timbertop P.B., Mansfield, 3722.

5. General Business. 6. Correspondence.

Monday, 11 October — Dr. D. M. Churchill.
Monday, 8 November — Presentation of Medallion.

GROUP MEETINGS
(8 p.m. at National Herbarium unless otherwise stated).

Thursday, 9 September — Botany Group. Mr. I. Morrison will show slides of Australian Proteaceae.
Wednesday, 15 September — Microscopical Group.
Monday 20, Tuesday 21, Wednesday 22 September — NATURE SHOW. Please give our Show as much publicity as possible and also personal assistance. Setting up Sunday, 19 September.
Friday, 24 September — Junior meeting at Hawthorn Town Hall at 8 p.m.
Monday, 4 October — Marine Biology and Entomology Group meeting at 8 p.m. in small room next to theatrette at National Museum.
Wednesday, 6 October — Geology Group.
Thursday, 7 October — Mammal Survey Group meets at 8 p.m. in Arthur Rylah Institute for Environmental Research, 123 Brown Street, Heidelberg.
Thursday, 14 October — Botany Group. Miss M. Lester will speak on "Coastal Plants".
Friday, 15 October — Montmorency and District Junior F.N.C. meeting in Scout Hall, Petrie Park, at 8 p.m.

F.N.C.V. EXCURSIONS

Sunday, 12 September — Botany Group excursion. Meet at Dandenong Station at 9.55 a.m. Bring one meal.
Sunday, 19 September — Setting up the Nature Show in the Melbourne Lower Town Hall. Because of this there will not be a general excursion in September but there will be two excursions in October, one a day trip and one weekend.
Sunday, 3 October — Beaconsfield district, led by Mrs. I. M. Naylor and other local residents. The coach will leave Batman Avenue at 9.30 a.m. Fare $1.60, bring two meals. Members travelling by car can meet at Starling Road, Officer, opposite the Officer Sports Ground, approximately 10.45 a.m.
Saturday, Sunday, 16-17 October — Weekend excursion to Bendigo, led by J. W. Kellam of the B.F.N.C. Saturday there will be a visit to the fossil locality at Spring Gully. Sunday we will join in the Bendigo F.N.C. excursion to Toolleen State Forest. Fare, including dinner, bed and breakfast at the Shamrock Hotel will be $15.00 and this should be paid to the Excursion Secretary by the October General Meeting. The coach will depart from Flinders Street, outside the Gas and Fuel Corporation at 8.45 a.m. Bring a picnic lunch for Saturday and two picnic meals for Sunday.

The
Victorian Naturalist

Editor: **G. M. Ward**

Vol. 88, No. 9   8 September 1971

CONTENTS

Articles:

    The Mating of a Huntsman. By Densey Clyne   244

    F.N.C.V. Trip to Mt. Beauty. By Jean Zirkler   249

    Along the Ginap Track. By Victor Jacobs   254

Field Naturalists Club of Victoria:

    Questionnaire (Removable centre pages)

    Diary of Coming Events   242

Front Cover:

    This photograph of the head study of the White-breasted Sea Eagle was taken by John Wallis.

The Mating of a Huntsman

by Densey Clyne

Illustrations by Author.

Huntsman spiders are sometimes known as triantelopes, sometimes as tarantulas. No-one seems to know the origin of the word "triantelope". As for the name "tarantula", these spiders are not at all closely related to the two groups of spiders which traditionally bear this name, one European and one American.

The Americans give the name to their Mygalomorph spiders, which are cousins to our funnelwebs and trapdoor spiders.

The European tarantula, on the other hand, is or was a wolf spider (Lycosidae). Because they live in holes in the ground, our own harmless wolf spiders are often sent to scientific institutions as funnelwebs. But the wolf spider's hole is in the open, and never surrounded by the large, funnel-web sheets of web that give the funnelweb spider its name.

The name "tarantula" was used in Europe in the Middle Ages for a species of wolf spider, and from this name came the dance called the "tarantella"—or vice versa. In any case, according to tradition, the effects of the spider's bite drove its victims to dance hysterically until they dropped dead, a condition known as "tarantism". And all these words are derived from Taranto, the town in Italy where it is supposed to have happened.

Anyhow, it seems unfair to saddle our Huntsman with a name that has such unpleasant connotations. It is a docile spider in my experience, and probably does more good than harm by catching flies and other potentially dangerous creatures in the house.

The following are some notes on the behaviour of a pair of Huntsman spiders. The species is *Isopeda vasta*, common in Sydney gardens and houses. It is a large spider, flattened dorsoventrally as are all members of the Sparassidae family to varying degrees, with laterigrade legs, i.e., legs that lie on their sides and curve forward. These two features, the flattened body and laterigrade legs, enable the spider to move about freely in its natural habitat, under loose bark and in similarly restricted places. The colour is light brown to greyish, and the body and legs are hairy.

The male and female under observation were found resting close together on a banana plant in my garden, and were brought indoors and placed in a glass-walled cage for observation.

At daybreak on 1 April, the female had undergone her final moult, and her cast skin hung in the cage. It seems common for male spiders to

Plate 1.
Huntsman Spider (*Isopeda vasta*)

be attracted to females some time after the penultimate moult, to remain close to them until the final moult, and to mate with them directly afterwards.

At 10 a.m. the same morning the spiders were mating, the male's right palpal organ having been inserted in the female's epigynum. (The mating organs of male spiders are at the tips of the two feelerlike palps on the front of the head. The genital openings of both sexes are situated on the abdomen, but prior to mating the male transfers the sperm fluid from his genital opening to a special sperm web, and from this web he draws it up into his palps.)

The spiders faced in opposite directions, the male standing partly over the female. Her legs and cephalothorax were flat against the cage wall, her abdomen tilted upward at an angle, so that the ventral surface could be reached by the male's palp. The male's flexed right legs straddled the female's body. On the horizontal plane, the two spiders were angled slightly to each other so that his head was closer to her than hers was to him.

The first expansion of the male's right palpal bulb lasted for two and a half minutes. During this time its membranous part, the *haematodocha*, was inflated in stages, with slight pauses in between, like a balloon being blown up, until it was fully expanded and visible as a translucent, bright blue bubble.

Concurrently with each of these minor inflations, the male's leg spines, which normally lie flat, rose to an erect position, and then fell. These periods of alternate erection and pause increased in duration from about 6 secs. to about 12 seconds. The angle of the spines at the top of the rise increased until, towards the end of the expansion period, they rose at an angle of about 65 degrees to the surface of the leg.

(In the following notes I use the terms "expansion" and "rest" to refer to the complete, overall inflation of the *haematodocha*, accomplished in stages and lasting for several minutes; and I use the terms "erection" and "pause" for each of its component stages of inflation accompanied by the rise and fall of the leg spines. The erection of the leg spines is probably merely a result of the burst of muscular activity, but it is convenient to use the term here to differentiate between the total expansion of the *haematodocha*, and the stages of inflation that bring it about.)

At 12 noon, expansion periods were timed, and lasted approximately 3½ mins., followed by ½ min. rest. When expanded, the *haematodocha* was clearly visible extending on either side of the palpal sheath. (The technical term for this sheath is the *cymbium*; it is the more or less cup-shaped tip of the palp, which partly encloses or covers the palpal organ). During rest periods it subsided more or less, but was usually still visible, though much reduced. At times it subsided so completely that it was hidden by the sheath. On these occasions the thin, dark coil of the *embolus* (that part of the palpal organ that actually conveys the sperm to the female) could be seen relaxing and partly returning into the sheath.

As the *haematodocha* expanded, a pink, knobbed process emerged from under the top of the sheath on the inner side, and moved around below the palp to appear on the other side of the expanded bulb. (Further investigation might identify this process in this species as the conductor of the embolus.) The chelicerae were depressed each time the leg-spines rose.

During each expansion period, the tempo of activity increased from start to finish, the spine erections lasting longer and longer at the expense of the pauses; e.g. a typical series might be 3-5, 3-5, 7-3, 8-3, 15-1, 27 . . . the first figure being the time of spine erection and the second figure the time of pause, in seconds.

At 12.40 p.m. an expansion period of 9½ mins. was noted, finishing with three quick climaxes of 0-1 secs., the leg spines rising very high. This was followed by 3½ mins. rest.

At 1 p.m. an expansion period of 10½ mins. was noted, comprising 30 erections lasting from 5 secs. to 18 seconds. The palp was then removed. During removal, the long, dark spring of the embolus was uncoiled and grew taut before leaving the female's epigynum and springing back inside the male's palp.

Immediately, the female lowered her abdomen to the normal position. The male remained standing partly over touching her, and cleaned his right palp, "chewing" it with the chelicerae. He made intermittent "shuddering" movements, and now and then straightened out the palp while his leg spines rose a little. The female showed no aggression.

Palp cleaning proceeded for five minutes, then the male turned to stand over the female, facing in the same direction for a few seconds before moving away, shuddering, then returning to the mating position and inserting the left palp in the female's epigynum.

At 1.20 p.m. expansion periods were noted as follows:
2½ mins., 12 erections. Pause.
2½ mins., 9 erections. Pause.
2½ mins., 8 erections. Pause.
1 min., 5 erections.

Observation ended at 1.35 p.m., and recommenced 25 mins. later when a long period with erections up to 18 secs. duration was noted. This was followed by a similar long period of about 20 mins., followed by a rest.

At 3.22 p.m. the right palp was inserted for the second time. Expansion now consisted mainly of two quick erections followed by a pause, the sequence being repeated many times; the two erections together taking about 35 secs., and the pause 10 seconds. This was followed by an expansion consisting of 4 quick erections, and a rest.

An expansion period of 6½ mins. at 3.50 p.m. consisted of 28 erections, and was followed by 31 min. rest. The maximum erection period in this series was 24 secs., the minimum about 4 seconds.

At 4.30 p.m., after 88 erections in 28 minutes, the male removed his palp.

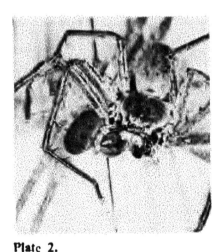

Plate 2.

Isopeda vasta mating. Female's abdomen is to left of picture. Note male's leg-spines in erect position.

Observation was discontinued at this point, and at 4.55 p.m. when the pair was next observed, the male had his left palp inserted. Erections and pauses were timed and noted, and after 2 hours 20 mins., the left palp was removed and cleaned.

A summary of the day's mating activities is as follows:

First insertion right palp —
3 hrs. 3 mins.
First insertion left palp —
2 hrs.
Second insertion right palp —
1 hr. 8 mins.
Second insertion left palp —
app. 2 hrs. 20 mins.

After this, mating took place nearly every day, usually for periods of 9-10 hours. The spiders were fed on moths, praying mantids and grasshoppers, and water was available in the cage.

The pair was watched daily. On 14 April the male's energy seemed to reach its peak. The periods of expansion and rest became progressively longer. At the climax of the longer periods of expansion, the leg spines during the final erections rose sharply to nearly 90 degrees, and snapped down again rapidly. The *haematodocha* disappeared from sight during resting periods, and did not relax much from full inflation during active periods, bulging well out beyond both sides of the sheath.

During one of the male's palp cleaning periods, the female was a little restless, but did not move from the male, who more or less encircled her, holding her with his legs and facing in the same direction. His first pair were in front of her, his second pair were between her second and third pair, his third between her third and fourth, and his fourth behind her fourth pair. During palp cleaning the male made the customary intermittent shuddering movements.

On 21 April mating was observed again, after a period of several days with little activity, although the spiders were not under constant surveillance during this time. There was no mating on 22 but on 23 at 11.25 the male inserted his right palp. This palp appeared to remain in operation until 4.15 p.m., when it was removed. At 4.35 the left palp was inserted, and at 8.55 mating ended.

No further mating was seen until 6 May. During this time the female fed on moths, cockroaches, blowflies and one bulldog ant. After capturing her prey she would bite and manipulate it, then circle around it, lifting the spinnerets high and touching them to the cage wall on either side of the insect to place strands of silk across it. When the prey was fully wrapped and still, she lifted it, freeing it to some extent from the cage wall before settling down to feed. Once she accepted a blowfly and fed without wrapping it. During

Plate 3.

Male and Female *Isopeda vasta* mating. Female's tilted abdomen can be seen on left of picture.

this period the female increased considerably in size, and by 6 May she was very large. Mating took place again on that day, commencing at 9.22 and ending at 4.45 p.m.

I have no note of mating after this period.

On 18 May the female made her white, lens-shaped egg-sac, about 1 inch in diameter, and attached by silk lines to the cage wall. When it was completed, she rested astride it, the male close by and facing her.

On 4 June I opened the cage and allowed the male to escape. I cut away a portion of the egg-sac, and found the eggs to be bright green. The egg-sac was now more or less attached to a piece of dead banana leaf in the cage, but the female invariably rested clasping it to her, her legs curved around the edges.

On 20 June I returned from some weeks' absence and found the egg-sac still intact, attended by the female.

No detailed diary notes were kept from this period on. The female remained with the egg-sac until the spiderlings emerged in spring, and then stayed quietly while they clustered close around her for some weeks. They appeared to feed on the prey which she caught, clustering close around her head and mouth while she held it in her *chelicerae* and predigested it.

The female and young spiders were released in my garden in late spring.

Plate 4.

Isopeda vasta. Male's left palp has just been withdrawn from female's epigynum, and *embolus* can be seen extending from palp to epigynum just prior to removal.

Plate 5.

Isopeda vasta female with egg-sac.

F.N.C.V. Trip to Mt. Beauty
26 December 1969 – 4 January 1970

by Jean Zirkler

When we left Melbourne at 9 a.m., weather conditions were very much in our favour and were to remain so throughout our trip. The first important stop was for lunch at Lake Nillahcootie, a new reservoir that gives water to the Broken River valley. We travelled on through Benalla, Glenrowan and Milawa (where the vineyards and winery were not unnoticed), and along the Ovens Valley through Myrtleford. Here we passed by neat farms growing tobacco and hops, then on to Bright for a rest stop and walk among the lovely trees that characterize this pretty township. Passing by walnut orchards we made our way up to Tawonga Gap, where the expansive view compels a stop; but here we had more than the scenery to force a temporary stay, for while we all took in the wonders of the valley from this vantage point and filled our lungs with the clear mountain air our driver, Ray Hicks, discovered that there was trouble lurking in our brand-new coach. He soon diagnosed the trouble, and we reached Mount Beauty Chalet, where he made arrangements for a different coach to be sent up to Euroa, from where he would collect it.

Our schedule for Saturday was upset, but in this magnificent country one is never put out: there is always so much to see. The morning was free, so 45 "Field Nats" were let loose over Mt. Beauty township and surrounding areas. Some found much pleasure and interest in watching birds in great numbers in the town square, with Noisy Friar-birds feeding in the Silky Oaks (*Grevillea robusta*), Pied Currawongs and several species of honeyeaters; others by walking along the track up the nearby mountain. A picnic lunch on the spacious lawns at the Chalet was all the more enjoyable because of the company of Blue Wrens and more honeyeaters.

Saturday afternoon was spent in good company too—that of Mr. Max Howell, of S.E.C. Tourist Information Office, a local naturalist who proved an excellent guide. First he allowed us to browse around the Administrative Building, where an interesting display enabled us to gain a splendid understanding of what we were later to see of the actual Kiewa Scheme, with the help of fascinating relief map models. Max also made informative comments on exhibits, including specimens of rock cores, flora and entomology. The surroundings of these administrative buildings have been attractively planted with a variety of native shrubs and trees, including *Eucalyptus pauciflora* (a Snow Gum that had been transplanted about 25 years ago), *E. camphora*, *E. viminalis*, *E. birostata*, and *E. chapmaniana*, all native to the region, the last species named comparatively recently after Brigadier Chapman and represented in Maranoa Gardens: it is known as the Bogong Gum from the place of its discovery during the time of Kiewa construction work. Most helpful were the brochures on the scheme issued to us by Max, and lists of plant and bird species and a summary of the geology of the Kiewa Works Area.

On Sunday morning (Ray having successfully exchanged coaches) we set out for the Bogong High Plains, eagerly expectant, and we were by no means disappointed. *Eucalyptus chapmaniana* was noticed along the roadside on the way up to Bogong village; at the foot of a very beautiful specimen of this tree there is a plaque commemorating the discovery of the new Bogong Gum. Buffalo Wattle (*Acacia kettlewelliae*) is quite plentiful between the 1400-3000 ft. altitudes, and Red-stem Wattle (*A. rubida*). Alpine Ash (*Eucalyptus delegatensis*) and *Pimelea axiflora* were easily distinguishable as we passed along. We crossed what is known as Sassafras Gully—now a misnomer, as Sassafras (*Atherosperma moschatum*) no longer exists there, for reasons unknown.

Royal Grevillea (*G. victoriae*) was there in abundant glory, but we couldn't stop to search for things of interest as the road is very steep in many places; to stop a large vehicle such as ours would be a danger to others. We noticed Derwent Speedwell (*Veronica derwentia*) everywhere in full flower. Passing the Clover and Junction Dams, we reached Bogong township, where we met Max Howell again and drove to the State Electricity Commission Laboratory, with more specimens of geology, entomology and flora of the Alps. Good research work is being done on pests such as the Sirex wood wasp, St. John's beetle and Queensland fruit fly. A vegetation type map was on display, made from 18,000 ft. up of the vegetation of the Alps.

We were now at an altitude of 4,000 feet above sea level and amongst many more sub-alpine and alpine plants, with birds prominent, willy wagtails, blue wrens, galahs, currawongs, white cockatoos, black-backed and white-backed magpies all being seen. Mountain Shaggy-pea (*Oxylobium alpestre*) gave a golden blaze of colour through the bush. Passing through Falls Creek we had a passing glance at the ski village, much grown in size since the club's earlier excursion based here. We climbed to nearly 6,000 feet and the Bogong High Plains, where some were rather surprised to find cattle still being grazed; the sheep that were pastured there in the early days are no longer seen and we understood why sorrel and other introduced pasture weeds are so common.

Lunch time (at last!) in the picturesque setting of Rocky Valley dam, in perfect weather, and all around us interesting plants like *Richea continentis*, *Drimys lanceolata*, *Epacris petrophila*, *Prostanthera cuneata*, *Celmisia longifolia*, *Hovea longifolia*, *Craspedia uniflora* (of a bright orange colour), *Pimelea ligustrina*, Cascade Everlasting (*Helichrysum thyrsoideum*) in great display, *Leucopogon nanus*, Sky Lilies, Alpine Marsh Marigold (*Caltha introloba*) and Mountain Gentian (*Gentianella diemensis*). Our Brown Edelweiss (*Ewartia nubigena*) is a mat plant of extreme beauty in flower.

We paid a visit to Basalt Hill, where basalt rock and road material was quarried during the building of Rocky Valley dam. The basalt is columnar in formation, like great pillars stacked erect side by side. One wonders about glaciers on the Bogong High Plains, but it is known that glaciation never took place, though evidence of periglacial features has been found. The Kiewa area is structurally ideal for our hydro-electric development but is poor in mineral wealth. Gold has been mined in the West Kiewa valley, some reefs fabulously rich but small in size. Tantalite a rare and valuable mineral, occurs on Mt. Bogong, and wolfram near Mt. Nelse, but neither of these minerals is present in economic quantities. The brown coal, which is of relatively high quality, is restricted to very thin seams with an overburden of hard basalt.

The following day we returned to this area for a short visit to the "Ruined Castle", then continued on to Pretty Valley. We passed the highest point on the High Plains. The road here is the highest official road in Victoria, 5,890 feet (any higher are firebreaks or Commission service roads). Our lunch stop was notable for the presence nearby of *Pimelea alpina*, *Astelia alpina*, *Boronia algida*, *Orites lancifolia*, *Drosera arcturi*, *Grevillea victoriae*, *Bulbine bulbosa*, *Veronica nivea*, *Euphrasia gibbsii*, *Veronica derwentia* and our first orchid so far, *Prasophyllum alpinum*.

On the return journey a short stop was made, and most of us took a walk along the road, taking in the clean mountain air after a thunderstorm and heavy rain. Along here we came across *Dianella tasmanica*, and to our delight lyrebirds were heard in the thickly wooded gullies, obviously enjoying the calm after the storm as we were. We had now come down again into the vegetation of the lower mountain regions, and while little streams of water rushed noisily from everywhere to join a larger stream as though in eagerness, like the birds we watched flitting through the trees, we enjoyed the beauty of the Soft Water Ferns and Fishbone Ferns along the little waterways. Down on our lists went *Daviesia latifolia*, *Notelaea ligustrina*, *Billardiera longiflora* and Mountain Grass. So ended, for us, another delightful day.

Next day (fore-warned of a six-mile hike in store) we set off by coach for Bundarrah Creek, located somewhere behind Mt. Jim, and travelled high up the mountain, always observing the birds, plants, trees and shrubs which the "experts" were usually able to identify. There was wondrous beauty in the wide views across streams and valleys, sometimes the reservoirs were in sight and of course the mountains near and far. The walk along the track, bordered by the water race, was most enjoyable, and the effort was rewarded by the final objective: a brown coal seam in a steep bank of Bundarrah Creek, where examination revealed leaf fossils showing up distinctly. Birds seen included Eastern Rosellas, Gang Gang Cockatoos, kookaburras, ravens and hawks, and among the plants were *Caladenia lyallii* (not recorded here previously), *Leucopogon gelidus* and *Eriostemon myoporoides*, to name only a few of those seen.

To begin another day we met Mr. Ken Mills, by arrangement, who is an official guide for the S.E.C. We set off for Mt. McKay power station, where we were taken by lift 250 feet down to be shown over the station. This station generates for peak periods and for boosting the load during winter months for country areas as well as the city of Melbourne. We learned of three helicopter pads in the mountains, useful in various ways. Mr. Mills pointed out an outcrop of large rocks where he has watched swifts resting, and firmly believes that this is where they rest before taking off on their long migration.

This day we stopped for lunch on Mt. McKay, 6,049 feet above sea level, at the fire lookout tower. Later we made another visit to the "Ruined Castle", as the first was spoilt by wind and rain. This outcrop of rocks is evidence of a basalt flow extruded 20-40 million years ago. Cooling of lava caused cracks which divided the rock into closely fitting columns, usually hexagonal in cross section. About one million years ago an uplift took place, forming the Bogong High Plains, followed by denudation and erosion. "Ruined Castle" is one of the few remaining examples of columnar jointed basalt in this area.

On our way back down to Mt Beauty we had a little time to spare, so we made our way on foot up the track to the beautiful Fainter Creek Falls, which are almost completely hidden by treeferns. Along the steep and rugged path we found Bird Orchids (*Chiloglottis gunnii*), and an unusual greenhood that so intrigued Dick Morrison and Tom Sault that they later went back to photograph it (and were caught in heavy rain and had to negotiate the return trip along the trail in semi-darkness!). There were very few orchids to be seen throughout this trip.

Our next day was disappointing. The coach took us a long journey to Mount Hotham summit, but as we reached there the whole area became enveloped with cloud, which made visibility impossible and strong cold winds made us rather uncomfortable. Most of us stayed in the coach to eat lunch, but those who braved the icy blasts found some very interesting plant life, with Snow Gums in flower, and acres of *Celmisia*. Three forms of *Celmisia* occur in the Bogongs: *C. longifolia* and its variety *latifolia* are both widespread on the High Plains, while *C. sericophylla* is found along creek banks on the High Plains with occasional occurrences down to 4,000 feet. These large white Silver Daisies and Silky Daisies, often interspersed with the vivid orange Billy Buttons (*Craspedia uniflora*), were in such profusion that they gave us natural gardens that were unforgettable.

Next day's planned trip across the Omeo Road had to be abandoned because of the heavy rain and reports of snow over that way, but the substituted trip to Lake Hume was great compensation, because it proved to be a wonderful day for birdwatching. We passed through Tawonga and along the Kiewa River valley. At Kergunyah the obvious richness of bird life demanded a stop, which proved to be a lengthy one, as there was so much to see. Dorothy Dawson, a prominent member of the Bird Observers' Club, was excited to see her first Dollar Bird, which, though not uncommon in the north-east of Victoria, is rarely seen in the coastal districts. Through the glasses we could clearly see the silvery-blue patches on the wings that give it its common name, and noted the typical flight pattern that justifies its other name of Roller. In a swampy paddock White and Straw-necked Ibis were feeding, and other water birds observed were spoonbills, egrets, herons and black ducks; while other birds seen included galahs, magpies, white-plumed honeyeaters, and several species of parrots, among them Eastern Rosellas and grass parrots.

At Sandy Creek there were black swans with cygnets, musk ducks, black ducks, crested grebe and plover. On reaching Lake Hume we were able to have a barbecue lunch as the weather had cleared sufficiently. Later we made a stop to have a closer look at a number of Fairy Martins and a few Welcome Swallows, in spite of more rain. Further along pelicans were seen perched high in dead trees near the shore.

On leaving Tallangatta we made a stop near a swamp to observe more birds, including Nankeen Night Heron, white cockatoos, cormorants, moorhens, Sacred Kingfisher and two Whistling Eagles, soaring above.

The next day was wet again, but this deterred only a few from the outdoor searching and observing. Some went up Mount Beauty itself and along the Cranky Charlie track, led by Max Howell. A great many species of plants were noted, one of special interest being Scented Rice-flower (*Pimelea curviflora*). Those who preferred indoors took the opportunity of getting together to identify and discuss the few specimens collected by those who have a permit to do so (in the interest of knowledge). Others willingly keep strictly to the law which makes the picking of wildflowers, except on private property with the owner's permission, illegal in Victoria, and the Kiewa area is a sanctuary for all bird, animal and plant life. (This does not include introduced species, but one such exotic, *Verbena bonariensis*, with rich purple flowers, Cluster-flowered Vervein, known locally as Squareweed from its stems, which are square in cross section, was common and colourful in the town of Mt. Beauty.

We did not see many wild animals, and in this party we didn't have the night searchers who like to go out and study the small creatures in their own habitats.

On Sunday, 4 January, we left for home via the Ovens River valley to connect with the Hume Highway. As always with these enjoyable excursions, thanks must go to all the local naturalists and officials of the S.E.C., State Rivers and Water Supply Commission and other departments who gave up much of their own time to make our trip a memorable one, and also to our excursion secretary, Miss Marie Allender.

F.N.C.V. PUBLICATIONS AVAILABLE FOR PURCHASE

FERNS OF VICTORIA AND TASMANIA, by N. A. Wakefield.
  The 116 species known and described, and illustrated by line drawings, and 30 photographs. Price 75c.

VICTORIAN TOADSTOOLS AND MUSHROOMS, by J. H. Willis.
  This describes 120 toadstool species and many other fungi. There are four coloured plates and 31 other Illustrations. New edition. Price 90c.

THE VEGETATION OF WYPERFELD NATIONAL PARK, by J. R. Garnet.
  Coloured frontispiece, 23 half-tone, 100 line drawings of plants and a map. Price $1.50.

  Address orders and inquiries to Sales Officer, F.N.C.V., National Herbarium, South Yarra, Victoria.

  Payments should include postage (11c on single copy).

Along the Ginap Track

by Victor Jacobs

Illustrations by Author

Although I was supposedly to head homeward on the 6 September 1968, after a North western tour of Victoria, I was tempted to make a diversion, whilst looking over my route maps on the previous night.

The eastern end of Wyperfeld National Park had always seemed so remote from Wonga Hut; but the road home passed Yarto, the nearest place-name to this part of the Park.

Temptation gave way to decision, and it was beside the railway line at Yarto that I parked the caravan.

Beyond the concrete silo and few houses that was Yarto, lay the dark fringe of Mallee eucalypts and the low undulations of the sand hills; and so after a short but "matey" chat with some of the "locals", I took my gear and headed across the grassy paddocks towards the beckoning Mallee.

Mediterranean Turnip, as elsewhere in the Mallee that year, was abundant. There was plenty of Evening Primrose and *Hibbertia stricta* was colonising the open area very well. I stepped over a fence; the drifting sand having left only a couple of inches of timber showing. I entered the Mallee heading due west, but had difficulty in maintaining a direct course, for the trees were quite close together. I was about to return when I came across a ploughed contour furrow which I followed, hoofing my heels in deeply so as not to slip. Wallowa (*Acacia calamifolia*), washed the bush with a pale gold, while the darker Hakea Wattle (*A. hakeoides*) was seen less often. Grey Mulga (*A. brachybotra*) was still in tight bud. A White-eared Honeyeater was escorting me, calling at intervals. When I had followed this furrow for some hundreds of yards, it ran into a fire access track (The Ginap Track I realised later) running approximately east and west. I ambled along this track in a westerly direction seeing many plants not seen in the eastern sectors of the park, but not many birds at all. After a mile or two I about turned and headed back, retracing my own, and the only human footprints visible. Instead of returning along the contour furrow I followed this fire access track to where it ended at the park margin. Here, standing amidst some fine flowering bushes of Dwarf Emu-bush (*Eremophila glabra*) was a notice which stated clearly that this was a fire access track and not to be used by any unauthorized vehicles. It was the Ginap Track. Taking a careful bearing I noted that to find this track from the Yarto Silo one has to head generally South West.

Back in the van I checked on my notes, using Ros Garnet's book; and later, having sent plants to the Herbarium, I was able to complete my plant-list. A lovely patch of Desert Heath-myrtle (*Baeckea crassifolia*) dense stands of Broombush (*Melaleuca uncinata*), Pink Purslane (*Calandrinia calyptrata*), Pygmy Sunray (*Helipterum pygmaeum*), Pink Velvet-bush (*Lasiopetalum behrii*), Fringed daisy (*Brachycome ciliaris*) Common Billy Buttons (*Craspedia glauca*), Tangled Burr-daisy (*Calotis erinacea*) on the dunes, Pale Turpentine-bush (*Beyeria leschenaultii*).

F.N.C.V. Members' Questionnaire
[These pages to be removed.]

The council of the F.N.C.V. considers that a survey of opinions of members on the objectives and functioning of the club will assist in the planning for the future. Would you please fill in the answers to these questions and return either by post to the Secretary, or by handing the page to him at a meeting.

1. Number of general meetings attended per year..........................

2. Reasons for your attendance at general meetings are —

    (a) To keep up to date with club affairs ☐
    (b) To see exhibits and discuss subjects of interest with other members ☐
    (c) To hear a speaker on a particular topic ☐
    (d) To hear a speaker on any topic ☐
    (e) Any other reasons —

3. Reasons for not attending general meetings are —

4. I would like to hear speakers on the following topics —

    Geology: ☐
    Any particular aspect?

    Marine Biology ☐
    Any particular aspect?

    Botany: ☐
    Any particular aspect?

    Mammals: ☐
    Any particular aspect?

    Entomology: ☐
    Any particular aspect?

    Microscopy: ☐
    Any particular aspect?

    Reptiles: ☐

5. Have you any suggestions for change of format or agenda at general meetings?

6. Do you attend group meetings regularly?   Yes/No
   Do you attend meetings of more than one group?   Yes/No
   What aspects of group meetings attract you?

7. When joining a new group have you felt welcome, and found it easy to become involved in the activities?

   If you no longer attend group meetings would you indicate reasons.

8. Are you interested in reading reports of group activities in the *Naturalist*?   Yes/No
   Do you consider that reports of group meetings and excursions should be written up in the *Naturalist*?   Yes/No

9. Would you care to comment on the *Naturalist* according to any of these aspects?

   Feature articles:

   Other articles:

   Reports of General Meetings:

10. If you consider that the F.N.C.V. excursions once per month are not providing what you wish in the study of natural history, then what particular objectives do you think they should have?

11. Do you consider that the excursions once a month by subject groups provide the necessary activities?　　　　　　　　　　　　Yes/No
    If not, what activities would you prefer?

12. Do you think that we have enough groups to cover all activities? Yes/No
    If not, what other groups do you consider necessary?

13. If F.N.C.V. groups plan to work in conjunction with other groups or organizations outside the F.N.C.V. on research projects extending over months or years, would you find this more interesting than uncorrelated studies of various areas?　　　　　　　　　　　　Yes/No
    Would you be able to be a fairly consistent attender of such group meetings and excursions if such projects were undertaken?　Yes/No

14. Do you need transport to the Herbarium for meetings?　　Yes/No
    If you do need transport or are able to give transport to other members please give your name, address and telephone number and say whether it is to give or obtain transport.

15. Are there any other aspects of the F.N.C.V. activities about which you would like to make recommendations?

It is stressed that all members should fill in this questionnaire, and return it as soon as possible.

New members should not feel that a short acquaintance with club activities limits the usefulness of their answers. We need their views and comments for future plans.

Further Comments

Common Fringe-myrtle (*Calytrix tetragona*), New Holland Daisy (*Vittadinia triloba*), Slaty Sheoak (*Casuarina muelleriana*), Wedge-leaf Hop Bush (*Dodonea cuneata*), *Helipterum jesseni*, Erect Guinea-flower (*Hibbertia stricta*), Tiny Sunray (*Helipterum demissum*). The fascinating Wheel Fruit (*Gyrostemon australasicus*), Scrub Pines, Fringed Heath-myrtle, Golden Pennants, Weeping Pittosporum and most exquisite of all the silvery Desert Phebalium (*Phebalium glandulosum* var. *bullatum*) looking under a hand lens as if a silversmith had spent careful moments on the artefact-like scaly stems.

The birds noted were White-browed Babblers, White-winged Choughs, Galahs, Mallee Ringnecks, an inquisitive Southern Scrub-robin, and a Chestnut Quail-thrush. I also heard Magpies and a Butcher-bird.

Having made a very small impression on this track I had a feeling that I would not be satisfied until I had walked its length leisurely and carefully. If it had not been used regularly, then its flora and fauna may have been worth seeing.

This germ of an idea gave way to the writing of letters to Mr. H. E. Tarr, Chairman of Wyperfeld National Park Committee of Management, and to Rudd Campbell, Ranger of the Park. The former had walked from Yarto to Eastern Lookout in company with Ron Falla, John Landy and Ian Moroshke some years ago; scrub-hashing a "direct" line by means of compass. This activity took twelve hours from 5.30 a.m. The answer from Rudd gave advice which mainly hinged on choosing a day not too hot, and he estimated the distance as about eighteen miles. With this information I gradually evolved a scheme. December and January would be too hot, and bird activity not so good as in the spring. The September vacation would not allow the twelve hours needed to do the walk

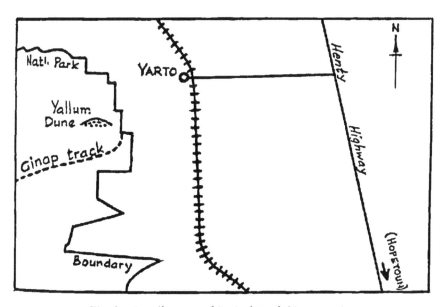

Fig. 1. Locality map of beginning of Ginap Track.

in daylight; therefore I planned to do the walk in September and spend one night out. Camera, binoculars, and compass would be necessary and so would food and water for a couple of days. As well as bedding, this made the rucksack sufficient for my back.

There the idea lay, semi-dormant until late in the second term of 1969, when Allan Thomson suggested a spell in the best Victorian National Park. I felt much the same.

We planned to meet there on Monday, 25 September, but having taken the same route as earlier, we travelled through Avoca and St. Arnaud, and overtook Allan at Warracknabeal where he had stopped for the night. After accepting his hospitality for afternoon tea we headed towards Rainbow in a flaming sunset that bathed road, sky, passengers and paddocks in a ruddy glow. As we drove up the gradual slopes the bitumen crests wore rose halos. At Rainbow, which had a 24-hour petrol service, we filled up our tanks so that we would have a plentiful supply for reaching Wonga Hut Clearing, driving along the tracks in the park, and reaching a petrol supply again.

It was dark when we reached the ornamental gate, and were agreeably surprised to find bitumen that gave us a smooth ride to our camp site. We parked in our usual spot beside the Black Box with the distorted branches, noting that the park had quite a population, and most of them caravanners. The time was 8 p.m.

Next day I was awake at dawn, Magpies, Red Wattle-birds, Galahs and Pardalotes could be heard. I walked up the Brambrook track as far as the fringe of River Red Gums around Lake Brambrook. The 1968 pest proportions of Mediterranean Turnip had fortunately not been repeated. Twiggy Guinea-flower and Fringed Heath-myrtle were plentiful.

but the Tea Tree in most cases though it had come into full bloom a short time before, had a rather dishevelled, withered appearance. Five big kangaroos made off looking quite chestnut in the morning sun. It is not surprising that some visitors report to the Ranger that there are "Big Reds" in the park. Variable Groundsel made gay splashes of gold, and looked at its best this year. The Devil's Pools were hardly that. One had a few square inches of water and the other a few square feet. There were no bee visitors this time. I went over the next hill — two more kangaroos lied. Two Pink Cockatoos landed, to breakfast in a Scrub Pine, but a Magpie chased them away and out of sight. A scuffle behind a tea tree attracted me, and I did a half circuit. The presumed bird did the same. I repeated my action. So did it.

Six times we went round the 'mulberry bush" without a sight, then he flew a dozen yards and was lost. I gave up and regained the track.

There, a dozen yards away, was my elusive bird watching me — a Chestnut Quail-thrush. And so we stood regarding each other till I stirred and the handsome bird flew away with a now familiar sound. I headed back to camp, noting visually or aurally, Red-backed Parrots, Grey Butcher-bird, Ringneck Parrots, White Cockatoos, Grey Thrushes, Kookaburras, Rufous Whistlers, Tree Martins, a Fantail Cuckoo, Ravens, Crested Pigeons, Galahs, Brown Thornbills and a family of Black-backed Wrens. The male was a superbly clad beauty, and would be better named the Blue-throated Wren, and except for his silvery-blue side-whiskers his blue was more of a lilac.

Allan and his wife arrived about midday and we walked across Mt Mattingley to join the Lake Brambrook track and reach Devil's Pools

On the way we added more birds to our list. Brown-headed Honeyeaters, Striated Pardalotes, Chestnut-tailed and Red-tailed Thornbills, Grey Currawong, White-browed Babbler, Peregrine Falcon (maybe the one that uses the disused Wedge-tailed Eagle's nest in the big gum tree at the foot of Flagstaff Hill), White-eared Honeyeater, Brown Hawk, Spiny-cheeked Honeyeater, Bronze Cuckoo, Gilbert Whistlers and Regent Parrots.

We approached the rise above Devil's Pools and saw for us, an unknown bird feeding beneath a Slender Pine. We watched it for more than five minutes as it fed at the base of the tree and flew to perch on a low branch. By the time it had flown to a distant dead tree, and was again watched, it was fairly accurately observed. That evening we searched unsuccessfully through all of the bird books including Brigadier Officer's new one on Flycatchers that had been brought to our notice that morning. I must admit that in spite of our notes we could not agree on which bird we had seen. I thought it had a brownish-grey back; but Allan was adamant that it was grey.

During the night, made bright by a full moon, I added Nightjars, Mopokes and Spur-winged Plovers to my tally.

On 26 August, the magpies had been calling since 4.30, but it was rather later than that when we headed for Eastern Lookout where we would walk and look for Lowan nests. On the way we saw a Chestnut-breasted Shelduck take off in a great hurry. Near to the Lookout there seemed to be such a hullabaloo from Pallid Cuckoos that we stopped the car to investigate. There were at least three of the birds in the area, and this displeased the Yellow plumed Honeyeaters. The latter having prodded one of them out a particular tree desisted their attacks and the cuckoo settled down to a breakfast of large, juicy, hairy caterpillars. Brown Treecreepers shrilled, Scissor Grinders grated gutturally, and Red-backed and Mulga parrots enlivened the scene. A pair of Brown Flycatchers sang melodiously. Here we could have watched the birds all day with hardly an interruption but we kept to our plan and followed that fire break track called Lowan Track. We covered about five miles without sighting a mound or a hint which eventually proved to be our own fault. A new plant sighted was *Prostanthera aspalathoides* with its attractive scarlet flowers growing on the low dark green bushes.

It was a glorious day: blue sky, little cloud and a temperature in the low eighties "I could take a lot of this" said Allan, I agreed.

Under Allan's tuition I had learned the call of a Crested Bell-bird. We both heard one call at the same moment and decided to track it. We left the track going North with the sun at 10 o'clock and taking frequent compass bearings (for being lost once was enough). If you take children to these places it would be a sound scheme to teach them this sort of practical path-finding. We walked at least a quarter of a mile apparently following that faint, elusive and possibly ventriloqual call. Then the call stopped and we turned about.

On the way back a couple of gruff grunts directed us to look and see a Mallee Fowl. It did not wait for an introduction but made off, although in no great haste. Near to the track a Yellow-tailed Pardalote called powerfully, standing on tiptoe for each "Tu Tu", and stretching so vigorously as to make a crest appear on the top of its head. We were back at the track and soon met up with the girls. Simone was investigating a bird making a throaty "Chrung Chrung". We joined the search and soon found the caller — a White-eared Honeyeater. This differed from the usual "Quilp Quilp" that we knew. During our visit this former call was heard many times in different parts of the park. On our way back to the car we found our original objective — a large domed working Lowan mound. Its diameter was 30 feet and its height 3 feet 6 inches.

The mound had been opened by the birds, digging a moat so as to leave a central smaller mound three feet in diameter. The moat was one foot deep and one foot wide. As we watched, the heat of the sun dried the sand causing it to trickle into the moat. The mound was only ten paces from the track. Someone wondered if there were eggs in it. This raised quite a disturbing thought in my mind. Our society has for years been a "Keep off the grass", "Don't touch" society where the eyes alone have been used for most observational learning. The new science course soon to be introduced into Victorian Primary Schools stresses participatory activity as a part of learning, which includes touching. This may be good instruction to make a person totally aware of his or her environment, but the result would be catastrophic if some person took this literally and tore a Lowan mound to pieces in an effort to discover whether or not it contained eggs. What a shocking thought! The new approach to science will have to show that man is not all important and that all creatures have their rights, and if one wants to discover the presence of eggs in a Lowan mound there are really much more scientific and patient ways of doing this than tearing it to pieces.

As we walked back a bird with a clockwork-like flapping flight flew across the track and perched within view. Two pairs of binoculars focused on it. This was the bird that we had seen yesterday near Devil's Pools — white eyebrow and black ear streak quite visible. It was a Black-eared Cuckoo. I had to admit that this bird was more grey than

Plate 1. Working Lowan mound near track to Eastern Lookout.

brown but later, by examining skins in the National Museum, it was found that all the skins had a brownish sheen. It is a great pity that we have no bird book which is completely reliable for our Australian Birds. One day, perhaps, one will be produced with illustrations and text to make a field observer's work much easier than it is now. That evening on Mt. Mattingley, Allan heard its call and saw the cuckoo again. Mt. Mattingley was pretty with Tea Tree, Fame Heath (*Styphelia behrii*) and Brush Heath (*Brachyloma ericoides*) in flower. The Sheoakes were tinged with a dainty brown.

That evening I packed my rocksack and checked that I had what I needed for the Ginap Track walk. Tomorrow morning should find me at the Yarto end of the track.

The Magpies called me just before dawn and I obediently rose and walked the short distance to Mt. Mattingley. As it grew light Black-backed Wrens became active and Yellow-winged Honeyeaters fed on the banksias. I heard the Black-eared Cuckoo call. A few drops of rain fell. The sky was quite light so I took the slight shelter offered by the mallees expecting it to stop. It became a sharper shower and soon was setting down. It was too wet to watch — spectacles and binoculars were blurred with rain. I went back to the van to change; a wise if belated decision, as the rain continued till past midday, soaking the area with 79 points. Quambatook had more than two inches and all of the state had its share. Several vans decided to leave, but only reached the gate, for the slight grade from the gate was a quagmire. The drivers sensibly turned around and returned to Wonga Hut to wait, but Allan who had an urgent appointment at Mildura took off without his caravan for twenty miles of sliding, skidding, swerving and probably cursing until the bitumen began.

The projected walk would now have to be done alone, and then not until the area had dried a little, and the sky had become less threatening.

By early afternoon the sun was shining, and I decided to investigate a new fire break track which had recently been graded. It led west into "no man's land", then north, and finally east to reach Pine Plains Gate. This was no car track, especially in its present condition, and gumboots were needed for walking. Birds were very active. There were White Cockatoos and Black-faced Cuckoo-shrikes, and more churring Brown Treecreepers in the Black Box section than I had previously heard. A puzzling small brown bird with a white wing margin and a rusty patch at the front of the head had me confused until later at the museum, it matched with the skin of a Red-capped Robin of indeterminate sex. Possibly a brighter female or an immature male. After a mile of open Black Box and River Red Gum where kangaroos browsed, Striated Pardalotes called, and Regent Parrots perched aloft, the track gradually climbed on to a plateau of tight mallee and banksias. The ground was alight with much Variable Groundsel. In the trees were Ringnecks, Black-backed Wrens and Pallid Cuckoos. I decided to turn back having seen a glimpse of a glorious "garden". I had commented previously that the dune vegetation west of Black Flat was becoming hazardous, being head high and trackless. Here there was a good remote track and one could leave it safely as the new growth had not yet reached four feet in most cases. I planned to see this area at dawn on the morrow.

September, 1971

259

By dint of an early rise, an early breakfast, and a drive through the misty pre-dawn, I was able to be walking through the box and gum sections in bright moonlight. The small box leaves made a fine tracery against the sky while the dark trunks looked like black stits in the faint light. Birds already calling included Pallid Cuckoos, Willy Wagtails, Magpies, Parrots, Wrens, Kookaburras and others not recognised. I slowed to a stop when a powerful whistling close by was answered by a low "bu-hoo". I waited until it was light enough to see an immaculate Golden Whistler still vibrating with song, and still being answered by his demure mate.

The sun was now up but not visible, as the moisture from the drenched ground had risen to produce a blanketing mist. The fine globules of moisture clothed the grass, leaves and banksia cones and festooned the many orb webs. It was difficult to avoid destruction of these delicate concentric patterns when leaving the track. Some were slung in the forks of branches, while others by means of horizontal lines extended flag-like from vertical sticks. Another web which resembled a space-frame consisted of numerous right-angled triangles of web tying in various planes, and dipping into a funnel that led to a hole. One half-inch hole contained a brownish wolf-like spider that came to his front door to study me with large eyes, but refused to come out.

The moon, still a glow in the mist, had its halo replaced by a semicircle of light—almost a moonbow. The area was like a newly planted, well stocked garden. Muntries spreading, Desert Banksia in flower, young mallee eucalypts with fresh tips, much Callitris regeneration, *Baeckea crassifolia*, Shrub Violet, Flexile Hakea, Guinea Flowers and Blue Boronia. A kangaroo on the track sat patiently until I had almost screwed the tele-lens on the camera, then bounded away. Wrens appeared close by and then a well camouflaged striated bird with a wren-like plume of a tail jumped up and down in a tangle of banksia stems, singing as energetically as it moved—a Striated Grass Wren. Chestnut-tailed Thornbills flew into the low mallee, while a flock of Tawny-crowned Honeyeaters "pip-pipped" and chased hither and thither, now near, now far, but never in one place long enough to have themselves photographed.

One is always loth to turn back in such an area knowing that so many treasures remain to be discovered, and that a few more steps may provide the clues leading to them. However, I turned back down the track to the car, and on to the van where Simone and Heather were working in a "dedicated" manner on matriculation studies. That afternoon I persuaded the studious ones to put aside their books and come as far as Round Lake. We walked half the nature trail then tracked across the bare dry lake where Pipits walked jerkily and a Spur-winged Plover called scratchily as it searched the lakeside vegetation. A track had been worn by many feet over the next dune and as far as the first high dune to the west. Here I noted Redbacked Parrots, White-rumped Miners, White-eared Honeyeaters, Chestnut-tailed Thornbills, Ringneck Parrots, and Red Wattle-birds. On the way back a solitary Pink Cockatoo flew overhead.

We completed the nature walk, but it was not the same as it had been. Human interference had caused a real loss of its natural integrity. In previous articles in this magazine 81 (11) 1965, and 85 (10) (11) 1968, I had commented on the value of these trails. Even then I felt that there may be some disadvantages.

This Round Lake trail in our remote park, cleverly as it was designed, has suffered from tramping feet and plucking fingers. Although only a few feet tramp in each instance, and only small pieces are plucked, the overall result has been spoilage, with the disappearances of many of the specific items marked in the booklet. Perhaps if the trail was re-routed every so often, and the markers moved to fresh vantage points, then the present trail would regenerate: but this would involve a great amount of work, too much to expect of any Ranger or Committee of Management. The present car route has not suffered to the same extent, as most observation is done visually without leaving vehicles, and my comments do not apply to it. On Mt. Mattingley a few pegs may indicate that someone is thinking of putting in a further walking trail. There are rare plants and animals in that area so easily reached on foot from Wonga Hut, and a Nature Trail here may destroy exactly what it sets out to preserve. Walking trails, while worth a try, are not proving effective and are really not essential. Armed with Ross Garnet's book, and National Park literature; and helped by the Ranger and other more knowledgeable campers, the newcomer will soon pick up a working knowledge of the flora and fauna.

The Ginap Track

At 9 a.m. on 29 August, I was at last within sight of the Yarto Silo to the North East. The map (Fig. 2) indicates the route which I followed, and differs from the one at the back of "The Vegetation of Wyperfeld National Park". The first left-hand fork is not the track that leads south to Eastern Lookout. This track leads to Lunar Clearing. Just outside the park Rufous Song-larks were singing, and as I moved west I heard Willy Wagtails, Pallid Cuckoos and Grey Thrushes. A Chestnut Quailthrush walked slowly across the track. I found the purple patch of Baeckea seen last year. Golden Pennants in heavy bud gave promise of beauty ahead. A dead Galah in a colourful splash of feathers gave no clue to its untimely end. New Holland Daisy was yet in bud. I passed dense communities of Broombush; smelled the heavy perfume of Wallowa in massed flower, with the soft tips waving in the breeze; and noted the tight green heads of Grey Mulga. Velvet-bush was similarly in bud. White-browed Babblers bounced, hid, then scattered cackling as I moved on. They were frequent on this walk. A large Tawny Frogmouth floated silently across my path. A pair of Yellow-plumed Honeyeaters was seen; the first of many on this trek. "Quitp, quilp" called a White-eared Honeyeater. A silent bird which sat immobile proved later to be a Horsfield Bronze Cuckoo, lacking the metallic sheen often illustrated.

Towards midday a large patch of flowering Hakea Wattle came into view and near it I saw my first pair of Purple-backed Wrens. They were busy feeding in bushes of Moonah. As I watched, a sudden clamour in a Scrub Pine was caused by a group of Babblers (apparently the resi-

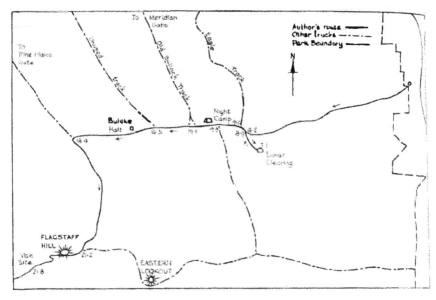

Fig. 2. Map of Ginap Track.

dents), and a pair of Yellow-plumed Honeyeaters which were investigating their ownership. A Grey Thrush also flew in. It, as most of the Grey Thrushes here, seemed larger, greyer, and appeared to have an aggressively thrust out beak as it called in a loud voice.

Ten minutes later I walked through a zone of Porcupine Grass and Manna Acacia (*A. microcarpa*) which was about to flower.

During the next half hour as I moved along, I was impressed with the large flowered *Olearia pimeleoides* with its spoon shaped leaves and the beautiful shrubs of yellow-flowered Phebalium. Here too were many Weeping Pittosporums and one Red Wattle-bird.

I reached the first turn off which led south to Lunar Clearing. At the corner, a movement accompanied by a twittering led my binoculars to a pair of tiny birds with wren-like tails and blue faces. One had a blue throat too, and a distinctive rufous crown. I had never seen them before and later, in examining the books, concluded that they were more like the Rufous-Crowned Wren than the Mallee Emu-wren. I waited until I saw the skins and found that once again the pictures were unreliable. The colour difference between crown and back is quite distinct.

The next sighting was a female or an immature Rufous Whistler and another three of the stalwart Grey Thrushes. I was following the track to the Lunar Clearing, and on the left of the track noted the scattered remains of a Lowan Mound. Many Turpentine Bushes were here. Willy Wagtails were heard distantly and then quite near. The track looped like a huge question mark and I was in the clearing.

Magpies and Willy Wagtails had taken advantage of the open habitat. The little dam was quite full, and a party of Spiny-cheeked Honeyeaters was having a frolic in the water and tussing amongst the branches of nearby trees. As I ate lunch I saw my first parrot, a Ringneck perched in a pine; and then a second, a Regent, liying over like a black and yellow dart. A few spots of rain made overlapping circles on the surface of the dam.

At 2 p.m. I left Lunar Clearing, and when back on the main track, I noted more Turpentine Bush and masses of Early Nancy. Another Horsfield Bronze Cuckoo was seen. The day was still good for walking, with the sun's rays obscured by thick grey clouds.

Soon a Scrub Robin emerged from a Porcupine Grass tunnel and advanced along a branch to within a few feet of me. The tunnel's mouth was framed with a mass of Clubmoss Daisy-bush.

Walking steadily, but investigating any sight or sound, I passed a belt of thick Scrub Pine where wrens were twittering but refused to show themselves.

As I passed the junction with Eagle Track, a White-eared Honeyeater signalled my approach. Two big kangaroos crashed off through the scrub with a thump from their tails.

Two minutes later I passed another track, this one leading south to Eastern Lookout. Black-faced Cuckoo-shrikes called from the thick trees and I saw one of what sounded like a large number. The Miners were more obvious, heralding my moves with cries of alarm and flying backwards and forwards across the track.

By mid-afternoon, I reached Balak Clearing on the right of the track, which was just an empty green space with no animal activity apparent.

As I walked on, a piping sound was heard by the track. Its origin was tracked to a small well-hidden bird. The parents, a pair of Yellow-plumed Honeyeaters, were agitated and showed this in flight and voice, but in spite of these remonstrations the young one refused to fly even when I approached to within two feet. It had probably only just left the nest and was in the middle of flying lesson number one.

A hundred yards further on I came to Ringneck Clearing which was similar to the last clearing. In the mallee by this clearing, a Lowan grunted, walked, ran, sprinted, and taking off whirred through the trees like a rocket. Mosquitoes were becoming active. Two Willy Wagtails

Plate 2.

Clearing
Wildflowers

scolded me and a couple of Kangaroos crashed away through the scrub.

By 4 p.m., the largest clearing lay before and behind me: Nightjar Clearing also spread on both sides of the track. Red Wattle-birds were in the trees on the margin of this clearing and the open area was not deserted, for a Magpie was in residence.

Just after leaving the clearing, a vociferous cacophony ahead of me directed my binoculars at a group of parrots sweeping head high towards me and energetically engaged in conversation. Whether the conversation was casual or a device for direction finding and formation preservation I do not know, but I had the distinct sensation that very shortly I would be impaled on a number of beaks and prepared to duck. At about twenty yards the formation split, and without losing a syllable, swept right and left to rejoin behind me and continue their airborne discussion as they pursued their clear track to the east.

I saw four more Yellow-plumed Honeyeaters. Two of them were youngsters but their flying had reached a higher standard than the perched one seen before. They were making use of this skill to pester their parents by continued pursuit and loud calls for food.

An elusive Crested Bellbird was heard quite clearly but faintly. There was always the temptation to try and sight this bird. This however was not the time or place.

I came to Emu Clearing; smaller than Nightjar Clearing, but larger than the other two, it was totally different from the three of them. It was flat, damp, and desolate, compared with the cheerful mallee. Where the track sloped into the clearing, and later where it rose out of it, were fringes of Weeping Pittosporum and seedlings in abundance of that species. The clearing had a few of these Pittosporums and some small Turpentine bushes but large damp patches were devoid of all vegetation. I was glad to move on again and into the Mallee, even though the track had begun to climb. Another pair of Purple-backed Wrens was noted as well as a kangaroo on the track ahead. Ten minutes later and still climbing, I entered a dense zone of Moonah with patches of Porcupine Grass. In one of these patches stood a small mallee, and glimpsing a slight movement, was lucky to aim the binoculars at the right spot and focus on a Shy Heathwren. Like most of these small wrens it did not stay in sight for any great length of time, but in my glimpse the white wing patch was quite discernible.

At 5.00 p.m. a Grey Thrush sang his evening solo, and taking his advice I unslung my pack and flexed my aching shoulders. Choosing a clean patch to the leeward side of a dense low-branched Moonah, I behaved like a Lowan and swept together all the adjacent leaf litter to create a six inch mattress. The wide sheet of polythene I carried was long enough to cover this mattress, make a small vertical windbreak, and then told back in lean-to fashion to serve as a cover for the sleeping bag. As I finished that job it was put to the test as light rain fell for half an hour. The gear remained dry. As I supped a Scrub Robin called. Two Regent Parrots perched across the track for a few moments before deciding to roost elsewhere.

By 6.30 it was nearly dark, and with daylight almost twelve hours away I wondered how much sleep I could absorb. I put my head down.

I had just shone my torch on my watch and having been under the impression I had slept for hours was quite surprised to find it was only 8 o'clock. I nibbled a couple of biscuits and put my head down again.

At 10.30 p.m. I was wide awake, so had a snack and a hot drink of coffee. Having had more than half my usual amount of sleep, I let my thoughts wander over a whole range of ideas. A persistent thought dealt with the comments of friends and acquaintances when they became aware of my intentions to "go it alone" in the bush. The frequent comment was, "What? By yourself?" and also, "I could never do that!" Apparently there was something unnatural about my behaviour. Ever since man stood up on all fours and began to develop his eyes and brain at the expense of his other senses he has not been satisfied to look out and enjoy the myriads of visual opportunities. I knew why some people wondered how I could stand the darkness and the silence. To so many, silence reverberates with the internal mutterings and imagined roars; while the darkness is lit with the bright flashes and coloured flak of fevered imagination. To me, man is an extension of the animal kingdom, one of them: Different in degree but not in kind and still a part of all natural evolution. So as a part of this natural kingdom I lay there in the darkness content to be a part of this great system and still eager to learn more of its creatures.

At 3.00 a.m., although still dark and a little chilly, I had a night's sleep behind me; so decided to move off dark or not. The sun was well below the horizon and the moon far from bright behind the masking clouds, but the track was quite visible. A Pallid Cuckoo called. Moving along with the torch flashing occasionally, and reflecting from the yellow and white of Hibbertia and Micromyrtus, I disturbed two silent winged birds that moved away with a gruff barking sound.

About an hour later I passed a track to the right. This track leads to Meridian Gate, and is part of the old hallock track from Hopetoun to Pine Plains.

As Magpies started to call, I passed another track. This one was used by Rudd Campbell, the Ranger, before the better tracks were made.

By 4.30, I reached the most undulating country that I had traversed on the Ginap Track, and even by the waning moon I could see that the mallee scrub had given way to spaced bulokes on the hillsides. The kangaroos that I disturbed thumped away. For some time my eyes had been peering left and right at the dark outlines of the vegetation and I regretted missing part of what I had set out to see. The bulokes made a good windbreak, so I brewed a cup of coffee and breakfasted.

Good light was still more than an hour away, and I felt that to walk on would be adding to what I had already missed; and to sit and wait was not to my liking either. So I made my decision, parked my pack beside the buloke, and with camera and binoculars set off to retrace my steps. The sky was much less overcast now with the moon, stars and some clouds clearly visible.

Wrens began to twitter, and by 6.15 I had reached my sleeping place. Many birds were calling now in the pale dawn; two called across the track apparently to each other. They continued to call in the dimness but when it became light enough for me to use the binoculars, they became silent and eluded me.

Shortly, I came to that section of the track where my torch had picked up the yellows and the whites. The track was liberally dotted with Hibbertia and Micromyrtus and was an ornamental delight. The edges of this attractive section were picked out with the red stems of Flexile Hakea. Not far away a White-eared Honeyeater called.

With my second, but first daylight journey over, this section of the track was rewarding and thus justified by my second sighting of the Mallee Emu-wrens. If any readers make this trip and see nothing else but these miniatures of our avifauna you will certainly count it worthwhile.

At this time, I heard the faintest sound that has ever impinged upon my eardrums. In fact I thought at first that my ears were producing it in opposition to the silence of the area. But, no! These faint squeaks came from some bird that hugged the Porcupine Grass. I did have a glimpse of brown movement but not enough for identification.

Close by, a solitary Ringneck walking amidst the Guinea flowers was having a breakfast of gold and brown as it delicately nipped petals and capsules.

The first ray of sun for the day lit two White Cockatoos. In the next half hour many birds were seen. The best sightings were Pink Cockatoos and Rufous Songlarks. I watched a group of Black-faced Cuckoo-shrikes, while some Chestnut-tailed Thornbills worked over the tree beside me. Some good specimens of *Pimelea stricta* and *Dodonaea attenuata* were seen in this section. Presently I came to the Buloke hillsides with many plants of Austral Bugle still in bud. Amidst the buloke was a small patch of typical mallee with plenty of Burr-daisies but little grass. Here a bachelor group of Red-backed Parrots moved, foraging for food with one Ringneck at a short distance.

The next hillside was clothed in a dense scrub of Mallee Bitter-bush and, hidden from sight, many wrens called. They sounded like the Black-backed Wrens but I wanted to be certain. For fifteen minutes I hunted hither and thither, hearing but not seeing them. I had the feeling that they watched me when my back was turned but quickly slipped out of sight when I faced them.

Two Pallid Cuckoos called as they flew and perched aloft in a tall dead tree.

By 9.00 a.m., the cloud had diminished greatly, and the sun was warm. I was walking along a straight track now that was flanked by River Red Gums and Black Box. This must be the extreme end of the "flood" plain of Lake Bramhrook. Every quarter of a mile or so, separate flocks of White-winged Choughs gave their alarm calls, and still whistling, mounted to safe high perches. There were young Weeping Pittosporums here, and much Creamy Stackhousia. Rufous Whistlers, Butcher Birds and a small flock of zigzagging and apparently crazed Galahs added to the scene. I looked up as a Pink Cockatoo flew over, and then ahead to where the Group Track ran into a T-junction. Here I turned left, and south. From the scattered Gum and Box country, there was a change to undulating dunes with Mallee and Tea Tree. Spiny-cheeked Honeyeaters were abundant here.

A little further south I met two more walkers, who were photographing subspecies of plants. I stayed to chat awhile as a Nankeen Kestrel circled overhead.

The dunes dipped down here to a small plain with some very large Bulokes, and then climbed to a vantage point which allowed me to see friendly Flagstaff Hill. The rain of the other day had so invigorated the

Ten Trees that they had lost their withered look and the blossom swarmed with masses of bees and hover flies. From here to the main East-West track that connects Eastern Lookout and Wonga Hut, was but a short distance, and I disturbed seven large kangaroos. I had the telephoto lens at the ready, but was not interested in the small rear shots that presented themselves. I was on the main track now and only a mile from base when a kangaroo came from behind a Red Gum and sat facing me. I was able to take this pose as well as one more in full flight as he bounded away.

At 11.35 I reached the van and was greeted by the girls.

Two more incidents of the rest of the visit to Wyperfeld are very clear in my mind. The first was Rudd Campbell's wide grin when he saw that I had returned safely.

The second occurred when a half dozen emus were wandering over Wonga Hut plain very close to the camps, and a new caravan came to a halt. Hardly had it stopped, than two small children appeared, crusts in fists and heading towards the grazing emus. Risking the danger of interfering between parent and child and of giving the latter a traumatic experience that might persist into adulthood, I gave a yell, and suggested that the emus were doing a good job of feeding themselves, and that the human approach would repel rather than attract them. I then had a chat with the parents, and it transpired that they had expected Wyperfeld to be a sort of extensive Healesville or Tidbinbilly. The conversation was agreeably fruitful.

The next morning I took a short walk to Mt. Mattingley, where the early morning sun rays lit Flagstaff Hill; and the mallee for many miles was softly wrapped in a grey swirling layer of mist. Down from the crest I went with Yellow-winged Honeyeaters flitting from banksia to banksia and a loud-voiced Scrub Robin shouting as I headed for the van, breakfast, and the move home.

Wyperfeld had let me into a few more of its secrets, but there remain ever so many more to share.

Plate 3.

Black-faced Mallee Kangaroo in flight between Eastern Lookout and Wonga Hut.

Field Naturalists Club of Victoria

Established 1880

OBJECTS: To stimulate interest in natural history and to preserve and protect Australian fauna and flora.

Patron:
His Excellency Major-General Sir ROHAN DELACOMBE, K.B.E., C.B., D.S.O.

Key Office-Bearers, 1971-1972.

President:
Mr. T. SAULT

Vice-Presidents: Mr. J. H. WILLIS; Mr. P. CURLIS

Hon. Secretary: Mr. D. LEE, 15 Springvale Road, Springvale (546 7724).

Subscription Secretary: Mr. D. E. McINNES, 129 Waverley Road, East Malvern, 3145

Hon. Editor: Mr. G. M. WARD, 54 St. James Road, Heidelberg 3084.

Hon. Librarian: Mr. P. KELLY, c/o National Herbarium, The Domain, South Yarra 3141.

Hon. Excursion Secretary: Miss M. ALLENDER, 19 Hawthorn Avenue, Caulfield 3161. (52 2749).

Magazine Sales Officer: Mr. B. FUHRER, 25 Sunhill Av., North Ringwood, 3134.

Group Secretaries:

Botany: Mr. J. A. BAINES, 45 Eastgate Street, Oakleigh 3166 (57 6206).
Microscopical: Mr. M. H. MEYER, 36 Milroy Street, East Brighton (96 3268)
Mammal Survey: Mr. D. R. PENTON, 43 Duke Street, Richmond, 3121.
Entomology and Marine Biology: Mr. J. W. H. STRONG, Flat 11, "Palm Court". 1160 Dandenong Rd., Murrumbeena 3163 (56 2271).
Geology: Mr. T. SAULT.

MEMBERSHIP

Membership of the F.N.C.V. is open to any person interested in natural history. The *Victorian Naturalist* is distributed free to all members, the club's reference and lending library is available, and other activities are indicated in reports set out in the several preceding pages of this magazine.

Rates of Subscriptions for 1971

Ordinary Members	$7.00
Country Members	$5.00
Joint Members	$2.00
Junior Members	$2.00
Junior Members receiving Vict. Nat.	$4.00
Subscribers to Vict. Nat.	$5.00
Affiliated Societies	$7.00
Life Membership (reducing after 20 years)	$140.00

The cost of individual copies of the Vict. Nat. will be 45 cents.

All subscriptions should be made payable to the Field Naturalists Club of Victoria, and posted to the Subscription Secretary.

JENKIN BUXTON & CO. PTY. LTD., PRINTERS, WEST MELBOURNE

F.N.C.V. DIARY OF COMING EVENTS
GENERAL MEETINGS

Monday, 11 October — At National Herbarium, The Domain, South Yarra, commencing at 8 p.m.

1. Minutes, Reports, Announcements.   2. Nature Notes and Exhibits.
3. Subject for evening — "Some aspects of Reproductive Physiology in Plants": Dr. D. M. Churchill, Director Elect of Royal Botanic Gardens.
4. New Members.

   Ordinary:
   Mr. J. H. Taylor, 7/116 Toorak Rd. West, South Yarra, 3141
   Miss P. Holman, 10 Andrew St., Vermont, 3133.
   Joint:
   Mr. Donald J. Grant and Mrs. Winsley S. Grant, 6/27 Hanover St., Fitzroy, 3065.

5. General Business.   6. Correspondence.

Monday, 8 November — Presentation of Medallion.

F.N.C.V. EXCURSIONS

Saturday, Sunday, 16-17 October — Weekend excursion to Bendigo, led by J. W. Kellam of the B.F.N.C. Saturday there will be a visit to the fossil locality at Spring Gully. Sunday we will join in the Bendigo F.N.C. excursion to Toolleen State Forest. Fare, including dinner, bed and breakfast at the Shamrock Hotel will be $15.00 and this should be paid to the Excursion Secretary by the October General Meeting. The coach will depart from Flinders Street, outside the Gas and Fuel Corporation at 8.45 a.m. Bring a picnic lunch for Saturday and two picnic meals for Sunday.

Tuesday, 2 November — President's Picnic to Mornington Peninsula. Fare $1.50. Bring two meals. The coach will leave Batman Avenue at 9 a.m. Please note time.

Sunday, 26 December-Monday, 3 January — Falls Creek. Accommodation has been booked at Spargo Lodge which is run by members of our club. Tariff ranges from $55.00 to $60.00 per week, full board. Fare for coach including day trips $22.00. Full fare should be paid to the Excursion Secretary when booking and accommodation paid individually at the Lodge.

Group Meetings

(8 p.m. at National Herbarium unless otherwise stated.)

Thursday, 14 October — Botany Group. Miss M. Lester will speak on "Coastal Plants".

Friday, 15 October — Montmorency and District Junior F.N.C. meets at 8 p.m. in Hall at Petrie Park.

Wednesday, 20 October — Microscopical Group.

Friday, 29 October — Junior meeting at 8 p.m. at Hawthorn Town Hall.

Monday, 1 November — Marine Biology and Entomology Group meeting at 8 p.m. in small room next to Theatrette in National Museum.

Wednesday, 3 November — Geology Group. "The Ballarat Goldfields": Mr. Charles Goodall.

Thursday, 4 November — Mammal Survey Group meets at 8 p.m. in Arthur Rylah Institute for Environmental Research, 123 Brown St., Heidelberg.

The Victorian Naturalist

Editor: G. M. Ward

Vol. 88, No. 10   6 October, 1971

CONTENTS

Articles:

Victorian Non-marine Molluscs — No. 5. By Brian J. Smith	272
Aboriginal Stone-axe Quarry at Mt. William. By D. A. Casey	273
Two New Greenhoods. By David L. Jones	277
The Present Distribution of Some Mammals in the Furneaux Group, Bass Strait. By J. S. Whinray	279
Rocks Plucked by the Sea. By Edmund D. Gill	287
Safety Beach, Dromana Bay, Port Phillip, Victoria. By A. W. Beasley	291

Features:

Readers' Nature Notes and Queries	278
Reptiles of Victoria — No. 3. By Hans Beste	296

Field Naturalists Club of Victoria:

Group Reports	298
Diary of Coming Events	270

Front Cover:

John Wallis' fine photograph of the Dingo (*Canis familiaris*), which together with the seals, is one of Australia's two non-marsupial carnivores.

Victorian Non-marine Molluscs
No. 5

by

BRIAN J. SMITH*

Two species of large operculate freshwater gastropods are included in the Victorian list by virtue of their occurrence in the River Murray. Unlike most of our non-marine gastropods (which are advanced molluscs with a lung and are therefore termed pulmonates), these two species are more primitive in ancestry and more closely related to the marine gastropods than to the other freshwater forms. They are easily recognised by the presence of an oval disc of horny material on their "tail" which forms a sort of "front door" on the shell when the animal retreats into it. This is the operculum, and the animals can therefore be termed operculates. A series of small black operculates which inhabit coastal streams and lakes will be the topic for a later article. However, two large unrelated species inhabit the whole of the Murray-Darling River Basin and may extend into northern Victoria in some north-flowing streams.

Plotiopsis balonnensis (Conrad, 1850)

This is an elongate, spiral shell about 25 mm in length which varies in colour from olive green to a dark greeny brown with dark red or brown streaks and dots. It has distinct spiral sculpture on the shell, and in most specimens a series of raised ridges or nodules occur on the shoulder of each whorl. In large specimens the apex or protoconch of the shell is often broken off. Little is known of its way of life. It belongs to the family Thiaridae.

Notopala hanleyi (Frauenfeld, 1862)

This is a large, solid, globular shell with a length of about 25 mm. It has a dark green to brownish periostracum with a microscopic sculpture of granose lirae. The animal is common in the mud of rivers and inhabits the Murray-Darling Basin. Other species in the genus live in Queensland and the Northern Territory. This animal is unusual in that it keeps its eggs in its body where they hatch, the young being born alive. This species belongs to the family Viviparidae.

*Curator of Invertebrates, National Museum of Victoria.

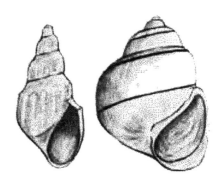

Fig. 1 *Plotiopsis balonnensis*.

Fig. 2 *Notopala hanleyi*.

(Drawn by Miss Rhyllis Plant)

The Aboriginal Axe-stone Quarry at Mount William near Lancefield

by D. A. CASEY*

We are indebted to the publishers of the *Bush Inn News*, Official Journal of the Gisborne and District Historical Society, for permission to reprint the following article which appeared in that journal for March-April, 1971.

The Aborigines have been in Australia for a very long time indeed, certainly for 20,000 years and probably for considerably longer than this. Originally their stone implements were simply pieces of stone with crudely chipped cutting or chopping edges. Later they made many finer and improved sorts of implements but all with working edges which were merely chipped to shape by trimming them with a hammer stone. Later still they adopted the technique of grinding them to shape by rubbing them against some hard abrasive stone so that it was possible to produce a much better and more efficient cutting edge. It is not known just when ground edge implements were first generally adopted in Australia, but it was in comparatively recent times. When Europeans first came to south-eastern Australia the ground-edge stone axe was the almost universal stone implement of the natives. These so-called axes were mounted on short wooden handles, which were bent around them, and to which they were fixed with resin and hair string. They were thus hatchets rather than axes and they were known to the early settlers by the American Indian name, tomahawk.

They were made from a wide range of different sorts of stone, but in Victoria at least most of them were made from diabase, a hard tough dark green stone which was very suitable for the purpose. Diabase is to be found in various places in central Victoria but the principal source of it which was exploited by the Aborigines was at Mount William. Every Aboriginal man would have, or would wish to have, a stone axe, and so the demand for this stone was very great. The outcrops of it at Mount William were extensively worked, and a considerable barter trade in this material was developed.

The stone was extracted and broken down and trimmed into pieces of convenient size for making into axes. The implements were not finished at the site, but the roughed out axe blanks were taken away for final trimming and grinding elsewhere. They were bartered with members of neighbouring, and distant, tribes who brought such things as — reed spears, animal skins, wooden implements, etc., to exchange for them. Actually no finished axes or stone implements of any sort have so far been found at the quarry site.

It is convenient to refer to the site as a quarry, but it would be perhaps more correct to describe it as a series of quarried outcrops. The outcrops occur over an area of 100 acres or more, and virtually the whole of this area is covered with chips and fragments of broken stone resulting from the quarrying operations. Much of this is now masked by soil and the growth of grass, but many heaps of debris indicate particular working

*From Tomahawks, National Museum of Victoria.

places adjacent to the outcrops. Amongst this debris many broken or imperfect axe blanks may be recognised.

Diabase ground edge axes have been found over most of central and western Victoria, and also well into the Riverina of New South Wales, and the south-eastern part of South Australia. The stone for these axes could only have come from central Victoria and no doubt much of it came from Mount William.

It is not clear just how the stone was extracted or broken out of the outcrops, although it has been stated that it was "dug out of the quarries with a pole of hard wood". It is notable that no hammer-stones have been found at the site such as might have been used for roughing out the stone blanks. It has been stated that the axes were finally trimmed to shape, prior to the grinding of the edge, by striking off flakes "with an old tomyhawk". This may well have been so, as battering is often observable on the butt end of axes as if they had been used for hammering. It is hardly likely however that "an old tomyhawk" would be adequate for the comparatively hard work involved in making the rough blanks. Possibly the stone was broken up and roughly shaped by striking it against any other large block or anvil stone.

There is no record of the quarrying operations having been seen by any European observer. The demand for the stone rapidly fell off with the decline of the Aboriginal population and the availability of iron, and the quarrying had ceased some time before 1854 when the site was visited by William Blandowski, who was the first to describe it. He wrote of it, amongst other things, that it "had an appearance similar to that of a

Plate 1. One of the working places amongst the diabase outcrops at Mount William

deserted gold-field". The existence of the quarry was however known before Blandowski visited it and he referred to it as the "celebrated spot which supplied the natives with stone for their tomyhawks". Its existence was known also to William Buckley, the escaped convict who spent 30 years amongst the Aborigines, although he never visited it.

Mount William was in the territory of the Wurundjeri tribe, sometimes called the Yarra tribe, which occupied the valleys of the Yarra and its tributaries, everywhere right up to the watershed. The quarry was in the country of a local group of the tribe whose area extended from the site of Melbourne, along the eastern side of the Maribyrnong, to the eastern end of the Macedon range. About the time of the first settlement of the Port Phillip district, the head-man of this group was Billi-billeri. He lived at the quarry site and conducted the quarrying operations and supervised the barter trade.

The quarry was tribal property, but the actual custody of it rested with a small group of related people who inherited their rights to it. Just how these rights came about, or what they were based upon, is however not clear. The group included at least one person who was outside the Wurundjeri tribe altogether. This was the husband of Billi-billeri's sister, who was one of the head-men of the neighbouring Kurnung-willam tribe and lived at the present site of Bacchus Marsh. His name was Nurrum-nurrum-biin, which is said to mean—Moss growing on decaying wood. At times when Billi-billeri was absent, his place was taken by Nurrum-nurrum-biin's son who on such occasions came to take charge of the quarry.

Plate 2. One of the heaps of debris amongst which many imperfect or broken axe blanks may be found.

Billi-billeri was a person of considerable character and personality. He conducted the business of the quarry strictly, and took good care to ensure the security of the enterprise. When a neighbouring, or a distant, tribe wished to obtain stone, emissaries were deputed to visit the quarry. If they did not loiter on the way and caused no trouble, they were freely allowed to travel unmolested, even through the territory of other tribes, and they were peacefully received. When they arrived at the quarry they camped close by, and indicated that they wanted to get stone. On one occasion Billi-billeri was heard to say, according to a native informant, "I am glad to see you and will give you what you want to satisfy you, but you must behave quietly and not hurt me or each other".

Sometimes stone was stolen from the quarry. In one case, probably in the late forties, stone had been taken by a man of the Wudthaurung tribe, and the story has been recorded in what are stated to be almost the exact words of an Aboriginal who had been present at the time. This simple naive description is nevertheless vividly descriptive and certainly has the ring of truth about it.

> "Billi-billeri sent a message to the Wudthaurung, and in consequence they came as far as the Werribee River, their boundary, where Billi-billeri and his people met them. At the meeting the Wurundjeri sat in one place and the Wudthaurung in another, but within speaking distance. The old men of each tribe sat together, with the younger men behind them. Billi-billeri had behind him Bungerim, to whom he "gave his word". The latter then standing up said, "Did some of you send this young man to take tomahawk stone?" The head-man of the Wudthaurung replied, "No, we sent no one". Then Billi-billeri said to Bunderim, "Say to the old men that they must tell that young man not to do so any more. When the people speak of wanting stone, the old men must send us notice". Bungerim repeated this in a loud tone, and the old men of the Wudthaurung replied, "That is all right, we will do so". Then they spoke strongly to the young man who had stolen the stone, and both parties were again friendly with each other".

Billi-billeri gained a reputation of being a good friend of the white men, and as a head-man of some standing he seems to have used his influence to prevent trouble between the Aborigines and the settlers. He was induced to help in the recruiting for the Native Police when that body was formed, under Captain Dana, in 1842, and he himself was the first to enroll in it. He was not however a success as a policeman. He would not ride a horse, and he refused to serve outside his own country. Also he soon tired of long hours of foot drill. Eventually it was arranged that he could retain his uniform and his musket, although doubtless without powder or shot, and to come on parade only when he pleased. He died in 1846.

REFERENCES

Thomas, W., 1854. *Brief Account of the Aborigines of Australia Felix.* Letters from Victorian Pioneers. 1898.

Blandowski, W., 1855. *Personal Observations Made in an Excursion to the Central Parts of Victoria.* Trans. Phil. Soc. of Vic. Vol. 1.

Brough Smyth, R., 1876. *The Aborigines of Victoria.*

Howitt, A. W., 1904. *The Native Tribes of South-East Australia.*

Guthridge, J. T., 1907-8. *The Stone Age and the Aborigines of Lancefield District.* Lancefield Mercury.

Mitchell, S. R., 1961. *Victorian Aboriginal Axe Stone.* Vict. Nat. 78 (3)

McCarthy, F. D., 1939. *Trade in Aboriginal Australia.* Oceania, Vol. IX, No. 4, and X, Nos. 1 and 2.

Stewart, A. J., 1966. *The Petrography of the Cohow Granite.* Proc. Roy. Soc. of Vic., Vol. 79, Pt. 2.

Two New Greenhoods for Victoria

by David L. Jones

Over the last three years, four species of orchid have been added to the Victorian flora. Two species of *Pterostylis* are the most recent of these being found during the current season (Autumn 1971) in the highlands around Wulgulmerang, in north-eastern Victoria. These are *P. laxa* J. A. P. Blackmore and *P. coccinea* R. D. FitzG.

Pterostylis laxa was the first species to be found. Its discovery followed a chance identification, by the author, of a solitary specimen amongst a box of *Pterostylis decurva* sent by Keith Rogers of Wulgulmerang. Following my request for further specimens, Keith contacted Cliff Beauglehole who was working in the area and the pair of them located it over a fair range of country. It is probable that in the past the species has been mistaken for a form of *Pterostylis decurva* although the two are not easily confused once the characteristics are known.

Pterostylis laxa is related to the Autumn flowering pair of *Pterostylis revoluta* and *P. reflexa*. All of these have acuminate labella and differ from *P. decurva* which has an obtuse labellum and conspicuously gibbous sinus. *P. laxa* differs from *P. reflexa* by having a shortly acuminate labellum and fleshy, lax sepals and from *P. revoluta* which has a much larger flower and a long narrowly acuminate labellum. Like these other two it grows in colonies and is often abundant.

Pterostylis coccinea is the most recent orchid to be added to the Victorian list. It was first found in 1970 once again by Keith Rogers of Wulgulmerang. At the time it was taken for a form of *P. revoluta*; however its distinctiveness was recognized in the bountiful 1971 season when he found it in great abundance on a couple of small peaks to the north of Wulgulmerang.

This species is certainly one of the most beautiful greenhoods in the genus. The flower is exceedingly large and gracefully curved. In colour it varies from a deep reddish green to a bluish green, all forms growing intermingled together. Botanically it can be recognized by the very large flower, exceedingly long labellum and by the markedly gibbous, scabrous sinus.

Collectors interested in orchids are especially asked to keep a lookout for both these species as there is a good chance their Victorian range will be extended considerably.

Flowers and Plants of Victoria in Colour

Copies of this excellent book are still available, and of course would make a wonderful gift. They are obtainable from the F.N.C.V. Treasurer, Mr. D. McInnes.

Readers' Nature Notes and Queries

Spider Earthworks

These interesting observations come from Mr. A. G. Fellows, our keen writer from Alabama Hill, at Charters Towers in Queensland.

On a bare spot in our lawn, I noticed a small funnel-shaped mound of earth composed of many rounded pellets of earth, each about 1" in size, and forming a circular wall which at ground level was 1" in diameter, but at its highest point was 1½" in diameter. The pellets were in a single layer, but seemed well stuck together, frail though the whole structure appeared. Late in the afternoon I visited the small cone again to see a tiny spider about a quarter inch across its leg-spread, rapidly running around the base of the cone seemingly in many directions.

After several futile attempts I at last made the tiny spider run on to a grass-stem which, with a quick jerk, dumped the spider into its funnel-shaped home. It immediately ran half way up the wall-side, down again, up and down again and then disappeared into the quarter inch opening at the base. The sun being now much lower, showed the finest of web material across the cone about half way up the wall. Also the sun now shone on a network of webbing around the cone outside. No wonder the spider had been busy when I first noticed its apparent haste. Two days later, the first cone has been partly broken—it would seem by a sharp shower that fell during the night.

A second tunnel had been erected one and a half inches away from the first, and the orifice to the first one had been carefully sealed with earth. Incidentally the bulk of each pellet used in the construction of the funnel would be a little larger than the spider's over-all size.

What an enormous amount of earth moving the small creature had accomplished and in such a short time. Before placing a temporary roof over the structure in case of another shower, I took two stereoscopic colour photos from one foot distance from the funnel to obtain detail as well as third dimension. A search over the lawn did not show any more such structures. The tiny spider was practically black in colour, with only one whitish spot near the base of the abdomen. The height of the funnel that replaced the first one was an inch and a half, its top diameter about the same, its base no more than half an inch.

The following day, early in the morning, I removed the "roof" from the two tunnels but they appeared the same.

However, in late afternoon, a third tunnel two feet from the first two, had appeared and was obviously very new. About the same size as the first two, it had been constructed in the few hours while I was absent doing other work. This was tantalising to say the least; to miss such a spectacle. Another "roof" graced this latter funnel also until the following day. I returned to the vicinity as often as possible, for surely more was to be learnt about these unusual structures. After twenty-four hours some very fine strands of web criss-crossed the third funnel, but no reappearance of the spider had taken place. About a foot from the third funnel a tiny trapdoor spider had opened its door for business, and that incidentally also "grew" in the night. It is hoped that as I watch for any change in the structures mentioned, I may learn more of these spiders' habits.

Platypus plural?

"Platypuses" is the most usually used plural, and is correct. It is acceptable too, to use the singular form for the plural also, as with "sheep" or "fish". However, if you must sound learned by using a classical plural, "platypi" is definitely wrong — for being a Greek word, the plural form is "platypodes" pronounced in four syllables by voicing the final "e".

(F.d.)

The Present Distribution of some Mammals in the Furneaux Group, Bass Strait, Tasmania

by J. S. WHINRAY*

INTRODUCTION

The main purpose of this note is to set out the present distribution of some native and introduced mammals in the Furneaux Group. It is compiled from my observations on various islands during the last six years. As well, information supplied by local residents, who include lessees and former lessees of islands, is used. Some museum specimens are mentioned. They are the only records of species for certain islands. When early literature records are available, the past and present distribution of species are compared.

The Furneaux Group occupies the south eastern part of Bass Strait and is made up of about fifty islands and islets (see Fig. 1). The islands mentioned in this note range from seventy five acres in area (Cat Island) to about half a million acres (Flinders Island).

MARSUPIALS

Red-necked Wallaby (*Wallabia rufogrisea*)

In 1828 Scott recorded this species for Flinders, Vansittart, Badger and Clarke's Islands (Scott, 1828: 1-2). The first record I could obtain for Cape Barren Island is from Mr. E. L. Maynard who began to snare on the island in about 1905. He was catching Red-necked Wallabies then (E. L. Maynard, pers. comm.).

There are Red-necked Wallaby populations on Flinders, Cape Barren, Babel and Badger Islands now. I have seen Wallabies on these four islands. The Babel Island population was introduced from Flinders Island in about 1965 (H. J. Hammond, pers. comm.). Only Red-bellied Pademelons were present on Badger Island in 1872 (Brownrigg, 1872: 22). Local informants insist that the present Red-necked Wallaby population was introduced to Badger Island by a former lessee (R. G. Morton Jnr., T. C. Morton, pers. comm.). There are no Wallabies on Vansittart Island (George Ross, pers. comm.) and I saw none on Clarke's Island during my visit.

Red-bellied Pademelons (*Thylogale billardieri*)

Red-bellied Pademelons were recorded in 1828 for East Sister, West Sister, Flinders, Babel, Prime Seal, East Kangaroo, Badger, Long, Great Dog, Vansittart, Preservation and Clarke's Islands (Scott, 1828: 1-2). Robinson recorded them for Cape Barren Island in 1831 (Plomley, 1966: 324). Red-bellied Pademelons are now found only on East Sister, West Sister, Flinders, Prime Seal, Cape Barren and Clarke's Islands. I have seen them on these six islands during the last five years. There were no Pademelons on East Kangaroo, Mount Chappell, Badger, Long, Great Dog, Babel or Preservation Islands when I visited them. Nor are there any on Vansittart Island (George Ross, pers. comm.).

Potoroo (*Potorous tridactylus*)

Clarke's Island — Five specimens of Clarke's Island Potoroos were purchased in 1923 by the National Museum of Victoria (J. M. Dixon, pers. comm.). None was caught by

*Flinders Island, Tasmania.

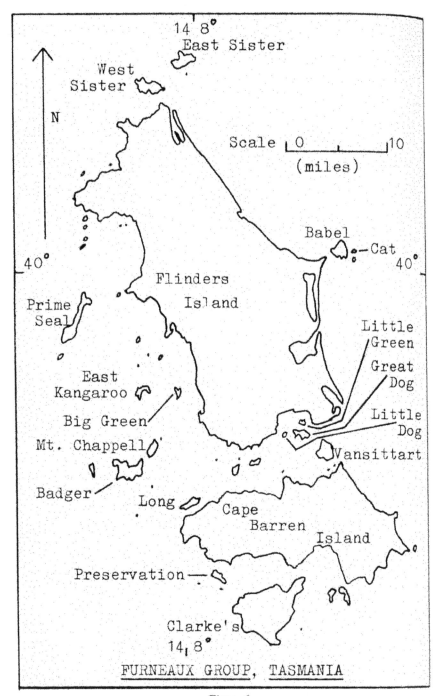

Figure 1

the caretaker during my visit in December 1966. He was setting gin traps for Rabbits and Pademelons.

Cape Barren Island — Kangaroo Rats were caught occasionally by snarers before the Second World War (R. E. Thomas, pers. comm.). These were probably Potoroos.

Flinders Island — I sent a Potoroo, killed near Flinders Island Aerodrome in July 1969, to the National Museum of Victoria. It is registration number C 8859 (J. M. Dixon, pers. comm.).

Short-nosed Bandicoot (*Isoodon obesulus*)

Flinders Island — The Bandicoot recorded by Backhouse (1843: 85) was probably this species; because the only Bandicoot bones in Mr. D. G. F. Smith's collections of bones from Flinders Island coastal blow-outs were from Short-nosed Bandicoots (D. G. F. Smith, pers. comm.). Hunters who worked the eastern coastal plain of Flinders Island between 1920 and 1939 did not catch Bandicoots (E. L. Maynard, G. A. Dargaville, pers. comm.).

West Sister Island — Two specimens from this island were lodged at the Australian Museum, Sydney, in 1928 (B. J. Marlow, pers. comm.) The National Museum of Victoria has two specimens which I collected on West Sister in 1966 and 1968. This population could be a relict one (Whinray, MS).

Southern Marsupial Mouse
(*Antechinus minimus*)

Prime Seal (Hummock) Island — The British Museum (Natural History), London, holds a specimen from this island (Thomas, 1888: 238).

Clarke's Island — A specimen from Clarke's Island was lodged at the Australian Museum, Sydney, in 1928 (B. J. Marlow, pers. comm.).

Flinders Island — Green (1969: 22) mentions two specimens from this island. They are held at the Queen Victoria Museum, Launceston, and were collected in 1961 and 1967.

White-footed Sminthopsis
(*Sminthopsis leucopus*)

Clarke's Island — The Australian Museum, Sydney, has one specimen from this island. It was received in 1928 (B. J. Marlow, pers. comm.).

West Sister Island — One specimen from West Sister was lodged at the South Australian Museum in late 1929 or early 1930 (P. F. Aitken, pers. comm.). The National Museum of Victoria has a specimen which was collected on the island in 1968.

East Sister Island — White-footed Sminthopsis were introduced to this island by a former lessee and were still present in 1966 (A. E. Blyth, pers. comm.).

Pygmy Possum (*Cercartetus nanus*)

Flinders Island — The Queen Victoria Museum, Launceston, has a specimen obtained from this island in 1965 (Green, 1969: 22). Another was sent to this museum from Flinders Island in August 1971 (R. H. Green, pers. comm.).

Cape Barren Island — "Possum Mice" were caught on the island in the 1950s by people cutting Grass Tree gum (S. F. Mansell, pers. comm.) They probably were *Cercartetus nanus*.

Ringtail Possum (*Pseudocheirus convolutor*)

Flinders Island — Gabriel (1894: 179) recorded this species for the Furneaux Group. This record was probably for Flinders Island. Ringtail Possums were common in the Strzelecki Scenic Reserve during my visits to it in 1965.

Cape Barren Island — Local residents report that the Ringtail is the only large possum on the island (John Mansell, C. B. Mansell, pers. comm.). It is the only species I have seen on Cape Barren Island.

Brushtail Possum (*Trichosurus vulpecula*)

Flinders Island — The Australian Museum, Sydney, received a specimen from this island in 1928 (B. J. Marlow, pers. comm.). My sightings of Brushtails during recent years suggest that the species is widespread on the island.

Prime Seal Island — Mr. Frank Jackson introduced Brushtails to this island in the 1920s (Frank Jackson, pers. comm.). They were present still in 1966 (J. W. Wheatley, pers. comm.).

East Sister Island — Brushtails were introduced to this land in the 1920s and were still present in 1966 (A. E. Blyth, pers. comm.).

Wombat (*Vombatus ursinus*)

Flinders Island — Wombats were recorded for this island in 1801 (Lee, 1915: 91). My sightings during the last few years suggest that Wombats are widespread on Flinders Island.

Cape Barren and Clarke's Islands — Flinders recorded Wombats for these two islands in 1798 (Rawson, 1946: 82). There were no Wombats on Clarke's Island by 1910 (W. A. Riddle, pers. comm.) and none on Cape Barren Island by 1908 (Kershaw, 1908).

MONOTREMES

Tasmanian Spiny Anteater (*Tachyglossus setosus*)

Flinders Island — Murray obtained a Spiny Anteater from this island in 1801 (Lee, 1915: 91). I have seen a number of Anteaters in the central and northern parts of this island in recent years.

Cape Barren Island — Flinders found Spiny Anteaters on this island in 1728 (Flinders, 1814: CXXXV). I have not seen any on the island but local residents report that they are present (John Mansell, C. B. Mansell, pers. comm.).

Vansittart Island — Spiny Anteaters were introduced to this island many years ago (John Mansell, pers. comm.). Constable D. Farrel saw one on the north point in late 1967 (D. Farrel, pers. comm.).

PLACENTALS

House Mouse (*Mus musculus*)

I have seen, or trapped, this species on Babel, Flinders, East Kangaroo, Great Dog and Cape Barren Islands. It is also on Little Dog Island (D. N. Brown, pers. comm.), Badger Island (A. C. Stackhouse, pers. comm.) and East Sister Island (A. E. Blyth, pers. comm.). There were House Mice on Prime Seal Island in the 1920s and 1930s (Frank Jackson, pers. comm.) and the species should, I think, be present there still.

The House Mouse appears to be one of the most successful introductions to the Furneaux Group and probably is more widespread than the records above indicate.

TABLE I

Table showing the distribution of marsupials and one monotreme in the Furneaux Group.

	E.	W.	F.	B.	P.	Bd.	V.	CB.	Ck.
Wallabia rufogrisea			x	i		i		x	
Thylogale billardieri	x	x	x	x				x	x
Potorous tridactylus			x					?	?
Isoodon obesulus		x							
Antechinus minimus			x		?				?
Sminthopsis leucopus	i	x							?
Cercartetus nanus			x					?	
Pseudocheirus convolutor			x					x	
Trichosurus vulpecula	i		x		i				
Vombatus ursinus			x						
Tachyglossus setosus			x				i		

Key to column headings:
(E. — East Sister; W. — West Sister; F. — Flinders; B. — Babel; P. — Prime Seal; Bd. — Badger; V. — Vansittart; CB. — Cape Barren; Ck. — Clarke's.
"x" indicates undoubted present occurrence of an apparently original population, "i" indicates a persistent introduced species; and "?" indicates uncertainty of either the present occurrence of a species or of the identity of the species present.

Ship's Rat (*Rattus rattus*)

Ship's Rats appear to be widespread on Flinders Island. As well, I have seen them on Big Green Island and trapped them on Cape Barren Island. On the latter island I even caught them in a fern gully high on the western slopes of Mount Munro in April 1967. One Ship's Rat was found dead in a birding shed on Great Dog Island in March 1967. It had apparently been killed by poison laid by the shed owner for House Mice. The rat was given to me on the island. When I mentioned it to the oldest birder on the island he told me that it was the first he had heard of on Great Dog in his forty-five years of mutton birding there (W. A. Riddle, pers. comm.). I caught no Ship's Rats during the month I spent on West Sister and Babel Islands.

Eastern Swamp Rat (*Rattus lutreolus*)

I trapped Eastern Swamp Rats at a swamp behind the western end of Killiecrankie Bay, Flinders Island, in late 1966. There were a number of burrows and an extensive system of tracks around the margin of the swamp. The National Museum of Victoria holds a specimen from Smith's Gully, Flinders Island. It is number C 9465 (J. M. Dixon, pers. comm.). On Cape Barren Island I trapped one Eastern Swamp Rat in a gully high on the western slopes of Mount Munro in April 1967. It was caught in an area of Common Bracken at the edge of a fern gully remnant. The specimen is held at Melbourne and is number C 9635 (J. M. Dixon, pers. comm.).

Eastern Water Rat (*Hydromys chrysogaster*)

Flinders Island — The National Museum of Victoria has one specimen obtained from this island in 1914. It is number C 4847. I saw this species twice on Flinders Island in 1965. The first sighting was by Whitemark Jetty and the second at Patriarchs' Inlet.

West Sister Island — The South Australian Museum has a specimen of Eastern Water Rat from this island. It was lodged at the Museum in late 1929 or early 1930 and is number M 2892 (P. F. Aitken, pers. comm.). The most recent record of this species for West Sister is for about 1960 when a fishing boat crewman caught two in gin traps set for Red-bellied Pademelons (Ivan Chamberlain, pers. comm.). Water Rats were seen recently on Cat Island (H. J. Hammond, pers. comm.) and were seen on Prime Seal Island during the 1920s and 1930s (Frank Jackson, pers. comm.).

Rabbit (*Oryctolagus cuniculus*)

Rabbits were introduced to Clarke's Island in about 1923 (W. A. Riddle, pers. comm.). They were common when I visited the island in December 1966.

Hare (*Lepus europaeus*)

I saw Hares on West Sister Island during a number of visits from 1966 to 1969.

Feral Cat (*Felis catus*)

Feral Cats appear to be one of the most successful introductions to the Furneaux Group. I have seen them on East Sister, Babel, Flinders, Little Green, Great Dog, Cape Barren and Clarke's Islands. They are reported from Little Dog Island (D. N. Brown, pers. comm.), Mount Chappell Island (R. G. Morton Jnr., pers. comm.) and they were very common on Prime Seal Island in the 1920s and 1930s (Frank Jackson, pers. comm.).

CONCLUSIONS

Details are given of the present range in the Furneaux Group of ten marsupials, one monotreme and seven placentals (five of which are non Australian species.) The presence of a monotreme on one island, and of marsupials on some islands, is shown to be the result of deliberate introduction.

Flinders Island, the largest island in the Group, is shown to have the richest marsupial fauna comprising eight species. East Kangaroo Island (341 acres) is the smallest island for which a marsupial, viz. Red-bellied Pademelon, has been recorded since the discovery of the Group. Red-bellied Pademelons have become extinct on seven of the thirteen islands for which they were recorded in the early nineteenth century. Bass Strait Wombats have, since the discovery of the Group, become extinct on the two smaller islands, viz. Clarke's (about 28,000 acres) and Cape Barren (about 110,000 acres), of their former range.

The widespread distribution of the Feral Cat and House Mouse suggests that these species have occupied previously unfilled niches in the islands' ecosystems. The capture of Ship's Rats on Mount Munro, Cape Barren Island, suggests that this species has also occupied an unfilled niche. These three non Australian species are probably more widespread than the records given above would indicate. As well, I think that they are likely to spread as opportunity offers, to any suitable islands which they have not yet colonized.

I consider that the range of the Whitefooted Sminthopsis, Southern Marsupial Mouse, Eastern Swamp Rat, Eastern Water Rat, Ship's Rat, House Mouse and Feral Cat will — with further work — be shown to be greater than the above records would indicate.

TABLE II
Table showing the distribution of some placentals in the Furneaux Group

	E.	W.	F.	Bb.	C.	P.	EK	BG	Ch	Bd.	LG	LD	GD	CB	Ck
*Mus musculus	x		x	x	?	x		x			x	x	x		
*Rattus rattus			x				x						x		
Rattus lutreolus			x										x		
Hydromys chrysogaster	?		x		x	?									
Oryctolagus cuniculus														x	
*Lepus europaeus		x													
Felis catus	x		x	x		?		x			x	x	x	x	x

Key to Column headings:
E. — East Sister; W. — West Sister; F. — Flinders; Bb — Babel; C. — Cat; P. — Prime Seal; EK. — Ied Kangaroo; BG — Big Green; Ch — Mount Chappell; Bd. — Badger; LG — Little Green; LD — Little Dog; GD. — Great Dog; CB. — Cape Barren; Ck.— Clarke's.

*indicates a non Australian species; "x" indicates undoubted present occurrence; and "?" indicates uncertainty about the present occurrence of a species.

REFERENCES

Aitken, P. F. Personal communication. C/- Curator of Mammals, The South Australian Museum, North Terrace, Adelaide.

Backhouse, J., 1843, *A narrative of a visit to the Australian colonies*. London.

Blyth, A. E. Personal communication. C/- Emita, Flinders Island. Tasmania.

Brown, D. N. Personal communication. C/- Cape Barren Island, Bass Straits, Tasmania.

Brownrigg, Canon, 1872. *The cruise of the Freak*. Launceston.

Chamberlain, Ivan. Personal communication. C/- White Mark, Flinders Island. Tasmania.

Dargaville, G. A. Personal communication. C/- White Mark, Flinders Island.

Dixon, J. M. Personal communication. C/- Curator of Vertebrates, National Museum of Victoria, Russell Street, Melbourne.

Farrel, D. Personal communication. C/- Lady Barron, Flinders Island.

Flinders, M., 1814. *A voyage to Terra Australis*. London.

Green, R. H., 1969. The birds of Flinders Island. *Records of the Queen Victoria Museum* No. 34. Also personal communication. C/- Queen Victoria Museum, Wellington Street, Launceston, Tasmania.

Hammond, H. J. Personal communication. C/- Lady Barron, Flinders Island.

Jackson, Frank. Personal communication. C/- Emita, Flinders Island.

Kershaw, J. A., 1908. Unpublished diary of the 1908 Australian Ornithologists' Union Bass Straits Expedition. In possession of R. H. Green, Queen Victoria Museum.

Lee, I., 1915. *The logbooks of the Lady Nelson*. London.

Mansell, C. B. Personal communication. C/- Cape Barren Island.

Mansell, John. Personal communication. C/- Cape Barren Island.

Mansell, S. E. Personal communication. C/- Lady Barron, Flinders Island.

Marlow, B. J. Personal communication. C/- Curator of Mammals, Australian Museum, College Street, Sydney.

Maynard, E. L. Personal communication. C/- Memana, Flinders Island.

Morton, R. G., Jnr. Personal communication. C/- 18 Helen Street, Launceston, Tasmania.

Morton, T. C. Personal communication. C/- Emita, Flinders Island.

Plomley, N. J. B., 1966. *Friendly Mission*, Hobart.

Rawson, G., 1946. *Matthew Flinders—Narrative of his voyage in the schooner Francis*, Golden Cockerel Press.

Riddle, W. A. Personal communication. C/- Lady Barron, Flinders Island.

Ross, George. Personal communication. C/- Lady Barron, Flinders Island.

Scott, Thomas. 1828. *Account of the . . . Furneaux Islands*. Manuscript held at the Mitchell Library, Sydney.

Smith, D. G. F. Personal communication. Mr. Smith's collections of bones from Flinders Island are held at the Western Australian Museum, Perth. C/- White Mark, Flinders Island.

Stackhouse, A. C. Personal communication. C/- Killiecrankie, Flinders Island.

Thomas, O., 1888. *Catalogue of Marsupialia and Monotremata in the collections of the British Museum (Natural History)*. London.

Thomas, R. E. Personal communication. The late Mr. Thomas spent much of his life on Cape Barren Island.

Wheatley, J. W. Personal communication. C/- Robertdale, Flinders Island.

Whinray, J. S. Manuscript. *Notes on West Sister Island, Furneaux Group, Tasmania*. (To be published soon in the *Victorian Naturalist*.)

ENTOMOLOGICAL EQUIPMENT
Butterfly nets, pins, store-boxes, etc.
We are direct importers and manufacturers,
and specialise in Mail Orders
(write for free price list)

**AUSTRALIAN
ENTOMOLOGICAL SUPPLIES**
14 Chisholm St., Greenwich
Sydney 2065   Phone: 43 3972

**BIOLOGICAL, GEOLOGICAL
AND ASTRONOMICAL CHARTS**

✦ ✦ ✦ ✦

STANDARD MICROSCOPE
7X, 10X, 15X eyepieces
8X, 20X OBJECTIVES $54.00
CONDENSER $22.00 extra

✦ ✦ ✦ ✦

**40X WATER IMMERSION
OBJECTIVES $18.00**

✦ ✦ ✦ ✦

A Wide Range of Plastic
Laboratory Apparatus
available at

**GENERY'S SCIENTIFIC
EQUIPMENT SUPPLY**
183 Little Collins Street
Melbourne: 63 2160

Rocks Plucked by the Sea

by Edmund D. Gill*

The surf zone is one of Nature's power tools. Waves oscillate then break, and surf rushes against the shore. Many of the resultant processes have been described, along with the effects of weathering, yet it is remarkable that although rocky shores have been studied for over a century, their processes of formation are still little understood.

One accessory process, that is sometimes mentioned but given little attention, is that of *plucking* by the sea. Many writers have described water pressure, the effects of compressed air (as seen dramatically in a blowhole), and the erosive power of sand and rocks swept by the water across the surface of a shore platform. Writers have likewise recorded how the sea quarries by working tirelessly along joint planes, bedding planes, zones of decomposed rock, fossil soils and such, so as to gradually remove massive boulders that tumble into the sea, or may be thrown landward during heavy storms.

Plucking is a different process, whereby pieces of rock are removed by tugging, sucking, and such like. It is closely allied to quarrying, but the difference will be apparent from the two examples described below that were observed on the coast of Victoria, Australia.

From Submarine Rock Floor

Plate 1 figure 1 shows the holdfast and adjacent part of the kelp *Durvillea* attached to a piece of Lower Cretaceous arkose. It was found on a rocky shore platform southwest of Von Mueller Creek on the Otway coast of Victoria. The kelp pulled a piece of rock from an outcrop at or below low water level, because that is where *Durvillea* grows. Such a piece of rock must have been more or less angular. The area of attachment is flat. The rock is now worn to the exact outline of the holdfast, and completely rounded. We can infer that:

1. The rock was rounded between the time it was plucked from the outcrop and when it was found on the shore on 12 June, 1971.
2. As the *Durvillea* was eroded but still alive, the time interval could not have been great, and therefore erosion of such rock in that environment must be rapid.

The brown colouration of the rock near the kelp is apparently due to organic stain from the alga. On the flat surface to which the holdfast was attached there is a series of faint roughly concentric fine ridges. Such I have not seen before, and presume that they are connected with the growth of the alga. In facetious mood I asked a fellow palaeontologist what fossil it was, but of course he had never seen one like that before!

In August 1971 a series of kelp fronds was examined on a beach a short distance south of St. George's River on the Otway coast near Lorne. It was noted that about half of the holdfasts had some rock attached. A

*Deputy Director, National Museum of Victoria

Plate 1

holdfast about 4 in. by 4½ in. had a number of small pieces adhering, including one of a few ounces weight. These were probably fragments of the rock originally attached, because some were broken surfaces, while the edge of the largest piece was eroded precisely to the curvature of the holdfast edge, and was rounded similarly to the piece illustrated. A small holdfast 2 in. by 2½ in. had a thin plate of rock attached. The side adhering to the holdfast was white calcium carbonate, seams of which are common in the rocks of the area. Over about half of the area was a layer of arkose, which was part of the rock intruded by the calcite band. The majority of the white carbonate surface was covered with a recent bryozoan, showing that,

1. The bryozoan preferred the carbonate as an attachment
2. Since the rock was plucked from the outcrop, and before it was cast ashore, the bryozoan had time to grow.

Some holdfasts at first glance appeared to have no rock attached, but closer examination revealed large numbers of small fragments. If the number of fronds pulling rock from the surface of the outcrops is as numerous as it appears by those heaped on the shore, then since the sea rose to its present level some 6,000 years ago, this process has supplied a significant armoury for the sea with which to abrade the shore platforms, and made a significant contribution to the sand supply for the beaches.

Edwards (1951) records holdfasts with rocks up to 20 pounds weight being thrown ashore. If the rocks plucked from a headland by kelp (the heavy blocks fallen to the sea floor plus those thrown ashore) total on average 100 pounds or so per year, then in the six millenia since the sea rose to its present general level, nearly 300 tons of rock have been removed by this process alone.

FROM ROCKY SHORE PLATFORM

Plate 1 figure 2 is a somewhat triangular piece of Lower Cretaceous arkose which the sea plucked from the cavity shown in the lower part of the photograph. The site is the surface of a supratidal shore platform on the bold headland called Point Sturt on the Otway coast of Victoria, Australia (see photos and section of Point Sturt in Gill 1971). The shape of the rock results from the fact that it is a loose block created by the coincidence of some lesser joint planes with a major one. The flat vertical face at the back of the cavity is the major joint plane. The flat surface of the plucked block that fits against that face is the thick side out of sight at the top of the picture. The rock was not levered by some person, or the bruises to the rock would have been evident. The block was found a short distance landward of the cavity, and was found to fit perfectly. The rush of

DESCRIPTION OF Plate 1

Fig. 1 Piece of Lower Cretaceous arkose (on right of photo) plucked by the sea from a rock outcrop, then worn to the shape of the attached kelp holdfast (on left of photo) before being cast ashore near Von Mueller Creek, Otway coast, Victoria, Australia. Ruler on inch scale.

Fig. 2 Block of Lower Cretaceous arkose from cavity shown in lower part of photograph plucked by fast-moving seas from shore platform at Point Sturt, Otway coast, Victoria. Rock made loose by convergence of oblique minor joint planes with major vertical one. Same three-foot ruler as used for scale in Fig. 1.

rapid waters across the top of the loose block must have plucked it out. The site is on the landward side of a break in a rampart originally formed by the same major joint plane. The sea forced through this channel at high speed could easily suck out this block in spite of its weight.

That the block was removed by the sea not long before the photograph was taken is shown by the fact that the hollow had not been filled in, the edges of the rocks were sharp, and the block removed was not far away. However, some of the very numerous marine gastropods (*Melaraphia*) found on the platform had crawled into the cavity, as the photograph shows. In other places, similar removal of loose blocks where converging joint planes free them, has been noticed. A week after the photograph was taken, the site was visited again but a storm had swept the area in the meantime. The block had been swept shorewards and was almost at the cliff, a distance of over a chain. I thought at first that some youth had hurled it there, but noted that there were no chippings or bruisings on it. Moreover, such a person would almost certainly have thrown it into the sea nearby to make a splash. Such blocks swept by forceful waters across the platform would act as a natural plane to reduce the surface of the rock.

The wedge of rock just described weighed 213 pounds. If five such rocks were removed each year from the platform at Point Sturt, and this continued for the past 6,000 years during which the sea has been at or near its present level, then some 300 tons would have been removed by this process. Thus although the process is a minor one, plucking is not without significance in the reduction of rocky coasts by marine erosion. This first attempt at quantification suggests that of the order of 600 tons of rock have been removed by this process from a typical exposed headland during the present phase of erosion.

REFERENCES
Edwards, A. B., 1951. Wave action in shore platform formation, *Geol. Mag.* 88: 41-49.
Gill, E. D., 1971. Ramparts on shore platforms *Pacific Geol.* In Press.

F.N.C.V. PUBLICATIONS AVAILABLE FOR PURCHASE

FERNS OF VICTORIA AND TASMANIA, by N. A. Wakefield.
The 116 species known and described, and illustrated by line drawings, and 30 photographs. Price 75c.

VICTORIAN TOADSTOOLS AND MUSHROOMS, by J. H. Willis.
This describes 120 toadstool species and many other fungi. There are four coloured plates and 31 other illustrations. New edition. Price 90c.

THE VEGETATION OF WYPERFELD NATIONAL PARK, by J. R. Garnet.
Coloured frontispiece. 23 half-tone. 100 line drawings of plants and a map Price $1.50.

Address orders and inquiries to Sales Officer, F.N.C.V., National Herbarium, South Yarra, Victoria.

Payments should include postage (11c on single copy).

Safety Beach, Dromana Bay, Port Phillip, Victoria

by A. W. BEASLEY*

The sandy shore of Dromana Bay, from the southern limits of the rock-cliffed coast of Mount Martha to the point where the Nepean Highway meets the coastline at Dromana, is known as Safety Beach. The beach has gradually deepening water offshore, and is a safe and popular place for recreation. It measures 2 miles in length and ranges up to 90 feet in width. The shore gradient is 5 to 6 degrees and, since the tide range is small (3 feet for spring tides at Dromana Jetty), the foreshore is not wide; this is unlike the foreshore in the Rosebud-Rye area where the gradient is very low (Beasley 1969). Much of the ground surface immediately behind the beach has been levelled by man, but for some distance the beach is bordered by a belt of low sand ridges which are fixed by vegetation.

Most of Safety Beach is sheltered from the north by the high ground of Mount Martha which extends north west to Martha Point. However, it is exposed to attack by waves generated by strong winds coming from westerly directions. Such attack occurs mainly in the winter months when powerful waves erode sand off the beach and deposit it in the offshore region; however, smaller waves generated by weaker winds later transport sand back to the beach.

Two small streams of little significance enter Dromana Bay along Safety Beach. These streams, known as Brokil Creek and Dunns Creek, are sluggish and transport little detrital material to the sea.

The geology of the Safety Beach-Dromana region is shown in maps by Keble (1950). The Mount Martha Granodiorite forms coastal cliffs that extend north from Safety Beach to Balcombe Bay, and from there for about 11 miles the coastal sections are composed largely of ferruginous sandstones known as the Baxter Sandstones. The shore west of Dromana consists mainly of sandy beaches bordered by sand ridges, with rock-cliffed sections and headlands of dune-limestone (aeolian calcarenite) occurring from near Rye to Point Nepean.

Sand covers the floor of Dromana Bay, and Beasley (1966) has shown that it extends out for a considerable distance, nearly to the 10-fathom line. The near-shore region is shallow, and the 5-fathom line is nearly 1 mile from the coastline.

To obtain information about the nature of the Safety Beach sand and to enquire into its origin, mid-tide samples were collected from both ends and from midway along the Beach (see map). A size analysis of each sample was carried out using British Standard sieves. A cumulative frequency curve and a histogram were constructed for each sample, and the median diameter and sorting coefficient (Trask 1932) were determined. Size fractions were examined with a binocular microscope, recom-

*National Museum of Victoria.

Plate 1. Safety Beach from N. end, looking SW. across Dromana Bay.

Plate 2. N. end of Safety Beach, showing cliffs of decomposed Mount Martha Granodiorite.

bined and treated with dilute 1:2 hydrochloric acid to determine the weight percentage of acid-soluble (mainly carbonate) material in each sample.

The sand from sampling station 1 at the northern end of Safety Beach, adjacent to coastal cliffs of decomposed "soft" granodiorite, is relatively coarse, having a median grain size of 0.83 mm. At station 2, midway along the beach, the median grain size is 0.66 mm, and at station 3 it is 0.55 mm. Thus, there is a continuous decrease in median grain size southward along Safety Beach. The decrease from station 1 to station 2 is particularly great, but that from station 2 to station 3 is less great. This indicates that appreciable southward drift of the larger sized particles does not occur.

Adopting Trask's (1932) classification, the sand at station 1 shows only moderate sorting, having a sorting coefficient of 2.56. Sorting improves markedly along the shore away from the cliffs of decomposed granodiorite. At station 2 the sorting coefficient is 1.20, and it is very similar (namely, 1.27) at station 3. The marked improvement from station 1 to station 2, suggests that the direction of drift there is south west along the shore.

In the histograms, weight percentage of material greater than 4 mm in size is indicated by vertical lines, since it represents material retained on the coarsest sieve used. The sand from station 1 is bimodal but the other two are unimodal. Sand from station 1 has a primary mode in the very coarse sand size-grade and a lesser mode in the fine sand. Its large secondary maximum and spread indicate an immature condition of sorting. Two sources for this sand are suggested by the nature of the histogram: there appears to be loading with coarse material supplied from the nearby granodiorite. A very conspicuous maximum size-grade occurs in the coarse sand grade of the samples from stations 2 and 3, and fine proximate admixture exceeds coarse proximate admixture in both of them.

The weight percentage of acid-soluble material in the sand at stations 1, 2 and 3 is 0.4 per cent, 0.4 per cent and 0.5 per cent. This very low content contrasts with that of the sands west of Dromana (Beasley 1969) which have a relatively high content of acid-soluble material. Microscopic examination indicates that whole shells and shell fragments made up almost all of the acid-soluble content, and that this material is present mainly in the coarser size-fractions of the sand. It seems to have come mainly from organisms indigenous to the nearby seafloor.

Fragments of granitic rock and hornfels occur in the sand at station 1. Apparently they have come from local sources: hornfels occurs around the Mount Martha Granodiorite as well as at The Rocks, Dromana (Baker 1938, Kehle 1950). Much of the granitic rock is of pebble size but the fragments range down through granule to coarse sand size; most of the smaller fragments exhibit a fairly low degree of roundness and it is clear that they have not been subjected to much transportation. The amount of granitic rock at stations 2 and 3 is considerably less than at station 1.

The Safety Beach sand consists mainly of quartz. Most of the quartz grains are colourless and subangular to subrounded. Small amounts of opaque reef quartz occur, and feldspar particles are relatively common at station 1. Some of the quartz grains in the coarser size-fractions are angular; this points to a short

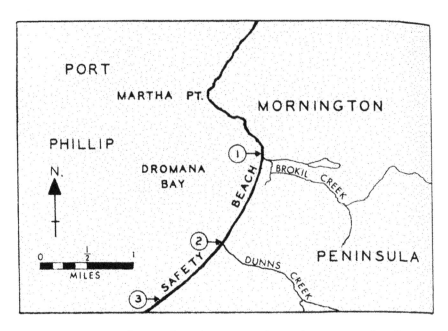

Fig. 1. Map showing location of sand sampling stations along Safety Beach.

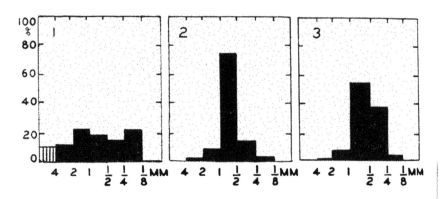

Fig. 2. Size-analysis histograms of Safety Beach sands.

detrital history, since rounding of larger size particles occurs at a relatively rapid rate.

Heavy minerals are not common in the sand, but local concentrations occur at the north end of the beach. Presumably the heavy minerals, transported by wave and current action, have become more concentrated there partly because of the configuration of the coastline; they are prevented from moving northward by the headland of Mount Martha Granodiorite which has steep cliffs and relatively deep water near-shore. More than 60 per cent of the heavy mineral assemblage is composed of black opaque oxide particles (mainly magnetite and ilmenite). Tourmaline, zircon, leucoxene and biotite make up most of the remainder. Much of the heavy mineral content has apparently been derived from the nearby granitic rocks, and many of the grains do not have a high degree of roundness. Magnetite grains frequently show edges of crystal faces, while a large number of the zircon grains are euhedral and subhedral, and tourmaline occurs commonly as prismatic crystals showing only slight abrasion. Some grains of zircon, tourmaline, rutile and ilmenite are rounded and well-rounded, and apparently have experienced a longer detrital history; they may have come from the Tertiary sandstones exposed in coastal cliffs to the north.

The fact that the sandy shore at Safety Beach is relatively stable, and that coastal erosion does not present great problems, make the locality popular and scenically attractive. There is plenty of material available to nourish the beach at all times. It is clear that the Mount Martha Granodiorite is an important source rock for the beach sand.

REFERENCES

Baker, G., 1938. Dacites and associated rocks at Arthur's Seat, Dromana. *Proc. Roy. Soc. Vict.* **50**: 258-278.

Beasley, A. W., 1966. Bottom sediments of Port Phillip Bay. *Mem. nat. Mus. Vict.* **27**: 69-105.

———. 1969. Beach sands of the southern shore of Port Phillip Bay, Victoria. *Mem. Nat. Mus. Vict.* **29**: 1-21.

Keble, R. A., 1950. The Mornington Peninsula. *Mem. geol. Surv. Vict.* **17**: 1-84.

Trask, P. D., 1932. *Origin and environment of source sediments of petroleum.* Gulf, Houston.

NATURAL HISTORY MEDALLION

At the November meeting, the award of this Medallion to Mr. Cliff Beauglehole will be made by Mr A. Dunbavin Butcher, Chairman of the Natural History Medallion Award Committee.

Mr. Beauglehole will speak on "The East Gippsland Flora Survey"

Geology Group Excursion

Sunday, 7 November — Geology Group. Trip to Sovereign Hill, Ballarat, guided tour by Mr. C. Goodall. Details to be arranged at Geology Group Meeting, Wednesday, 3 November, 8 p.m.

reptiles of victoria – 3

by HANS BESTE

Plate 5

Amphibolurus muricatus — Tree Dragon or Jacky Lizard.

A handsome dragon which is rarely found away from forested areas.

Length: to 12 inches.

Head long, narrow much more pointy than that of *A. barbatus*. Ear-opening large. Mouth and tongue yellow to pale orange. Five fingers and five toes. Hind foot and toes slender and long. Colour greyish, with several rows of spinous scales forming vertebral crests. Two rows of spots on upper surface which may run into each other. Usually pale yellow but indistinct grey in western parts of the state. Under pale grey with faint grey pattern. Throat often black.

Found: on tree stumps, fences or rocks, while sunning itself, throughout state.

Best distinguishing features — slim build, vertebral crests.

Plate 6

Amphibolurus diemensis — Mountain Dragon.

A drab dragon of the high country.

Length: to 9 inches.

Head, short and rounded, distinct from body. Snout blunt. Mouth lilac. Legs well developed, five fingers and five toes. A fast lizard, expert at catching flies. Colouration grey to brownish-grey, with indistinct pattern. Texture of skin rough, like sand paper.

Found: in alpine areas, also in Grampians.

Best distinguishing features—blunt round head. Lilac coloured mouth and gums.

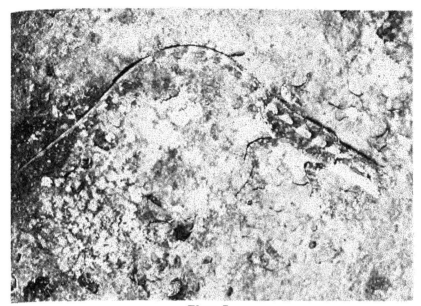

Plate 5.
Amphibolurus muricatus — Tree Dragon. Photo: H. Besle

Plate 6.
Amphibolurus diemensis — Mountain Dragon. Photo: H. Besle

Field Naturalists Club of Victoria

Marine Biology and Entomology Group Meetings
5 April, 1971

Mr. Condron chaired the meeting of 16 members. The members held a discussion on exhibits shown under their microscopes. Messrs. K. Strong and D. McInnes spoke of their observations of the gall wasps and it was agreed that Mr. K. Strong would speak of his further observations of these wasps at a later date this year. Much discussion followed relating to the T.V. programme from Channel 2 on the development of antivenene for the sting of the "sea wasp", or "box jellyfish". Mr. F. H. Baxter, of Commonwealth Serum Laboratories, was in charge of this research. Mrs. Z. Lee advised that she would ask Mr. Baxter to address the group.

Exhibits

Mr. B. Condron exhibited Butterflies caught recently, and moths caught in a service station and upon which magpies were feeding.

Miss White showed a geometer moth species to be identified.

Mrs. Z. Lee: A moth, *Hippotion celerio*—a hawk moth taken at Wattle Park. This moth, of the Sphyngidae was released in the room in order to observe the wing colouring and the creature's habit of hovering. Mrs. Lee also showed larvae of the golden eye lace wing with its stack of empty husks of its victims on its back.

Dr. Brian Smith exhibited a bivalve shell *Phragmorisma watsoni*. This specimen was dredged from 26 fathoms, 20 miles from Lakes Entrance. Dr. Smith told the meeting that this rare shell would be the first and only one for the Melbourne Museum, and that only 5-6 specimens have ever been taken; the first of which was by the "Challenger Expedition" last century from off an island in Bass Strait. The South Australian Museum have one taken in 1957 at an island off South Australia. The specimen was in alcohol owing to the very fragile nature.

Mr. Condon exhibited three stages of a small beetle; Ladybird larva, pupa, and adult. They were feeding on aphids on a swan plant.

Mrs. K. Strong exhibited Froghopper eggs and "cuckoo spittle" on a eucalyptus leaf collected on F.N.C.V. excursion to Brisbane Ranges; also the wasp from a gall on eucalyptus leaf, the body length of which was 2.25mm, collected on the same excursion.

3 May, 1971

Mr. Condron chaired meeting of 18 members. Miss Jenny Forse spoke on her work on Bull Ants and illustrated it with black and white slides of ants of different casts in an "anturium", and colour pictures of her work in the field. She explained how the young were cared for and the function of a queen and the founding of a new colony. Members kept Miss Forse answering many questions showing great interest in a very lucidly presented address.

Exhibits

Mr. H. Bishop showed under the microscope moths which had emerged from a gall on an acacia.

Mr. McInnes showed galls on leaves caused by bugs not yet identified. Mr. McInnes showed an introduced slug found at East Malvern. Mrs. Z. Lee showed an ichneumon wasp taken at Reservoir.

Mr. McInnes referred to what had been taken for an oil slick on water at Glenelg which measured approximately 50 yds. in width, and which, on examination of a jar full, turned out to be diatoms. Mr. K. Strong advised that the same phenomenon had been observed at Brighton clinging to seaweed.

Dr. Brian Smith exhibited a book by Cox on "Australian Land Snails" published in 1868.

7 June, 1971

President, Mr. R. Condron, chaired meeting of 21 members. A short note from Mr. John Strong, secretary on leave, telling of his limpet collecting etc., in the U.K., was read to the meeting.

The guest speaker, Mr. E. H. Baxter of Commonwealth Serum Laboratories, then addressed the meeting on the work done on the production of an antivenene for the sting of the "sea wasp"

Chironex fleckeri". He advised that 60 deaths from the sting of this jellyfish were recorded, but explained of the possibility of much larger figures, as owing to the rapidity with which the victim could die from a lethal dose of the venom, as short as two minutes, it is likely that reported drownings would in reality be caused by this stinger. Slides showed the two species of jellyfish and pictures of the extent of the scars and wheals resulting from a victim's contact with the stinging nematocyst.

Mr. Baxter advised on first aid, by pouring alcohol in form of methoated spirits, whisky, etc., over the affected area. This dehydrates the gelatinous strands and the stinging capsules are possibly, inactivated, so that together they may be peeled off the flesh with a soft dry cloth or tissue.

He warned against the use of rubbing with wet sand, as moisture appears to activate the nematocysts. A tourniquet, if possible, and artificial respiration is most important. The possibility of immunisation for those persons whose livelihood exposes them to the dangers of the sea wasp is at present receiving the attention of the research workers, whom Mr. Baxter heads. The public in general should not enter the waters off our coast, frequented by the jelly fish during the months of November to March. The breeding place had not yet been located, and it is supposed it could be in some inlets to the north of our country. The method of making the anti-venene was explained in detail, but it was explained that it was not by any means near completion. The method of catching the jelly fish, and obtaining the venom, was dealt with, and the extremely high cost of an amount which was sufficient to work with, was only hinted at. The attentive audience asked many questions, and at the conclusion Mr. Peter Kelly moved a vote of thanks to Mr. Baxter.

Campout

A campout is being planned for October during a week-end, probably to the Moonlight Creek Valley, and those wishing to join should give their names to either Mrs. Burleigh or Miss Piper at the October meeting, or else by 'phone to 89 8981.

Photoflora '72

The Native Plants Preservation Society is now planning Photoflora 72, its seventh biennial competition and exhibition featuring colour slides of Australian wildflowers.

Entry forms are now available from the Competition Secretary, Miss B. C. Terrell, 24 Seymour Avenue, Armadale, 3143. Public screenings have been arranged in 19 suburban and country centres during March and April, 1972. Details of these will be published later.

Field Naturalists Club of Victoria

Established 1880

OBJECTS: To stimulate interest in natural history and to preserve and protect Australian fauna and flora.

Patron:
His Excellency Major-General Sir ROHAN DELACOMBE, K.B.E., C.B., D.S.O.

Key Office-Bearers, 1971-1972.

President:
Mr. I. SAULT

Vice-Presidents: Mr. J. H. WILLIS; Mr. P. CURLIS

Hon. Secretary: Mr. R. H. RIORDAN, 15 Regent St., East Brighton, 3187. (92 8579)

Subscription Secretary: Mr. D. E. McINNES, 129 Waverley Road, East Malvern. 3145

Hon. Editor: Mr. G. M. WARD, 54 St. James Road, Heidelberg 3084.

Hon. Librarian: Mr. P. KELLY, c/o National Herbarium. The Domain, South Yarra 3141.

Hon. Excursion Secretary: Miss M. ALLENDER, 19 Hawthorn Avenue, Caulfield 3161. (52 2749).

Magazine Sales Officer: Mr. B. FUHRER, 25 Sunhill Av., North Ringwood, 3134.

Group Secretaries:

Botany: Mr. J. A. BAINES, 45 Eastgate Street, Oakleigh 3166 (57 6206).
Microscopical: Mr. M. H. MEYER, 36 Milroy Street, East Brighton (96 3268).
Mammal Survey: Mr. D. R. PENTON, 43 Duke Street, Richmond, 3121.
Entomology and Marine Biology: Mr. J. W. H. STRONG, Flat 11, "Palm Court", 1160 Dandenong Rd., Murrumbeena 3163 (56 2271).
Geology: Mr. T. SAULT.

MEMBERSHIP

Membership of the F.N.C.V. is open to any person interested in natural history. The *Victorian Naturalist* is distributed free to all members, the club's reference and lending library is available, and other activities are indicated in reports set out in the several preceding pages of this magazine.

Rates of Subscriptions for 1971

Ordinary Members	$7.00
Country Members	$5.00
Joint Members	$2.00
Junior Members	$2.00
Junior Members receiving Vict. Nat.	$4.00
Subscribers to Vict. Nat.	$5.00
Affiliated Societies	$7.00
Life Membership (reducing after 20 years)	$140.00

The cost of individual copies of the Vict. Nat. will be 45 cents.

All subscriptions should be made payable to the Field Naturalists Club of Victoria, and sent to the Subscription Secretary.

Magnificent stand of While Mountain Ash, *Eucalyptus regnans*, in the Marysville State Forest

FORESTS COMMISSION VICTORIA

. . . . . . *preserving the beauty of our forests for your enjoyment.*

The
Victorian Naturalist

Editor: G. M. Ward

Vol. 88, No. 11
3 November, 1971

CONTENTS

Articles:

Identification of the Black-headed Snakes within Victoria.
By A. J. Coventry 304

Penguins. By Myfanwy Beadnell 306

Distribution and Habitat of the Broadtoothed Rat in Victoria.
By J. H. Seebeck 310

Victorian Non-marine Molluscs — No. 6. By Brian J. Smith 325

Book Review:

"Australian Butterflies" 324

Field Naturalists Club of Victoria:

Report on September General Meeting 328

Western Victorian Club Reports 329

Diary of Coming Events 331

Front Cover:

The Spotted Marsh Frog (*Limnodynastes tasmaniensis*) is caught in "thoughtful mood" by the photographer, John Wallis.

Identification of the Black-headed Snakes (Denisonia) within Victoria

by

A. J. COVENTRY*

The elapid snake genus *Denisonia* (sensu Boulenger, 1896) contains over twenty Australian species, including a number which bear a strong superficial resemblance in that they are relatively small (seldom exceeding 700 mm), have light brown bodies, and black heads. For this reason these species are often confused. However, there is no doubt that the similarities of some are due to convergent evolution, and that the group is polyphyletic.

This article discusses the Victorian species of black-headed snakes belonging to the genus *Denisonia*, and conservative taxonomic conclusions are made in an attempt to aid the identification of the species and help resolve some of the confusion surrounding them. It is recognised, however, that an Australia-wide revision is necessary before final taxonomic conclusions can be made. It should be pointed out that juvenile specimens of the Brown Snakes, *Demansia textilis* and *Demansia nuchalis*, which also have light brown bodies and black heads occur throughout Victoria. These can be separated very easily from the species under discussion, in that *Demansia* has divided anal and subcaudal scutes, while the black-headed *Denisonia* species have single anal and subcaudals.

Rawlinson (1971) published a checklist of the reptiles known to have been collected in Victoria. This list includes five species of black-headed *Denisonia*, viz. *D. brevicauda* Mitchell, 1951, *D. flagellum* (McCoy, 1878), *D. gouldii* (Gray, 1841), *D. nigrostriata* (Krefft, 1869), and *D. suta* (Peters, 1863).

Recent sorting and checking of specimens in the National Museum of Victoria revealed that the Victorian snakes previously identified as *D. brevicauda* and *D. nigrostriata* are conspecific (referable to *D. brevicauda*) and that Victorian specimens referred to *D. gouldii* belong to *D. dwyeri*. Thus it appears that there are only four species of black-headed snakes of the genus *Denisonia* in Victoria, and these are *D. brevicauda* Mitchell, 1951, *D. flagellum* (McCoy, 1878), *D. dwyeri* Worrell, 1956, and *D. suta* (Peters, 1863).

KEY TO THE SPECIES FOUND IN VICTORIA

1. Black head patch divided — *flagellum*
   Black head patch undivided — 2
2. Scales in 19 rows — *suta*
   Scales in 15 rows — 3
3. Dark vertebral stripe present — *brevicauda*
   Dark vertebral stripe absent — *dwyeri*

*National Museum of Victoria

Denisonia flagellum (McCoy, 1878)
Little Whip Snake

This is readily distinguished from other Victorian species by the head pattern, which has two distinct dark patches. The first of these begins on the rostral scale, and extends back onto the nasals and internasals, while the second extends from the anterior border of the frontal, back over the entire head and nape for some six or seven vertebrals. This leaves a pale band across the snout in the region of the prefrontals, which immediately separates this species from any other black-headed Victorian *Denisonia*. Although listed in the literature as having 17 scale rows, ten of the seventy specimens counted had only 15 rows at mid-body.

Distribution: The Museum has specimens from the Eyre Peninsula, S.A., through southern Victoria to the Melbourne area, and then northwards, on the western side of the Great Dividing Range to the A.C.T

Denisonia suta (Peters, 1863) Curl Snake

This species can be identified by both scalation and colour. It always has 19 scale rows at mid-body (15-17 in the other species under discussion), and its head colouring is distinct in that the pre- and post-oculars are pale, and separated from the upper labials which are also pale, by a dark lateral stripe which extends on each side from the temporals forward, to meet on the rostral. In older specimens the black hood on the head tends to fade, although the lateral head stripe remains prominent.

Distribution: Confined to the north-central and north-west within Victoria, and extending into South Australia, Northern Territory, south-west Queensland and New South Wales.

Denisonia dwyeri (Worrell, 1951) Dwyer's Snake

This species has been confused with Gould's Snake, *D. gouldii* which is confined to Western Australia. It differs from *D. gouldii* in either lacking, or having a much paler reticulated pattern over the body, having a flatter head, and a shorter, heavier body. The black head patch covers the entire head, except for the pre-

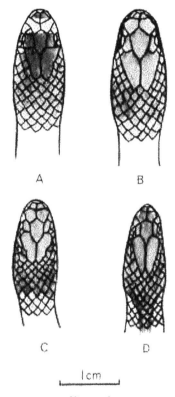

Figure 1

Dorsal aspect of heads of:
A. *D. flagellum* N.M.V. Regd. No. D16019
B. *D. suta* N.M.V. Regd. No. D13355
C. *D. dwyeri* N.M.V. Regd. No. D3613
D. *D. brevicauda* N.M.V. Regd. No. R11119

Drawings by Miss Rhyllis Plant.

ocular, upper labial and rostral scales, which are a creamy colour. The head patch extends back for 4-6 vertebral scale rows behind the parietals. In one specimen examined there is a faint sign of a vertebral stripe, which is less prominent than found on *D. brevicauda*. Like *D. brevicauda*, this species has 15 scale rows at midbody.

Distribution: Within Victoria, southwards from the Murray River through the central regions and the western foothills of the Great Dividing Range to the Seymour district. The Museum has one early specimen labelled "Frankston", but some doubt exists as to its provenance. *D. dwyeri* also occurs in similar habitats from southern Queensland through New South Wales.

Denisonia brevicauda Mitchell, 1951. Mitchell's Short-tailed Snake.

This species has a similar head shape and pattern to *D. dwyeri*, and can be best separated from it by the dark vertebral stripe as well as a different habitat. Originally described as a subspecies of *D. nigrostriata*, a long-tailed species from north-eastern Australia, these species can be readily separated from one another by the length of the tail, which in *D. brevicauda* has 23-29 subcaudals (Mitchell, 1951) as against 50-64 (Boulenger, 1896) in *D. nigrostriata*.

Distribution: The warmer drier areas in the north-west of the State, from the Little Desert northwards to the Murray River. It also occurs in the adjacent areas of South Australia.

Acknowledgement: I thank Mr. P. A. Rawlinson of Latrobe University for offering constructive advice.

REFERENCES

Boulenger, G. A., 1896. Cat. Snakes in the British Museum 3: 343-4.
Krefft, G., 1869. Snakes of Australia: 70.
Mitchell, F. J., 1951. *Rec. S.A. Mus.*, 9: 550-1.
Rawlinson, P. A., 1971. Victorian Year Book, 85: 11-36.
Worrell, E., 1956. *Aust. Zool.*, 12: 202-5.
Worrell, E., 1963. Reptiles of Australia: 135-6.

Penguins

by Myfanwy Beadnell

Those funny little men in evening suits! Everybody loves penguins. There are about eighteen varieties of penguins, depending on how some of the variations are classified, ranging from the huge Emperors, standing 48 inches high, to our own 14 inch Fairies, properly known as Northern Blues. Not all penguins are black and white; Emperors and Kings have a lot of yellow and gold, some have greyish or bluish feathers, and many have crests of various sizes, shapes and colours. Nearly all chicks are brownish or greyish.

Penguins are not, as many people think, degenerate birds that have lost the art of flying, but are amongst the most highly specialised of our fauna, and are perfectly adapted to the arduous conditions under which most of them breed. They spend most of their time at sea and have very powerful pectoral muscles and strong flippers to push against the dense medium of water. Their feet are placed far back on the body, and thus together with the tail, act as a rudder when swimming. The tail also acts as a prop or support on land. This and the fact that the hips cannot rotate on the backbone, gives the birds a waddling, pompous gait.

Other adaptations to marine life are firstly, the ability to drink salt water. Penguins have a filter-gland between the nose and the skull, which takes in sea-water and discharges strong brine out through the nostrils and transfers nearly pure water to the digestive system. Secondly, they do not get "the bends" on diving. Before diving they empty the lungs of air and shut them off from the rest of the body. This means that there is no problem with dissolved nitrogen appearing as bubbles when normal pressure is regained on surfacing, but it also means that the birds must swim and create energy without using oxygen. To do this the metabolism is slowed down and the heart only beats about 4 or 5 times per minute, about one-twentieth of its normal rate. On surfacing the bird must rapidly make good the back-log of oxygen lack and get rid of accumulated waste products, which it does very efficiently; birds swimming near the surface can be seen to "porpoise" along, rising every few minutes to breathe deeply and reactivate the system.

Keeping warm is the most basic problem of all. To begin with, there is a special arrangement of the feathers, which are small (averaging 70 to the sq. in.), curved and oily at the tips, overlapping like tiles on a roof, from the shaft of the feathers numerous fibres grow down, forming a kind of matted, fibrous undershirt. Then there is the thick layer of blubber, which not only serves as an insulator, but as a source of food during the long fasts which all penguins undergo. Keeping cool is also a problem as the birds need to cool down rapidly when coming out to sit on a sunny rock or sheet of ice after a strenuous bout at sea. Because of the feathers and the thick blanket of blubber, a special mechanism is required. The bare patches, the feet and the inner surface of the flippers are very rich in blood vessels and can be seen to glow pink like radiators as the hot blood comes to the surface and the heat dissipates into the atmosphere.

Most penguins are very gregarious and very noisy; some "arpp", some shrill and some bray like donkeys. New Zealanders living in remote coastal areas frequently complain of the noisy braying under their houses at night. Penguins eat an enormous quantity of fish, the stomach taking up almost the whole body area of the bird. A large colony will eat as many fish as 70 modern trawlers could catch in a day. The crop is so equipped that a penguin can suspend digestion when carrying fish home to its chick by secreting a thick wall of mucous; penguins have been known to starve to death with a full crop if prevented from feeding this to the chick.

Most penguins have a very strong attachment to their nesting site and will return to the same nest each year, the tendency being most marked in the male, who returns first each season and starts scraping about and adding to the last season's nest. The female returns soon after and joins him. There is a great deal of recognition ceremony, posturing and necking, scraping at and adding to the nest. Occasionally the wrong female arrives first and settles in, the male not objecting. If the proper mate then arrives the interloper is soon ejected.

minus quite a few feathers. If the true mate is very late, and the new wife firmly established, it may not be possible to evict her. However, the late-comer is not usually long in finding an unattached millie instead. Most penguins lay two eggs, though often only one is reared. Some lay one only, some three, and a very few four or more. Usually both parents incubate the eggs and tend the chicks in turn, the other bird being at sea fishing. Recognition ceremonies are repeated at each change of shift. Chicks and parents recognise each other by sound only, or possibly by sound and smell, but not by sight; sight seems specially adapted for distance vision.

The antarctic summer is very short and for the larger penguins it is a problem to get the chicks launched in time to take advantage of the seasonal time of food bounty. The two large varieties, Emperors and Kings, lay only a single egg. Neither builds a nest but lays directly onto the ground or the sea-ice. It is immediately transferred to the feet of the male where it is enfolded and kept warm and safe in a pouch of downy skin. In the case of the Emperor, the male does all the incubating, which takes about nine weeks. The egg is laid early in winter and when born the chick is fattened up very rapidly, its growth being accelerated so that it can be launched in mid-summer. This means that with the pre-nuptial period of five weeks, and journeys of at least a week before and afterwards to the open-sea feeding grounds, hundreds of miles away, the male bird fasts for almost five months, through the desperate cold of an antarctic winter. During this time he drops in weight from 95 to 45 pounds. The female, having laid her egg, goes off to sea to recover and gain weight, coming back with an enormous load of fish just when the chick is due to be hatched. If she is late, the mate can carry on; he can secrete in his throat a kind of milk-mucous which will keep the chick alive until the fish arrives. When the hen arrives, the male goes off to fatten up. After launching the chick the parent birds fatten up again, then starve for a second period during the annual moult, then fatten up again for the next breeding season.

The King, also a large bird (37 ins. high) takes 7½ weeks to incubate its egg. The male starts incubation but both take turns, transferring the egg from the feet of one to the feet of the other with their beaks. This is possible for Kings breed in the sub-antarctic islands, where the sea is not so inaccessible. The King has a different arrangement to beat the season: she lays in summer and the chicks are kept alive but almost starving during the winter, their growth being slowed down so that they can be launched in spring. During this long period both parents are away fishing and the chicks huddle together in enormous rookeries, keeping each other warm. The parents return about every three weeks, and cram the chick as full as it will go before leaving it again. This slow development means that the breeding season takes about nine months, and therefore Kings can only breed twice in three years, breeding in rotation, early in the season, late, and not at all.

For the Adelie penguin the season begins around mid-October, the birds walking 200 or 300 miles across the sea-ice to their nesting site. The males return first, in long straggling lines of 100 or so, and are soon followed by the females. Great recognition ceremonies and nest building activities take place, even though the nest is only a pile of stones. In Adelie also, the male does all the incubating, presumably because the enormous distances to the open sea and fish makes shift work impracticable. Sometimes the bird will be caught in a blizzard when sitting on his nest, and he buried right up to the tip of his beak in the snow. The two chicks are launched, able to fend, but not fully grown, in the spring, when the parents go off to sea to fatten ready for the moult.

Most penguins live in large colonies, the Adelies often being 5 or 6 million strong. This is partly because they can only nest in those areas where there is a break in the ice-shelf surrounding most of the continent, and they are able to scramble ashore. At Cape Hallett carbon dating shows that the colony is over 1000 years old.

The species that live further north have a kinder climate, and most build either in tussocky grass or in burrows, often burrows vacated by other animals. The Snares Island penguin roosts in the lower branches of scrubby trees.

The main enemy of the penguin is the Leopard seal, and to a much smaller extent other seals and sea-lions. On land, Skuas and Giant Petrels attack eggs and chicks, but rarely adults unless these are wounded. Man is now becoming the penguins' worst enemy, forcing him to drown in oil and depriving him of his home through industrialisation; DDT and other poisons have been found in their tissues.

Because penguins look so comically human, people are apt to collate their behaviour with that of humans, but of course it is instinctive, involving no mental process, but having evolved through trial and error throughout the ages, and also through the oft-repeated experiences of individual birds, and the social ritual of the colony. Older birds tend their eggs and chicks much more efficiently than younger couples.

Much study is being done and much information is appearing about these birds, perhaps the most fascinating and lovable of all our avian friends.

F.N.C.V. PUBLICATIONS AVAILABLE FOR PURCHASE

THE WILD FLOWERS OF THE WILSON'S PROMONTORY NATIONAL PARK, by J. Ros Garnet.
Price $5.25, (discount to members); postage 20c.

VICTORIAN TOADSTOOLS AND MUSHROOMS, by J. H. Willis.
This describes 120 toadstool species and many other fungi. There are four coloured plates and 31 other illustrations. New edition. Price 90c.

THE VEGETATION OF WYPERFELD NATIONAL PARK, by J. R. Garnet.
Coloured frontispiece, 23 half-tone, 100 line drawings of plants and a map. Price $1.50.

Address orders and inquiries to Sales Officer, F.N.C.V., National Herbarium, South Yarra, Victoria.

Payments should include postage (11c on single copy).

Distribution and Habitat of the Broadtoothed Rat, Mastacomys fuscus Thomas (Rodentia, Muridae) in Victoria

by J. H. Seebeck*

The broadtoothed rat has been recorded from few localities in Victoria, the most recent published records being Loch Valley (Warneke 1960) and Leongatha (Lyndon 1960). Earlier locality records have been summarized by Warneke (1960) and Calaby and Wimbush (1964). Since 1960 several other localities have been discovered and the known modern range extended. The continuing presence of the species in the Otway Ranges has been confirmed.

This paper presents this additional distributional data and some details of the habitat associations of the species. Figure 1 shows the present distribution of *Mastacomys fuscus* in Victoria. Specimens reported in this paper are in the collections of the Fisheries and Wildlife Department or the National Museum of Victoria. Table 1 lists registration numbers of *Mastacomys* in those collections.

NEW LOCALITY RECORDS

1. *Kalorama*

In February 1965 a preserved specimen of *Mastacomys fuscus* was received from Dr. G. Ogilvie, of Jasper Road, Kalorama, in the Dandenong Ranges, 40 km (25 mls) east of Melbourne (lat. 37° 48'S, long. 145° 18'E) at an altitude of about 500 m (1620 ft) above sea level. It was a young non-parous female which had been captured by the family cat. Subsequent extensive trapping in the area was carried out in the ensuing weeks but no further specimens of *Mastacomys* were collected, although a large number of other small mammals were captured.

In September 1968, further exploratory trapping was carried out on the opposite (southern) side of the ridge from Dr. Ogilvie's home, some 120 m (400 ft) away). Three specimens of *Mastacomys* were trapped (20, 22, 23 September). These were an adult male, an adult parous female and a juvenile female. Further trapping in the area failed to yield any further specimens. Subsequently, two further specimens of *Mastacomys* (an adult female and a juvenile male) were found dead, near to this trapping zone. Both had apparently been taken by a cat.

The area in which these recent specimens were trapped was at the western end of Erith Road, on a block of privately owned land which has now been cleared of most of the undergrowth. Before April 1968 the scrub on this block was contiguous with other uncleared private land, but most of the adjacent blocks were bulldozed in late April 1968 to remove the scrub as a fire precaution measure. Figure 2 shows the location of the capture site and the bush remaining at the time of writing (July 1970).

Soils in the area are derived from hypersthene dacite, and in this situation the parent rock is close to the surface, being exposed in a number of places. The altitude and south westerly aspect result in a high rainfall; the average for Mt. Dandenong 3.2 km (2 mls) south-east is 147.3 cm per annum. This rainfall, together

*Fisheries and Wildlife Department and Ryluh Institute for Environmental Research, 123 Brown St., Heidelberg, Victoria, 3084

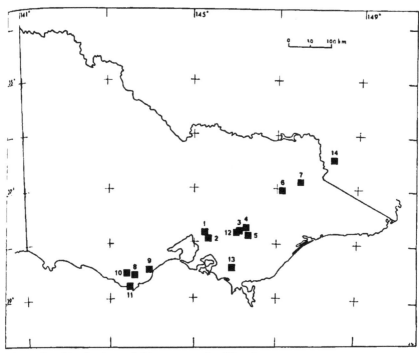

Figure 1. Distribution of *Mastacomys* in Victoria.

■ Localities mentioned in text
1 Kalorama, 2 Belgrave, 3 Pennys Saddle, 4 Upper Thompson, 5 Mt. Baw Baw, 6 Mt. St. Bernard, 7 Falls Creek, 8 Aire Valley, 9 Benwerrin, 10 Carlisle River, 11 Glen Aire, 12 Loch Valley, 13 Leongatha, 14 Mt. Kosiusko.

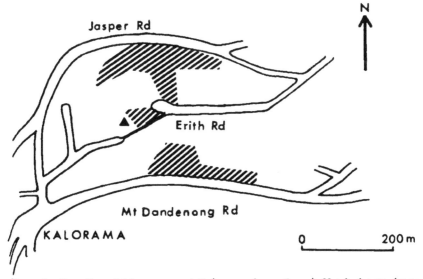

Figure 2. Location of *Mastacomys* at Kalorama shown thus ▲ Hatched area shows remaining bushland between Jasper and Mt. Dandenong Roads.

with the soil characteristics, has resulted in a typical wet sclerophyll vegetation which has however been modified by fire and mechanical clearing. This modification has resulted in the vegetation being composed of mature or over-mature fire damaged eucalypts with a low, dense shrub layer of regenerating shrubs and trees. The tree layer was tall with a very open canopy, and composed mainly of *Eucalyptus obliqua* with some *E. radiata* and a few isolated *E. regnans*, all about 30 m high. There was a small grove of mature *Acacia melanoxylon* about 20 m tall in the south-western corner of the block, but these have mostly been felled.

The understorey, of average height 3-5 m, with some eucalypt saplings 6-8 m tall, was dominated by *Pomaderris aspera*, but also present were *Bedfordia salicina*, *Acacia melanoxylon*, *A. verticillata*, *Olearia argophylla* and *Cassinia aculeata* with scattered *Hedycarya angustifolia*, *Olearia lirata*, *Pimelea ligustrina*, *Goodia lotifolia* and *Indigophora australis*. Lower shrubs included *Pimelea axiflora*, *Helichrysum dendroideum*, *Lomatia fraseri* and young plants of most of the taller shrub species present.

A dense field layer to 1.2 m formed thick cover over most of the area but there were some open areas of grass, mainly tussocks of *Poa australis*. The field layer was composed of *Pteridium esculentum*, *Lepidosperma elatius*, *Polystichum proliferum*, *Chryanthemoides monilifera*, *Dianella tasmanica*, *Tetrarrhena juncea* and two species of *Rubus*. Climbing plants such as *Clematis aristata* and *Rubus triphyllus* (= *R. parviflorus*) were scattered throughout the area. Low herbaceous plants completed the dense ground cover,

including *Senecio vagus*, *S. linearifolius*, *Geranium solanderi*, *Stellaria flaccida*, *Acaena anserinifolia*, *Centaurum minus*, *Cynoglossum latifolium* and *Carduus tenuiflorus*.

Many tunnels and runways were present at ground level. Two of the recent *Mastacomys*, the male and the juvenile, were caught at one trap station on a well defined runway near a fallen log. The adult female was caught at the entrance to a tunnel in a dense patch of *Tetrarrhena juncea*.

The most abundant small mammal in the area was the bush rat *Rattus fuscipes assimilis*, but the swamp rat *Rattus lutreolus*, black rat *Rattus rattus* and house mouse *Mus musculus* were also trapped in small numbers. *Antechinus stuartii* (brown phascogale) and *A. swainsonii* (Swainson's phascogale) were also present, as was the short-nosed bandicoot, *Isoodon obesulus*. It is possible that *Mastacomys* will become extinct at this locality, as much of the suitable habitat has been cleared. This habitat type is, however, fairly widespread in the Dandenongs and section 2 records the presence of a further Dandenong Ranges colony.

2. Belgrave

In mid-July 1970, N. A. Wakefield and A. Brugman, of Monash Teachers College, located a small colony of *Mastacomys* in wet schlerophyll forest near Belgrave, in the Dandenong Ranges (lat. 37° 55' S, long. 145° 22' E). Field studies are in progress on this colony and the results of these studies will be published in due course. Wakefield (pers. comm.) has reported that the colony consists of about twenty-five individuals, and that four specimens (1 ♀, 3 ♂), which were accidental deaths in this colony during the initial phases of this study, are lodged with the National Museum of Victoria.

3 *Pennys Saddle*

Pennys Saddle, an open plain 12-16 hectares (30-40 ac) in extent is situated on a ridge separating Loch Valley from the Toorongo Valley, at an altitude of 733 m (2600 ft) A.S.L. about 19 km (12 mls) north-north-east of Noojee, on the Noojee-Matlock Road (lat. 37° 47' S, long. 146° 04' E). This locality is about 6.5 km (4 mls) from Loch Valley Pine Plantation, the site of Warneke's (1960) rediscovery of *Mastacomys* although it is some 300 m (1000 ft) higher in altitude. In June 1968 two adult male *M. fuscus* were trapped in a shallow gully at the eastern end of the saddle by the author. Both animals were trapped under thick, bushy *Nothofagus cunninghamii*.

Soils at Loch Valley are deep friable reddish clays derived from metamorphosed Silurian sediments and Devonian granites of the Toorongo Range (Warneke 1964). It is likely that the soils at this present site are mainly derived from the Toorongo granites. The Atlas of Australian Soils (1962) classifies them as Brown Friable Porous Earths.

Meteorological characteristics of Pennys Saddle are probably similar to those at Loch Valley. Warneke (1964) has collated the climatological data for that station. These show that the average annual rainfall is 143.3 cm, mainly of winter incidence. The driest months are January, February and March, each of which has a mean of about 7.6 cm. July and August are the only months that normally receive snow falls, although snow has been recorded for all months except January and February. Summers are warm to hot with mean maximum temperatures usually below 27°C and mean minima about 7.8°C. January is the hottest month. Frosts are common, and winter temperatures often fall below 0°C, although mean maximum in winter is 14.4°C with mean minimum about 1.7°C in July, the coldest month.

The vegetation of the Saddle itself is short grass, sedges, rushes and low bracken merging with regenerating *Eucalyptus regnans* forest, with associated shrubs. The gully in which *Mastacomys* was trapped was very wet at the time, with a small stream developing in it. The floor of the gully was densely covered with *Carex appressa* the dominant species. Fallen trees, tree fern stumps and occasional outcrops of granite broke the continuity of the ground cover. Tussocks of *Gahnia radula*, *Juncus* spp. and *Senecio linearifolius* were scattered throughout the sedge, as were the ferns *Polystichum proliferum*, *Blechnum nudum* and *B. procerum*. Small ground plants such as *Acaena anserinifolia*, *Viola hederacea*, *Hydrocotyle* sp. and *Geranium* sp were also present. Rather open shrub cover to about 3 m (10 ft) was provided by *Olearia phlogopappa* with scattered plants of *Cassinia aculeata* and *Nothofagus cunninghamii*. Near the gully edge, saplings of *Acacia dealbata*, *A. melanoxylon* and *Eucalyptus regnans* appeared, and as the gully deepened down the slope and tree cover of these three species became more complete, treeferns were common. Ground cover was less, and was composed of small ferns, e.g., *Blechnum fluviatile*, herbs such as *Australina muelleri* and many species of moss.

Further trapping at this locality has yielded only bush rat, *Rattus fuscipes assimilis*, brown phascogale, *Antechinus stuartii* and Swainson's phascogale, *Antechinus swainsonii*. In addition, Warneke (1964) records the presence of swamp rat (*R. lutreolus*) and long-nosed bandicoot (*Perameles nasuta*) at Loch Valley, and the Fisheries and Wildlife Department has records of *Trichosurus caninus*, *Petaurus breviceps*, *Gymnobelideus leadbeateri*, *Pseudocheirus peregrinus*, *Wallabia bicolor*, and *Vombatus ursinus* in the area.

4. Upper Thompson River

On 22 May 1971 N. A. Wakefield trapped a three-quarters grown male *Mastacomys* near the headwaters of the Upper Thompson River at a site about 20 km (12.5 mls) north-east of Noojee at an altitude of 1020 m (3350 ft) A.S.L. (lat. 37° 15′ S, long. 146° 12′ E). The specimen is lodged with the National Museum of Victoria, and the occurrence of the species in this area will be reported in full in due course (Wakefield, pers. comm.).

5. Mount Baw Baw

Mount Baw Baw Alpine Village is close to the summit of Mount Baw Baw, 114 km (71 mls) east of Melbourne, and at an altitude of about 1460 m (4800 ft) A.S.L. (lat. 37° 51′S. long. 146° 17′E). Most of the area is a fairly dense woodland of *Eucalyptus pauciflora* with a low shrub understorey, but close to the village the East Tanjil River runs through a wide flat valley, where the vegetation is dominated by *Richea continentis*. Within the forested area there are many outcrops of granite (see Plate 1). It was close to one such outcrop that P. N. Homan of the Mammal Survey Group, F.N.C.V., trapped an adult male *Mastacomys fuscus* in March 1969. The animal was trapped in snow grass (*Poa* sp.) under mountain pepper (*Drimys aerophila*) under snow gum (*Eucalyptus pauciflorus*). The area was visited by the author in May 1970. Trapping was carried out in both available habitat types (*E. pauciflora* woodland and *Richea* swamp) but no further specimens of *Mastacomys* were taken.

The Baw Baw plateau is composed of upper Devonian granite and the soils are mainly alpine humus soils (Organic Loamy Soils) although in the creek valleys they appeared to be much deeper and richer in organic matter and probably approached a bog peat. Among the extensive granite outcrops soil was thin and restricted to the fault lines. (Anon. 1964, Costin 1962, Neilson 1962).

There are no detailed meteorological records for the Baw Baw plateau, but Hogg (1970) indicates that rainfall is in excess of 150 cm per annum, with heaviest falls in winter and spring. Summer rainfall is about 30.5 cm, rather higher than that recorded at Hotham Heights. Snow falls usually occur between May and October, with persistent snow lying during the winter and spring. Winter mean maximum temperatures are about 7.2°C, and mean maximum temperatures in summer probably do not exceed 18°C.

The dominant plants in the area were *E. pauciflora* growing to a height of 6-9 m, with a mid-dense canopy. Interspersed with the multi-stemmed living trees were many standing fire-killed slags. The low shrub layer varied in height from 0.5-1.5 m dependent on aspect, and in density from open to very dense. It was dominated by *Drimys aerophila*, *Leucopogon maccraei* and *Pultenaea muelleri*, with scattered bushes of *Olearia phlogopappa* and taller herbs. *Wittsteinia vaccinacea* often formed dense cover, particularly near tree bases or under the taller shrubs. The ground layer was usually quite dense, but dependent on the shrub cover was locally absent. Commonly encountered species included *Poa australis* (dominant ground cover in places), *Stylidium graminifolium* and *Celmisia longifolia*, while the fern *Polystichum proliferum* was abundant at the bases of many granite outcrops. Much fallen timber provided additional cover, and herbs such as *Geranium potentilloides*, *Acaena anserinifolia*, *Haloragis tetragyna*, *Asperula gunnii*, *Orites lancifolia*, *Dianella tasmanica*, *Senecio quadridentatus*, *S. linearifolius* and *Veronica nivea* were occasional components of the ground layer, the latter three particularly near the disturbed road verges. The *Richea* heath area merged into the woodland, and several other shrub species were found in the ecotone. These included *Epacris petrophila*, *E. paludosa*, *Baeckea utilus*, *Olearia algida* and occasionally *Callistemon sieberii*.

Runways and diggings of small mammals were found in many areas, being particularly obvious under low shrubs such as *Drimys* and *Pultenaea*. A number of specimens of bush rat (*Rattus fuscipes assimilis*) and brown phascogale (*Antechinus stuartii*) were collected, together with one Swainson's phascogale, *Antechinus swainsonii*. Wombats (*Vombatus ursinus*) and brushtail possums (*Trichosurus* sp.) occurred in the area and utilized the ground for feeding and movement. Ringtail possums (*Pseudocheirus peregrinus*) were also present.

Plate 1. Habitat of *Mastacomys* in snow gum woodland at Mt Baw Baw.

Plate 2. Habitat of *Mastacomys* at Mt St. Bernard.

6. *Mount St. Bernard*

In February 1967 W. M. Bren, Fisheries and Wildlife Department, trapped an adult parous female *M. fuscus* adjacent to a creek near the Dargo High Plains Road, about 0.8 km (0.5 ml) south-east of the Alpine Way (lat. 37°01'S, long. 147°05'E). The altitude of the creek at this point is 1418 m (4650 ft) A.S.L. The creek is a tributary of the Dargo River and at the location of the *Mastacomys* capture runs in a steep-sided valley 60-90 m (200-300 ft) below the road. The valley floor is fairly level, with a maximum width of about 15 m and was densely covered with thickly growing *Carex appressa*. Plate 2 shows a typical section of this gully.

The soils in the area are alpine humus soils (Organic Loamy Soils) on a bed-oz'; of Ordovician slatey mudstones (Anon. 1964, Costin 1962, Neilson 1962).

Meteorological records have not been kept at Mount St. Bernard, but rainfall and temperature data are available for Hotham Heights, 6.5 km (4 ml) east-north-east distant and some 400 m (1300 ft) higher than this site. These data show that the area has a rainfall of 149 cm per annum mainly occurring in winter and spring. Snow may occur in all months of the year but usually falls between June and October, and is normally persistent, often forming deep drifts. Winter temperatures are low, with average daily minimums in July, the coldest month, being below freezing point. Daily maximums below -1.1°C are frequently recorded. January is the warmest month but average maximums are only about 16°C. Two excellent summaries of the climate of Victorian alpine regions may be found in Anon. (1962) and Hogg (1970).

Mingled with the *Carex* on the valley floor were ferns (*Polystichum proliferum* and *Blechnum pennamarina*), tussock grass (*Poa* sp., probably *P. australis*) and many herbs including *Hypochoeris radicata*, *Rumex acetosella*, *Haloragis tetragyna*, *Geranium potentilloides*, *Viola hederacifolia*, *Stellaria pungens*, *Acaena anserinifolia*, *Oreomyrrhis* sp., *Hydrocotyle* sp., and *Lotus* sp. Many fallen logs were present, together with isolated shrubs of *Oxylobium alpestre*, *Olearia phlogopappa*, *Helichrysum secundiflorum*, *Acacia obliquinerva*, *Drimys aromatica*, *Leptospermum grandiflorum*, *Leucopogon suaveolens* and *Daviesia ulicifolia*. The shrubs listed formed a rather dense low layer on the valley slopes. *Leucopogon gelidus*, *Tieghemopanax sambucifolius* and *Gaultheria appressa* also occurred within the shrub layer. The shrubs were rarely more than 1.5-2 m tall and were overtopped by *Eucalyptus delegatensis*, which rose to about 25 m, but rarely had a trunk diameter greater than 30 cm.

The *Mastacomys* was trapped on a runway in the sedge, close to the creek bank. These runways were abundant and criss-crossed most of the valley floor. They were probably constructed by the bush rat, *Rattus fuscipes assimilis* which is very common in the area. Trapping carried out by the author in May 1970 at this site yielded only *R. fuscipes* — 7 specimens from 21 traps. The only other ground mammal collected in the area was the brown phascogale, *Antechinus stuartii*. Wombats (*Vombatus ursinus*), brushtail possums (*Trichosurus vulpecula*) and ringtail possums (*Pseudocheirus peregrinus*) also occur in this locality.

Further evidence of the presence of Mastacomys in this part of Victoria is provided by a record of a damaged specimen collected in

October 1966 by N. A. Wakefield near the headwaters of the Dargo River under Mount Hotham, altitude 1667 m (5500 ft) A.S.L. This specimen is lodged in the National Museum of Victoria.

2. Falls Creek

Dixon (1971 a) reported the discovery of *Burramys parvus* from the Falls Creek area of the Bogong High Plains. During the course of further field work by that author at this locality an adult male *Mastacomys* was trapped on 24 May 1971 (Dixon, 1971 b). The specimen, now lodged with the National Museum of Victoria, was collected in an area of alpine shrubs and snow gums (*Eucalyptus pauciflora*) at an altitude of 1798 m (5900 ft) A.S.L., 7.2 km (4.5 mls) south-west of Falls Creek (lat. 36° 53'S, long. 147° 15'E).

4. Aire Valley

Aire Valley Forests Commission Plantation (lat. 38° 39'S, long. 143° 15'E), altitude between 330 and 530 m (1100-1750 ft) A.S.L. lies about 3.2 km (2 mls) south-west of Olangolah, where Brazenor (1934) collected *Mastacomys fuscus*. The plantation, which covers some 2870 hectares (7000 ac) was established in 1929 to restore to a productive condition some of the extensive partially cleared land which had been selected under early settlement schemes in the 1880s and 1890s and later abandoned (Central Planning Authority, 1957). The plantation area is now forested with conifers of several species, mainly *Pinus radiata* and *Pseudotsuga menziesii*. It previously carried a typical wet sclerophyll forest, dominated by *Eucalyptus regnans* and *E. bicostata* with associated dense understorey plants. Little eucalypt forest now remains in the plantation area, but regrowth scrub clothes creek gullies, some roadside areas and the valley of the Aire River. It was in such areas of native bush that exploratory trapping by the author in August 1962 and April 1968 yielded specimens of *Mastacomys*.

Soils in the area are podsolic—friable grey loams on decomposing Jurassic sandstones (Central Planning Authority, 1957) — and are listed in the Atlas of Australian Soils (1962) as Hard Acidic Duplex Soils.

The mean annual rainfall at Beech Forest (which adjoins the north-west boundary of the Plantation) is about 170 cm, with the heaviest falls coming in winter and spring, but frequent mists and fogs, coupled with heavy dews maintain high moisture content in the bush throughout the year. Temperature data are not available for this section but it is likely that average summer maxima do not exceed 18°C with February being the hottest month. In winter, the average maxima are probably below 4.4°C with June being the coldest month (Central Planning Authority, 1957).

The areas in which *Mastacomys* has been collected were dominated by medium height (4.5-6 m) shrubs of *Acacia melanoxylon* with other shrub species such as *Cassinia longifolia*, *Olearia lirata*, *O. argophylla*, *Prostanthera lasianthos* and *Hedycarya angustifolia* scattered throughout. The lower shrub layer was either characterized by other shrubs, e.g., *Pimelia axiflora*, *Bursaria spinosa* and the introduced *Hypericum androsaemum* or was dominated by the tree fern *Dicksonia antarctica*. The lowest levels of the undergrowth were normally a dense tangle of *Tetrarrhena juncea*, *Rubus* sp., *Hypericum* and ferns, mainly *Polystichum proliferum*.

Blechnum nudum and *B. procerum*. In more open areas *Pteridium esculentum* was present, with *Senecio linearifolium*, *S. lautus* and occasionally the fern *Hystioropis incisa*. (On the river flats and some cleared hillsides the dominant vegetation was *Pteridium esculentum*). In these more open areas the ground cover species included *Prunella vulgaris*, *Acaena anserinifolia*, *Geranium* sp. and *Hydrocotyle* sp. In very moist situations the small ferns *Blechnum fluviatale* and *B. aggregata* were occasionally present, together with sedges and rushes such as *Carex appressa* and *Juncus procerus*.

Several other species of small terrestrial mammal were present in the scrub zones within the plantation. The most abundant of these was the bush rat, *Rattus fuscipes assimilis*. The swamp rat, *R. lutreolus* was common in some situations, and the introduced rodents *R. rattus* and *Mus musculus* were in small numbers, particularly in the vicinity of the Forest Commission camp site at the Aire River. Two small dasyurid marsupials, *Antechinus stuartii* and *A. swainsonii* were fairly common. In 1962 the author trapped a potoroo, *Potorous apicalis* along one of the many unamed creeks which feed the Aire River, about 1.6 km (1 ml) from the forestry camp. Black wallabies, *Wallabia bicolor* and ringtail possums, *Pseudocheirus peregrinus* were also found in the area.

9. *Benwerrin*

Benwerrin, 35.5 km (22 mls) northeast of Aire Valley is at an altitude of about 600 m (2000 ft) A.S.L. (lat. 38° 29'S, long. 143° 56'E). In June 1970, members of the Mammal Survey Group, F.N.C.V., collected an adult female *Mastacomys fuscus* during exploratory trapping. The locality was about 4.8 km (3 mls) south of Benwerrin junction, on the Mount Cowley Road. The physical characteristics of the area — soil and climate — are probably similar to those at Aire Valley though the rainfall is somewhat less, probably between 130 and 150 cm per annum.

The vegetation was a fairly open canopied forest, dominated at higher levels by tall (25-30 m) *Eucalyptus obliqua*, with scattered *E. cypellocarpa*. To the south down a fairly steep slope *E. cypellocarpa* became more dominant. The forest has been logged in the past and most trees did not exceed 45 cm in trunk diameter.

There was a tall shrub layer present, ranging in height from 2.5-3 m. Composition of this layer varied with altitude and degree of disturbance by man. Near the road it was predominantly *Acacia verticillata* but further downhill areas dominated by *Cassinia longifolia* and *Acacia verniciflua* occurred. Under the *E. cypellocarpa* in the wetter gully area many tree ferns, *Cyathea australis*, were present. Other shrub species were scattered throughout—*Pomaderris aspera*, *Olearia argophylla* and *Zieria smithii* — and many saplings of the eucalypt species were present. Lower shrubs included *Spyridium parviflorum*, *Pimelia axiflora* and *Goodenia ovata*, but the dominant feature of the lower levels was *Tetrarrhena juncea*, which was generally about 1.2-1.5 m high and formed a more or less continuous field layer. Scattered throughout were a number of other low-growing species — *Senecio linearifolius* (often forming large clumps), *Lepidosperma* sp. *Dianella tasmanica* and the fern *Blechnum nudum*, *Pteridium esculentum* (which was more common towards the road), *Polystichum proliferum* and occasionally *Hystioropis incisa*. Ground plants were present in some open patches, e.g., *Geranium g., Haloragis tetragyna, Viola hederacea* and *Acaena anserinifolia*, and there were many fallen limbs and logs.

The lower, dense layers were riddled with runways formed by small mammals, and it was on such a runway that the *Mastacomys* was trapped. Other small mammals trapped in the area were the bush rat, *Rattus fuscipes assimilis* (in large numbers) and the brown phascogale, *Antechinus stuartii*. Diggings probably made by the long-nosed bandicoot, *Perameles nasuta*, were seen but no specimens were trapped. The black wallaby, *Wallabia bicolor* was fairly common in the area, and the only phalangerid encountered was the ringtail possum, *Pseudocheirus peregrinus*.

10. Carlisle River

This locality (lat. 38° 33'S, long. 143° 25'E), 17.7 km (11 mls) north-west of the Aire Valley and at an altitude of about 150 m (500 ft) A.S.L. is very different from other areas in Victoria from where *Mastacomys* has been collected. The predominant vegetation characteristics were those of a wet heath. The single animal collected here in September 1965 by K. G. Peisley, Fisheries and Wildlife Department, was trapped in heathland close to a *Melaleuca* swamp. The area was visited again in June 1970, and it was apparent that much of the heathland had been burnt in the intervening years. The swamps were largely unburnt. The surrounding heathland and the low scrubby forest on higher ground was regenerating strongly.

This vegetation has developed on soils derived from Tertiary clay sediments. They are acid grey sandy loams overlying light clay or sandy-clay subsoils, and contain appreciable amounts of organic matter in the surface layers (Central Planning Authority, 1957).

Meteorological data for the locality is not available. Rainfall records from Gellibrand River West, 12.3 km (7.5 mls) south-west of Carlisle River show that station to have an average rainfall of about 107 cm with most rain falling in winter. Carlisle River lies midway between the 100 and 130 cm isohyets (Central Planning Authority, 1957) and rainfall is probably slightly higher than at Gellibrand River West.

There are no temperature recording stations near this locality but Central Planning Authority (1957) gives, in general terms, the following information for inland areas of the region. February is the warmest month Average summer maxima are probably about 27°C, minima about 13°C Winter average maxima are about 13°C and minima 4.5°C or slightly less, with July being the coldest month.

Swamps at this locality were composed mainly of tall (3-6 m) *Melaleuca squarrosa* with scattered *Leptospermum juniperinum*. Free water was present at the time of the second visit. Around the perimeter of the swamps the vegetation was only about 0.5 m tall, a mixture of *M. squarrosa* and *L. juniperinum* with *Sprengelia incarnata*, *Epacris lanuginosa*, *E. impressa*, *Gymnoschoenus sphaerocephalus*, *Xyris operculata*, *Lepidosperma filiforme* and *Restio tetraphyllus* present, scattered throughout. Close to the ground were *Selaginella uliginosa*, *Stylidium graminifolium*, *Tetratheca ciliata*, *Calorophus lateriflorus* and *Patersonia* sp. The latter group were all less than 25 cm tall. On the rising ground *Xanthorrhoea australis* became a dominant feature and in places was extremely abundant (see Plate 3). Adjacent

Plate 3. *Melaleuca* swamp and surrounding *Xanthorrhoea* heathland at Carlisle River

Photo: J. Seebeck

to the swamp was a sandy rise bearing scrubby open *Eucalyptus baxteri* and *E. radiata*, with *Dillwynia glaberrima* and *Banksia marginata* widespread, together with many other typically heath species forming a low, dense shrub layer.

Other native mammals collected at Carlisle River in 1965 were the bush rat, *Rattus fuscipes assimilis* and the swamp rat, *R. lutreolus*. The latter seemed to be more abundant. Local property owners reported the presence of echidna (*Tachyglossus aculeatus*), bandicoot (probably *Isoodon obesulus*), grey kangaroo (*Macropus giganteus*) and black wallaby (*Wallabia bicolor*).

11. *Glen Aire*

Mulvaney (1962) reported the excavation of two rock shelters, which had been used by aboriginals, near the mouth of the Aire River (lat. 38° 48'S, long. 143° 29'E). Some mammalian bone remains were recovered from stratified layers in Shelter 2. These remains included *Mastacomys* material in the upper-most layer (Layer 1). This material has recently been examined by the author, and consists of six maxillary tooth rows and a single damaged right mandible; the remains probably represent three individuals.

Mulvaney considered that accumulation of Layer 1 probably began less than a century ago. Radio carbon dating of charcoal from Layer 4, 2 m deep, has given an age of 350 ± 45 years B.P.

Other mammal remains recovered from this layer included *Rattus lutreolus*, *Rattus fuscipes assimilis*, *Sminthopsis leucopus*, *Perameles nasuta* and *Trichosurus vulpecula*. All these species presently occur in the Glen Aire district.

DISCUSSION

The present distribution of *Mastacomys* is considered to be relict by Calaby and Wimbush (1964). This is supported by the observations of Green (1968) and those of the present author, though the distribution of the living species in Victoria is now shown to be much wider than previously reported. Wakefield (1967) and Green (1968) have also demonstrated wider distribution as a subfossil. It is probable that future investigation in Victoria, particularly in the alpine regions will show continuity with New South Wales populations, even though colonies may be disjunct.

In Victoria, the reported habitat for *Mastacomys* has been the dense undergrowth associated with wet sclerophyll forest, dominated by *Eucalyptus regnans* or *E. obliqua*, at altitudes below 700 m (2300 ft). The habitat at Kalorama, Belgrave, Pennys Saddle, Aire Valley and Benwerren is of this type. The micro-habitat in which *Mastacomys* was collected at Mount St. Bernard is wet sedgeland. This form of habitat has been reported by Green (1968) at Cradle Mountain, Tasmania, although that area lacks top cover as is provided by *E. delagatensis* at Mount St. Bernard. The Mount Baw Baw vegetation is different from other Victorian localities at which *Mastacomys* has been collected. The single specimen so far collected was taken in *E. pauciflora* woodland, but close to extensive areas of wet alpine heathland, similar to that described by Calaby and Wimbush (1964) for Whites River, Mount Kosiusko, New South Wales and alpine regions in Tasmania (Green 1968). The individual concerned may have been extralimital from the heathland vegetation association. The Falls Creek habitat is also similar to other high altitude associations reported.

The occurrence of *Mastacomys* at Carlisle River is of some significance. The habitat at this locality is structurally similar to alpine heathland but with a much lower rainfall. Wakefield (1963) has reported the presence of *Mastacomys* (as a subfossil) in drier zones of Victoria. It is possible that living *Mastacomys* may be present in isolated colonies in similar wet heath in the Grampians and south-western Victoria.

At all sites examined, the population density of *Mastacomys* has been low. Calaby and Wimbush (1964), Warneke (1960) and Green (1968) have all reported a similar situation, and Calaby and Wimbush point out that in all situations *Mastacomys* is associated with one or more species of *Rattus* in much greater abundance. This is also the situation with all localities so far reported in Victoria.

Acknowledgements

I am indebted to Mr. R. M. Warneke, Fisheries and Wildlife Department, Victoria, for permission to include valuable unpublished material and for his critical appraisal of the manuscript. My thanks also are due to the Mammal Survey Group, F.N.C.V., for permission to include unpublished material from their survey data, and Mr. N. A. Wakefield and Miss Joan Dixon for allowing me to include several locality records thereby permitting a more complete distribution record than would otherwise have been possible. The Director and staff of the National Museum of Victoria permitted me to examine the Glen Aire material. Mr. G. Barnes, Fisheries and Wildlife Department, Victoria, shared much of the field work with me, and Miss H. Aston, Mr. A. Court and Mr. A. C. Beauglehole of the National Herbarium kindly identified plant material.

TABLE 1

Mastacomys fuscus specimens in (a) Fisheries and Wildlife Department and (b) National Museum of Victoria collections.

Number	Sex	Locality	Date
(a) RF 2408	♂	Aire Valley, Maddens Hill Rd.	10 Aug 1962
R. 2954	♀	Kalorama, Jasper Rd.	5 Feb 1965
R. 3045	♂	Carlisle River, near Rifle Butts	30 Sep 1965
R. 3116	♀	Mt. St. Bernard, Dargo Rd.	14 Feb 1967
R. 3446	♂	Pennys Saddle	9 Jun 1968
R. 3447	♂	Pennys Saddle	9 Jun 1968
R. 3459	♀	Aire Valley	14 Apr 1968
R. 3527	♂	Kalorama, Erith Lane	20 Sep 1968
R. 3528	♀	Kalorama, Erith Lane	22 Sep 1968
R. 3529	♀	Kalorama, Erith Lane	23 Sep 1968
R. 4131	♂	Mt. Baw Baw	31 Mar 1969
5170	♀	Benwerrin	13 Jun 1970
5468	♂	Kalorama, Erith Lane	10 Oct 1970
5563	♀	Kalorama, Erith Lane	8 Nov 1970
(b) C. 9998	—	Mount Hotham	Oct 1966
C. 9999	♂	Belgrave	Jul-Aug 1970
C. 10000	♀	Belgrave	Jul-Aug 1970
C. 10001	♂	Belgrave	Jul-Aug 1970
C. 10002	♂	Belgrave	Jul-Aug 1970
C. 10003	♂	Upper Thompson River	22 May 1971
C. 10065	♂	Falls Creek	24 May 1971

REFERENCES

Anon. (1962). Victoria's Mountain Regions, pp. 43-67 in *Victorian Year Book No.* 76. Commonwealth Bureau of Census & Statistics, Melbourne.

Anon. (1964). Soils of Victoria, pp. 1-10 in *Victorian Year Book No.* 78. Commonwealth Bureau of Census & Statistics, Melbourne.

Atlas of Australian Soils (1962). Melbourne-Tasmania area, Sheet 2. Melbourne University Press, Melbourne.

Brazenor, C. W. (1934). A new species of mouse, *Pseudomys* (*Gyomys*), and a record of the broad-toothed rat, *Mastacomys*, from Victoria. *Mem. natn. Mus. Vict.* **8**: 158-61.

Calaby, J. H. & Wimbush, D. J. (1964). Observations on the Broad-toothed rat, *Mastacomys fuscus* Thomas. *CSIRO Wildl. Res.* **9** (2): 123-33.

Central Planning Authority, Victoria (1957). *Resources Survey, Corangamite Region*. Govt. Printer, Melbourne.

Costin. A. B. (1962). The soils of the high plains. *Proc. R. Soc. Vict.* **75** (2): 291-99.

Dixon, Joan M. (1971 a). *Burramys parvus* (Broom) (Marsupialia) from Falls Creek area of the Bogong High Plains, Victoria. *Victorian Nat.* **88**: 133-138.

———. (1971 b). The Broad-toothed Rat *Mastacomys fuscus* Thomas from Falls Creek Area, Bogong High Plains, Victoria. *Victorian Nat.* **88**: 198-200.

Green, R. H. (1968). The Murids and Small Dasyurids in Tasmania. Part 4. *Rec. Queen Vict. Mus. Launceston* **32**: 12-19.

Hogg, D. (ed.) (1970). *Guide to the Victorian Alps*. Melbourne University Mountaineering Club.

Lyndon, E. (1960). Broadtoothed rat in South Gippsland. *Victorian Nat.* **77**: 193-4.

Mulvaney, D. J. (1962). Archaeological Excavations on the Aire River, Otway Peninsula, Victoria. *Proc. R. Soc. Vict.* **75** (1): 1-15.

Neilson, J. L. (1962). Notes on the Geology of the High Plains of Victoria. *Proc. R. Soc. Vict.* **75** (2): 277-84.

Wakefield, N. A. (1963). Mammal subfossils near Portland, Victoria. *Victorian Nat.* **80**: 39-45.

Wakefield, N. A. (1967). Preliminary Report on McEachern's Cave, SW Victoria. *Victorian Nat.* **84**: 363-83.

Warneke, R. M. (1960). Rediscovery of the broadtoothed rat in Gippsland. *Victorian Nat.* **77**: 195-6.

Warneke, R. M. (1964). Life History and Ecology of the Australian bush rat, *Rattus assimilis* (Gould), in Exotic Pine Plantations. Unpublished M Sc. Thesis, University of Melbourne.

Photoflora '72

The Native Plants Preservation Society is now planning Photoflora '72, its seventh biennial competition and exhibition featuring colour slides of Australian wildflowers.

Entry forms are now available from the Competition Secretary, Miss B. C. Terrell, 24 Seymour Avenue, Armadale, 3143. Public screenings have been arranged in 19 suburban and country centres during March and April, 1972. Details of these will be published later.

Flowers and Plants of Victoria in Colour

Copies of this excellent book are still available, and of course would make a wonderful gift. They are obtainable from the F.N.C.V. Treasurer, Mr. D. McInnes.

book review

Australian Butterflies
by Charles McCubbin
206 pp. 13½" x 9½"
Published by Nelson. Price $25

In view of the widespread interest in Australian butterflies, it is remarkable that, until now, no definite book has appeared on the subject since Waterhouse and Lyall's "Butterflies of Australia" in 1914 and "What Butterfly is That?" by Dr. Waterhouse in 1932.

This superb book by Charles McCubbin illustrates and describes all the known Australian species, over 330, and many varieties. The text, which is written in a pleasant yet authoritative style, contains more information about our butterflies than has appeared in any one book before. A page and a half of acknowledgments and many more throughout the text testify to the fact that Mr. McCubbin has drawn heavily on the accumulated experience of his many friends, nevertheless much of the information is a result of his own painstaking observation.

As well as a description of each species and their known varieties, there is a great deal of information about their habitat, food plants, life history, range and habits. All this is written with a minimum of technological jargon and is thus fully comprehensible to the average reader but still accurate and scientific. The author has resisted the temptation to write a detailed entomological introduction but has given a concise and quite up to date outline of the taxonomy and morphology of butterflies in 12 pages. A comprehensive section on collection, mounting and storage, a short glossary of entomological terms, a most useful bibliography compiled by Mr. Arturs Neboiss and two indices one for butterflies and one for plants completes the text of the book.

It is not the text, however, that makes this such a distinctive book, but the illustrations. We do not find here a hundred pages of unadorned text followed by 50 or so plates of butterflies in neat rows as is so often the case but throughout the whole book are Mr. McCubbin's delightful watercolours. Every butterfly is depicted life size and with remarkable accuracy. In a great many cases they are shown with their food plant and in some cases with their larvae or other aspects of their life history. Many of the full page or double page paintings have a background of landscape depicting the region where the butterflies are found and such is Mr. McCubbin's art that these places are mostly instantly recognisable to the naturalist without reference to the text. A high degree of accuracy has been achieved in the rendition of colour as will be seen on comparing actual specimens with the illustrations and here praise is also due to the printer for the excellent quality of reproduction. This is particularly noticeable with the subtle blues and purples of many of the Lycaenidae.

All in all, this book, which would delight Sir Charles Snow in its effective fusing of "The Two Cultures", will be a valuable reference work for the scientist and student, an absolute must for all naturalists even remotely interested in Australian butterflies and a delight to all as a work of art.

Peter Kelly.

Victorian Non-Marine Molluscs — 6

by

BRIAN J. SMITH*

Freshwater Mussels—Family HYRIIDAE.

Three major groups of freshwater bivalves occur in Victoria, two being groups of small to medium-sized bivalves to be described later, and the group of large black-shelled molluscs known as freshwater mussels, which belong to the family Hyriidae. Seven species of freshwater mussels are described for Victoria but only four can be described as common. The shells are all large, fairly solid, and dark brown to black in colour. They can be found in rivers, creeks, lakes and dams, and live buried in sandy mud with only the posterior end protruding from the sand to filter food particles from the water. The species in this group are difficult to identify as a high degree of individual variation exists in most species. However, the mean characters of a sample from the population coupled with a knowledge of the locality should enable most mussels to be identified in Victoria.

Velesunio ambiguus (Philippi 1847)

This is a large, elongate-oval mussel, only slightly winged posteriorly and in many waters the beaks show a great deal of erosion in old specimens. This is widespread throughout the State, being found commonly in the coastal rivers both of Gippsland and Western Victoria, and in the Murray and its tributaries. It is also found readily in farm dams where it has been introduced by birds or on introduced fish. This species also occurs throughout much of N.S.W., Queensland and eastern South Australia. Average length 70 mm.

Alathyria jacksoni (Iredale 1934)

This species has a larger, heavier shell than the previous species and shows more pronounced wings posteriorly. However, in Victoria it is confined to a small area in the far east of the State and to the Murray area. It is not a common species but does occur throughout much of the Murray-Darling Basin. Average length 105 mm.

Genus Hyridella

Five species of this genus occur in Victoria, all but one being very similar to each other, belonging to the subgenus *Hyridella sensu stricto*. These are small to medium shells, reaching about 100 mm in length, and tend to be elongate in shape. Some are localized to one river system only while others are fairly rare.

Hyridella (Hyridella) australis (Lamarck, 1819)

This has a fairly large, oblong-ovate shell, with prominent posterior ridges and heavy beak sculpture. It is found in the coastal rivers of central and eastern Victoria and up the coast of N.S.W. into southern Queensland. However, it does not appear to be a common species. Average length 60 mm.

Hyridella (Hyridella) depressa (Lamarck, 1819)

This is more elongate than the previous species, but its most marked character is that the dorsal margin

*Curator of Invertebrates, National Museum of Victoria.

325

anterior to the beaks slopes away markedly. In Victoria it is confined to a few coastal rivers of east Gippsland, where it is not common, and extends up the coast of N.S.W. to southern Queensland. Average length 50 mm.

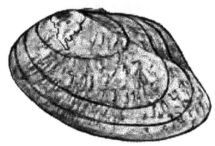

Fig. 1 *Velesunio ambiguus.*
Fig. 2 *Hyridella (Hyridella) drapeta.*
Fig. 3 *Hyridella (Protohyridella) glenelgensis.*

(Drawn by Miss Rhyllis Plant).

Hyridella (Hyridella) drapeta
(Iredale, 1934)

This species is intermediate between the previous two species but is separated from them by its elongate ovate shape, the flattened posterior ridges and the delicate beak sculpture. This is the commonest and most widespread species of the genus in Victoria being found fairly abundantly in most coastal rivers and streams of the central and eastern part of the State; also in some rivers in the Otways area. Outside Victoria it is also found in the coastal rivers of N.S.W. and southern Queensland. Average length 80 mm.

Hyridella (Hyridella) narracanensis
(Cotton and Gabriel, 1932)

This has a small shell with a distinctive almond-shaped outline and a small, very strong hinge. It is confined to coastal rivers of central and eastern Victoria, and the Tamar River system in northern Tasmania. Average length 30 mm.

Hyridella (Protohyridella) glenelgensis
(Dennant, 1898)

This highly distinctive species is placed in a different sub-genus from all other species of the genus in Victoria because of its small size, the strong shell sculpture, and exceptionally strong hinge. It is almond-shaped and is winged posteriorly with a prominent posterior ridge.

This species appears to be confined to the Glenelg-Wannon river system of far western Victoria. Average length 30 mm.

REFERENCE

McMichael, D. F. and I. D. Hiscock. 1958. A Monograph of the freshwater mussels (Mollusca: Pelecypoda) of the Australian Region. *Aust. J. Mar. Freshw. Res.* 9: 372-507.

SPECIAL MEETING
The Future of the F.N.C.V.

The December meeting will be devoted to discussing suggestions for improving and modernising the Club, arising from the recent questionnaire. If you have any constructive suggestions to make concerning the aims, activities, or operation of the Club, or if you would like to speak at the above meeting, please notify the Secretary in writing by 30 Nov. enclosing a summary of your ideas.

R. H. RIORDAN,
Hon. Sec.

15 Regent Street,
East Brighton, 3187.

GEOLOGY GROUP EXCURSIONS

Sunday, 7 November — To Sovereign Hill Historical Park, Ballarat. Leader: Mr. Bert Strange who will arrange demonstrations of gold panning, and use of the cradle and sluice box. Admission 50 cents.

Sunday, 5 December — To Lederderg Gorge with the Hawthorn Juniors.

Transport is by private car. Spare seats are usually available for those without their own transport. Excursions leave from the western end of Flinders Street Station (opposite the C.T.A. Building) at 9.30 a.m.

BIOLOGICAL MICROSCOPE OBJECTIVES
9X, 40X, 40X water immersion, 90X oil immersion

EYEPIECES
7X, 10X, 15X SUBSTAGE CONDENSER	$79.35
MECHANICAL STAGE	$27.60
BINOCULAR HEAD	$90.85
SUBSTAGE LAMP	$17.25

GRIFFIN & GEORGE BEAM BALANCE
2 mgm – 250 gm $30.00 WEIGHTS $11.33

8 dram plastic flip-top vial (ideal for field collecting) 80c. per dozen

PLASTIC PETRI DISHES 3½" diam.
90c. per packet of 10

Available from **GENERY'S SCIENTIFIC EQUIPMENT SUPPLY**
183 Little Collins Street, Melbourne, 3000
Phone: 63 2160

Field Naturalists Club of Victoria

Birds of Marsh and Lake
[Summary of talk given to the F.N.C.V., 13 September, by Miss Joan Forster.]

Miss Forster commenced with some of the lesser known items of bird anatomy and physiology. It is not always obvious which parts of the skeleton of the bird correspond to those in mammals; thus the thigh is short and the knee is hidden in the feathers, so that the first joint seen in the leg hare of feathers is the ankle. In the wing, the flight feathers are attached to the forearm and the main flight feathers to the hand. Large muscles attached to the breast bone pull the wings down. Heart rate is 120, doubles when flying, and is at an uncountable rate when frightened. The lungs contain interconnecting tubes through which the air may be passed back and forth, and in diving birds this may occur until most of the oxygen is absorbed.

Mounted specimens of water birds were on display as Miss Forster discussed each one.

The Grebe has feet set far back, no tail, and the wings are small and rounded. They have difficulty in taking off and prefer to dive for safety expelling air from the body to do so. They swim with periscoped head, and are quick in the water. The feet are lobed, there being flaps of skin on the toes and not webbing. The plumage is soft and velvety and is eaten by them and fed to their chicks. The nest is of soggy vegetation, and the four to six eggs are covered with decomposing material. The grebes are world wide and inhabit fresh water.

The Little Pied Cormorant is twenty-three inches long, black on the back and top of the head, with white starting above the eye and extending down the front. The feathers are gleaming and patterned. The beak is yellow with a hook. To this group belong the oldest fossil remains of any living order. The throat is a distendible pouch. The four toes are webbed and are used in propulsion as the bird swims and dives very fast. The wings are held open in the wind to dry. They perch with body upright and head held up. Their habitats are lagoons, ponds and rocky shores. They form colonies with other wading birds and lay from four to seven pale blue eggs.

The Herons are marsh birds with long legs, long necks, and stout bills. Some of the feathers are crumbled to a powder and are used for cleaning feathers when covered with slime or oil from an eel. The powder absorbs the oil and is then scratched out with a serrated claw on the middle toe. An oil for water proofing the feathers is obtained from a gland at the base of the tail. They wade in swamps and shallow water, some times up to their shoulders looking for their food of frogs, yabbies, and fish. They have a broad slow flight. The nest is a platform in a tree maybe up to one hundred feet. Two to four pale green blue eggs are laid in the spring. The back is a light chestnut, the front creamy white and the crown black with three long plumes from the back of the head.

The Black Duck inhabits fresh water and tidal estuaries and is found all over Australia. The colour is a rich dark brown, the feathers being edged with pale fawn. The bill is grey with a blue tip. These birds gather aequatic plants and animals from the mud. The eggs up to fourteen, are laid from July to December, and the nest may be far from water, or in a tree. How the fledglings get to the water is often far from certain. The female has a loud quack and the male a small voice.

The plumage of the Black Swan is black with a brownish tinge, and this bird was first seen in W.A. in 1697. Flocks of one thousand may be seen in pastoral lands in Victoria and there may be up to twenty thousand in one area. They frequent lakes, rivers, and shallow estuaries, and often move at dusk with trumpeting calls. Their food consists of aequatic plants and small water creatures, and it is thought that they can compete with stock for grass.

The Dusky Moorhen belongs to a group present in the Tertiary period, and a number of this group are now near to extinction. The colour is dusky brown and grey; bill and its base are red; the feet are red, and there is a scarlet garter above the ankle. The toes are attenuated, lobed not webbed, and the bird can run over lily leaves with wings outspread. As it walks it jerks its tail showing the white underneath. When swimming the head jerks back and still in time with the feet. It feeds on aquatic plants and animals in the shallow water of swamps and streams.

Miss Forster showed slides of some of these birds with young in the Botanic Gardens. These were followed by an excellent film of the bird life in an island lake where many species nest during the breeding season, and some remain during the drier months.

An appreciative audience expressed their thanks for this excellent survey of the subject of water birds given by Miss Forster.

Western Victorian F.N.C. Association
Report on Club Activities for year 1970/71

ARARAT

President: Mr. Stan. Kelly.   Secretary: Miss Z. Banfield.

Membership 17.   Meetings First Monday, during winter Saturday afternoon at C. of E. Hall. Excursions during the year proved interesting to members.

In October and November children from two Melbourne primary schools during their "Mobile Excursion" week, visited Ararat and were conducted along a Nature Trail in McDonald Park. Short talks were given to these children on orchids, trigger plants and local trees, etc. Three hundred children and teachers were in the two groups.

BALLARAT

President: Secretary: Mrs. E. Bedggood.

Membership 40.   Meetings First Friday at School of Mines.   Excursions Second Sunday each month, meetings and excursions well attended.

The Forests Commission has been approached re the extension of the existing reserve of 185 acres at Mt. Beckworth — final reply awaited.

Our club together with Geelong F.N.C. has arranged a meeting in the near future with the Forests Commission regarding the future of Mt. Cole. Members are listing the flora and fauna of the area.

Interest is being taken in the conservation of wet lands in the Ballarat and Western Districts.

COLAC

President: Mr. R. Missen   Secretary: Mrs. G. Skinner.

Membership 55.   Meetings Second Friday C.W.A. rooms. Excursion following day.

The year's programme has been varied. Local and visiting speakers have covered such topics as interstate trips, aquatic plants, Mallee parrots, Otways land surveys, non-marine mollusc census, Kruger National Park, South Africa. Excursions have been made to Grey River Reserve, local lakes, Beauchamps Falls and Gellibrand and Carlisle areas. Members are actively interested in the preservation of Turton's Track in the Otways and one-third of an acre has been developed near Barongarook in a tree-planting venture.

CRESWICK

President: Mr. L. M. Williams.   Secretary: Mr. H. L. Barclay.

Membership 30.   Meetings Third Thursday each month, Wesley Hall, Victoria St.

The year 1970 saw our Club grow in membership and with the added numbers greater interest and activities almost 100% attendance at meetings, and well over that per cent at excursions (this number swelled with interested friends).

Members joined in well at camp-outs usually extending their stays to enable further excursions in the various districts visited.

Our Club library is now operational with a good selection of books and periodicals available, to enable further additions members subscribe to a special fund for the purchase of recommended books.

Meetings have been well varied with subjects, two nights were set aside for members to have their say on own choice subjects, prepare and present a paper, these nights proved a highlight.

In common with so many other clubs we are watching our near bush-land being cleared for pine plantations, after much work and interviews we have now been assured that the main and choicest area containing Sluty Creek and tributaries are to be left intact. So we will still have a small piece of bush-land close at hand.

DONALD
President: Mr. H. J. Falla. Secretary: Mrs. B. K. Grewar.
Membership. Adult 28, Junior 7 Meetings First Friday, Primary School Centre.

Excursions were held to Yarb Plains, Maldon and with Stawell F.N.C. to Roses Gap and Mt. Langi Gherin, Delegates have attended each W.V.F.N.C.A. meeting Inspections of a small area suitable for a wildflower reserve have been made. Interest is maintained in matters of conservation. Guest speakers to meetings include Mr Barry Golding - "The Geology of the Donald District", Mr. J. Ros Garnet — "Australia's Nat. Parks", Mr. C. P. Whelan — "The Wimmera and Mallee Water Supply System", Mr. Ken Barket — "Introduced Plants". In addition three members' nights were held.

GEELONG
President: Mr. J. Wheeler. Secretary: Mr. Gordon McCarthy.
Membership 400 Meetings First Tuesday, McPhillimy Hall.

Excursions have taken place to a variety of areas and were well attended Three were by bus. Two weekend camp-outs were slated during the year and the Otways Survey Group within the club hold bi-monthly camp-outs.

Many conservation projects were supported.

HAMILTON
President: Mr. Murray Gunn. Secretary: Mr. John James.
Membership: 32 Adults, 6 Juniors Meetings Third Thursday, Gray St Primary School.

Meetings have been addressed by local and visiting speakers During Hamilton's "Yulunga Festival" Mr. Roy Wheeler addressed a large gathering. Other visiting speakers were Mr. Don. Taylor, Mr. Ralph Illidge and Mr. Jack Gillespie.

Excursions were held monthly and in October the club joined Portland F.N.C in a visit to the Smoky River at Hotspur.

Our club has members on various committees including Grange Burn Development, Western Victorian Conservation Committee, and work on a case for the declaration of Mt. Napier as a Nat. Park is under way.

A native shrubs and tree-planting project was undertaken at the Town Hall Car Park and members supported the Nature Show in Melbourne

HORSHAM
President: Mr. C. Kroker. Secretary: Miss S. Robertson.
Membership 54 Meetings Fourth Thursday, Horsham Tech. School.

Each meeting members are allowed 7 minutes to speak on a subject of their choice. This has been successful. The annual Club Essay Competition saw a record number of entries—450. A high standard was attained.

Local and guest speakers were varied — "A Field Naturalist in Tonga", "A Trip to Ouldea, W.A., to search for a rare prostrate Eucalypt", "Travel in the South Pacific", "Multiple Use of Land", "Fire Control and the work of the C.F.A", "Geology of the Grampians", and visitors and members viewed "Photoflora '70".

Excursions were held locally. Two were to the Little Desert and the Toolan scrub lands.

Members assisted in the formation of the now active Arapiles Conservation Committee.

The fencing project at the John Smith Memorial Sanctuary is now complete and debt free.

The remainder of this report will be published next month.

F.N.C.V. DIARY OF COMING EVENTS
GENERAL MEETINGS

Monday, 8 November — At National Herbarium, The Domain, South Yarra, commencing at 8 p.m.
 1. Minutes.   2. Announcements.
 3. Presentation of Australian Natural History Medallion to Mr. A. C. Beauglehole by Mr. A. Dunbavin Butcher, chairman of the Medallion Committee.
 4. New Members.

Ordinary:
 Miss Ana Maria Bailey, 7 Beacon St., Glen Waverley, 3150.
 Mr. Andrew J. Cope, "Nummurcena," Angus Ave., Ringwood East, 3135.
 Mr. Ian Maxwell Hall, 27 Anthony Ave., Doncaster, 3108.

Joint:
 Miss Lois Garland, 51 Grosvenor St., Balaclava, 3183.
 Mr. Kirwin L. Pickering, Mrs. Faye A. Pickering, 19 Callander St., Hughesdale, 3166.
 Mr. Desmond Spittall, Miss Denise Deerson, 24 Magnolia Rd., Gardenvale, 3185.

Country:
 Mrs. Judith L. Wood, Aradale Hospital, Ararat, 3377.

 5. Subject for evening: "The East Gippsland Survey" by Mr. Beauglehole.
 6. Correspondence.   7. General Business.

Monday, 13 December — Special Meeting. See page 327 of this issue.

F.N.C.V. EXCURSIONS

Sunday, 21 November — Healesville-Warburton area. The coach will leave Batman Avenue at 9.30 a.m. Fare $1.60. Bring two meals.

Sunday, 12 December — Beach excursion, subject Marine Biology. The coach will leave Batman Avenue at 9.30 a.m. Fare $1.70. Bring two meals.

Sunday, 26 December-Monday, 3 January — Fall's Creek. Accommodation has been booked at Spargo Lodge, tariff $55.00 to $60.00 per week, full board. Fare for coach, including day trips $22.00. The fare should be paid to the Excursion Secretary when booking, cheques to be made out to Excursion Trust. The coach will leave from Flinders St. outside the Gas and Fuel Corporation at 8 a.m. Bring a picnic lunch.

Group Meetings
(8 p.m. at National Herbarium unless otherwise stated.)

Thursday, 11 November — Botany Group. Some plants seen in the Flinders Ranges. Mr. I. Morrison will show his slides taken on the recent excursion supplemented by slides from other members. Will all members who have slides of botanical subjects make a list of the plants taken and bring a selection of those they would be prepared to show in time for the additional slides to be selected before the meeting starts.

Friday, 12 November — Montmorency and District Junior F.N.C. meets at 8 p.m. in Hall at Petrie Park.

Wednesday, 17 November — Microscopical Group.

Friday, 26 November — Junior meeting at 8 p.m. in the Hawthorn Town Hall.

Wednesday, 1 December — Geology Group.

Thursday, 2 December — Mammal Survey Group meets at 8 p.m., in Arthur Rylah Institute for Environmental Research, 123 Brown St., Heidelberg.

Monday, 6 December — Marine Biology and Entomology Group meeting at 8 p.m. at National Museum in small room next to theatrette.

Thursday, 9 December — Botany Group.

Field Naturalists Club of Victoria

Established 1880

OBJECTS: To stimulate interest in natural history and to preserve and protect Australian fauna and flora.

Patron:
His Excellency Major-General Sir ROHAN DELACOMBE, K.B.E., C.B., D.S.O.

Key Office-Bearers, 1971-1972.

President:
Mr. T. SAULT

Vice-Presidents: Mr. J. H. WILLIS; Mr. P. CURLIS

Hon. Secretary: Mr. R. H. RIORDAN, 15 Regent St., East Brighton, 3187. (92 8579)

Subscription Secretary: Mr. D. E. McINNES, 129 Waverley Road, East Malvern, 3145.

Hon. Editor: Mr. G. M. WARD, 54 St. James Road, Heidelberg 3084.

Hon. Librarian: Mr. P. KELLY, c/o National Herbarium, The Domain, South Yarra 3141.

Hon. Excursion Secretary: Miss M. ALLENDER, 19 Hawthorn Avenue, Caulfield 3161. (52 2749).

Magazine Sales Officer: Mr. B. FUHRER, 25 Sunhill Av., North Ringwood, 3134.

Group Secretaries:

Botany: Mr. J. A. BAINES, 45 Eastgate Street, Oakleigh 3166 (57 6206).
Microscopical: Mr. M. H. MEYER, 36 Milroy Street, East Brighton (96 3268).
Mammal Survey: Mr. D. R. PENTON, 43 Duke Street, Richmond, 3121.
Entomology and Marine Biology: Mr. J. W. H. STRONG, Flat 11, "Palm Court", 1160 Dandenong Rd., Murrumbeena 3163 (56 2271).
Geology: Mr. T. SAULT.

MEMBERSHIP

Membership of the F.N.C.V. is open to any person interested in natural history. The *Victorian Naturalist* is distributed free to all members, the club's reference and lending library is available, and other activities are indicated in reports set out in the several preceding pages of this magazine.

Rates of Subscriptions for 1971

Ordinary Members	$7.00
Country Members	$5.00
Joint Members	$2.00
Junior Members	$2.00
Junior Members receiving Vict. Nat.	$4.00
Subscribers to Vict. Nat.	$5.00
Affiliated Societies	$7.00
Life Membership (reducing after 20 years)	$140.00

The cost of individual copies of the Vict. Nat. will be 45 cents.

All subscriptions should be made payable to the Field Naturalists Club of Victoria, and posted to the Subscription Secretary.

Victorian Naturalist

Vol. 88, No. 12 DECEMBER, 1971

Published by the
FIELD NATURALISTS CLUB OF VICTORIA
In which is incorporated the Microscopical Society of Victoria

*Registered in Australia for transmission by post as a periodical.
Category "A"*

45c

Magnificent stand of White Mountain Ash, *Eucalyptus regnans*, in the Marysville State Forest

FORESTS COMMISSION VICTORIA

. . . . . preserving the beauty of our forests for your enjoyment.

The Victorian Naturalist

Editor: G. M. Ward
Assistant Editor: G. Douglas

Vol. 88, No. 12   8 December, 1971

CONTENTS

Articles:

Aboriginal Relics on the Mornington Peninsula. By A. E. Spillane .. 336
Further Notes on Tool-using by Birds. By A. H. Chisholm 342
Rock Fractures Called Joints. By A. W. Beasley 347
Winter Insect Collecting in Northern Territory. By J. C. Le Souëf .. 350

Personal:

Australian Natural History Medallion Award. By Margaret G. Corrick 344
The A. E. G. (Rudd) Campbell Memorial Appeal 349

Features:

Victorian Non-marine Molluscs — No. 7. By D. C. Long 357
Reptiles of Victoria — No. 4. By Hans Beste 358

Field Naturalists Club of Victoria:

General Meeting Report 360
Western Victorian F.N.C.s. 361
Diary of Coming Events 363

Front Cover:

The Tree Kangaroo (*Dendrolagus lumholtzi*) inhabits the north-eastern rainforests of Queensland; and was one of two species of tree kangaroo which crossed between New Guinea and Australia when Torres Strait was exposed as land. The other species is Bennett's Tree Kangaroo. Photo: John Wallis.

Aboriginal Relics on the Mornington Peninsula

by A. E. SPILLANE

(Illustrations by author)

The Mornington Peninsula consists of an area of about 300 square miles; made up mainly of gently undulating land, the highest peak of which is Arthur's Seat (1,031 feet) situated at Dromana. Geologically, the Peninsula is very interesting, as it contains many of the geological formations found in Victoria. The base rock is Ordivician and Silurian. The climate is quite temperate, with an annual rainfall of about 28 inches.

Matthew Flinders was one of the first explorers to find evidence of Aborigines on the Peninsula. In 1802 he entered Port Phillip Bay, and in the course of his investigations climbed Arthur's Seat; and at the summit noticed deserted camp fires and heaps of oyster shells which had been left by the natives. Looking towards the southern end of the Peninsula, he observed smoke from their fires.

The Bunurong tribe of Aborigines inhabited the Peninsula when Australia was discovered. Not a great amount is known about this tribe — but by going back through the records of Victoria's short history, and by studying the remains of camp sites, middens, and other relics; we are able to obtain some idea of how these people lived before the first settlers arrived.

The Bunurong people were mainly a coastal tribe; and their territory extended from the Werribee River, through the southern suburbs of Melbourne, the Mornington Peninsula, and the rest of the coast to Anderson's Inlet. The Bunurong in turn belonged to the Kulin Nation, which was made up of numerous tribes. These tribes occupied quite a large area of Victoria.

The Bunurong people met the natives of the Kurnai tribe at Anderson's Inlet, where they intermarried. The Kurnai tribe which was divided into five clans, occupied most of Gippsland.

As winters on the Peninsula are cold, the natives would have resorted to the use of kangaroo and possum skins for clothing, and the erection of mias mias made from strips of bark for shelter.

Food would have been no problem; as the records of early settlers provide ample evidence of an abundance of animal life, including kangaroos and emus. The seas surrounding the Peninsula would have supplied numerous species of fish, while the rocky sections of the coast carried a large range of shellfish.

Vegetable food would have been available; as the following plants are found in the area under consideration, and are known to have been eaten by Victorian Aborigines: The Native Cherry (*Exocarpa*); various types of fungi; the spikes and stalks of the Grass-Tree (*Xanthorrhoea*); gum exuded by various *Acacias*; the nectar from *Banksia marginata* and *Banksia integrifolia*. Grass and Acacia seeds were ground up and eaten; and tubers of the following orchids were consumed: *Dipodium*, *Prasophyllum*, *Microtis*, *Pterostylis*, *Lyperanthes*, *Caladenia* and *Glossodia*. The fern *Dicksonia antarctica* still occurs in

...nous gullies, and would have been useful source of food.

The Aboriginal stone tool industry [of] the Peninsula, was not as extensive [as] in some areas of Australia; nevertheless, the former occupants of this [part] of the Continent, showed great [skill] in the manufacture of the various [stone] implements needed in their [daily] life. The artifacts produced, [would] have been quite adequate for [the] fabrication of their spears, shields, [boomerangs], and the skinning of [animals] for food requirements and clothing.

Viewed collectively, an assemblage [of] implements from the Peninsula is [most] interesting, as it contains articles which range from very crude [flaked] pebble-choppers and knapped [flakes], to ground-edge axes and [microliths] which exhibit outstanding [workmanship]. Other types of stone [tools] used were hammer-stones, whetstones, knives, mills, and ground-edge chisels.

The ground-edge axe accompanied [the] natives wherever they went; and [was] used for tree climbing, the manufacture of spears, shields, canoes, and obtaining honey from trees. The handles were usually made from the wood of *Acacia deulbata*. It was [heated], bent over the axe-head, and [tied] together with the sinews from a Kangaroo; the head of the axe was [then] cemented into place with a [resinous] gum, sometimes obtained [from] the Grass Tree (*Xanthorrhoea*).

A person viewing the thriving city [of] Frankston today, would find it [hard] to imagine that about 135 [years] ago the area was a popular [haunt] of the Bunurong natives. The [evidence] of their former presence [was] fairly plentiful not so very long [ago], but unfortunately has now practically vanished due to the expansion of the city and other factors.

There was an aboriginal "workshop" about two and a half miles out of Frankston near Cranbourne Road; it has now disappeared due to soil being removed from the site. It was situated on a high sandy ridge from which a good view of the surrounding countryside would have been obtained. Microliths were collected at this camp; they were mostly made from fine grained quartzite and show a high degree of skilled workmanship.

There used to be a very fine example of a midden on the coast just south of Sweetwater Creek, but unfortunately it was destroyed when Nepean Highway at Oliver's Hill was made wider a few years ago.

Frankston has a gently sloping beach which is ideal for the night spearing of fish. As the Bunurong Aborigines were known to engage in these activities, using burning branches to attract fish, they undoubtedly used the beach for this purpose.

There are several creeks which drain into Port Phillip Bay between Frankston and Mount Eliza, and they would have supplied ample fresh water for the Aborigines when they were in the area. The rocky foreshore in this location must have supplied good quantities of shellfish such as oysters, mussels, limpets and periwinkles, as their remains are still in evidence on top of the cliffs.

Mount Eliza was a popular location with the Aborigines, as evidenced by the middens still to be seen on the coastline, but unfortunately they are gradually eroding away, and must disappear in the near future.

The writer has a very fine ground-edge axe-head which was found at Canadian Bay many years ago. It was made from diabase, evidently from the Mount William quarry near Lancefield, Victoria.

At the little inlet just south of Canadian Bay, a thick scum of oyster

Ground-edge Axe. Canadian Bay.

Microliths from The Blac Camp, Cape Schanck.

Whetstone (used for the sharpening of axes). Somerville site.

Shells left by the Aborigi Boags Rocks.

Plate 1

of shells may be observed. However, at Fossil Beach there are three middens, one in particular is quite large and it would be ideal for carbon dating as charcoal is plentiful.

There used to be an interesting camp site on the coast at Dromana. It was situated at the mouth of a small creek which had its outlet into the Bay just north of the present shopping centre. At this site the writer has found utilized pieces of different types of stone, and several microliths. Camps of this type are rare on Port Phillip Bay. Unfortunately, this area is no longer suitable for collecting, as filling has been spread over the former camping ground.

One of the best examples of a camp site on the Peninsula is situated at Portsea. It was discovered by Mrs. K Begg, on her eight acre property "Warrener", which is in Elizabeth Avenue on the Portsea Downs Estate The site is nicely positioned on a sandy well elevated piece of land, and there are several adjacent waterholes and lagoons, in all, an ideal camping ground.

The grounds of the property give ample evidence of past activities of the Aborigines, by the large amount of scattered shells, some of which are species found on the Bass Strait shoreline Over a period of about four years, Mrs. Begg has gathered together a very interesting collection of artifacts from this site. They include a pebble chopper, an anvil used in the manufacture of microliths, bondi points, burins, thumbnail scrapers, crescents, and triangles.

Archaeologically, the most interesting area of the Peninsula is the Bass Strait shoreline, and the country extending a few miles inland. It is an area of rugged and majestic beauty. A great part of the coast is made up of high basaltic cliffs, with their bases extending out into the Strait as reefs. From the evidence still remaining, it is evident that the main attractions of the Aborigines to this area were the sandy nature of the country, and the large amount of shellfish to be found on the rocky beaches. In the records of the early settlers, mention was made of the abounding wildlife to be seen, and this would have been an added inducement for the natives to frequent this locality. Inland, the country is mainly gently rolling hills, consisting of sand lightly held in place by grasses; and trees and shrubs which are mainly *Casuarinas*, *Leptospermums*, *Melaleucas* and *Leucopogons*.

Traces left by the Aborigines, may be observed along the coast extending from Point Nepean to Flinders. An example of these traces may be seen at Koonya Beach, Blairgowrie. At the eastern end of the beach there is a rocky headland jutting out into the Strait; on this platform are the remains of a large midden. Apparently this area was mainly used as a feasting ground, as no stone chips or implements are to be seen.

Between Rye Back Beach and Capri Beach, abundant evidence of the past presence of the Aborigines may be seen. The calcareous dune sand is drifting exposing shells in tremendous quantities. Stone chips and artifacts are very scarce indeed. About a quarter of a mile west of Rye, the writer found a scraper. A short distance east of Capri Beach, two microliths and a scraper were found — the microliths were very small and appeared to be very old.

South East of Capri Beach, are to be found some very interesting rock formations which are named Bougs Rocks. In between both ends of the Rocks, is the largest collection of Aboriginal shellfish remains on the Peninsula. Artifacts are extremely scarce, the writer has found only one large stone flake (no doubt used as a knife) in the area. Apparently the natives travelled long distances to collect the shellfish, and took most of the harvest back to permanent camps further inland.

There is a camp site about three-quarters of a mile inland from Gunnamatta Beach. A sand ridge composed of calcareous dune sand runs from Trueman's Road in a westerly direction, the camp is towards the west end. Abundant shellfish remains are scattered around, (mainly Turbos and Limpets) and a large blackened patch of sand can be seen where the natives evidently lit their fires. Stone material left by the former occupants is scarce in this locality; however, the writer has found several worked implements, mostly microliths, one of which is exceedingly small.

At Cape Schanck, there is a large midden on the top of the cliff which overlooks Pulpit Rock. A large amount of charcoal is mixed with the shells. The main stone material found here, is flint in the form of sharp flakes.

There was until recently, a large camp site just off the Cape Schanck Road, about half a mile from the sea. Wind blown sand had shifted, revealing cooking stones and artifacts just as the Aborigines had left them. Many beautifully made microliths have been gathered by collectors from this site. The camp has now been levelled, and planted with Marram Grass to hold the shifting sand.

The best known site on the Peninsula is the Blacks' Camp. It is situated at the intersection of the Flinders and Cape Schanck roads. It was ideally situated for a permanent camp. Sandy, well drained, it had plenty of water available from a large pond, and most importantly, it was sheltered from the strong sea breezes from Bass Strait. What makes this camp so interesting, is the fact that early settlers actually saw the Aborigines using it. Collectors have found a wide range of implements, particularly microliths, at this site over the years.

Signs of past Aboriginal activities on the coast of Western Port Bay are meagre. Where rocky beaches occur, small middens may be observed, but on the mangrove sections of the coastline, indications of their former presence are very rare.

The Bunurong natives showed a distinct preference for sandy country for the establishment of their camps. This is borne out by the fact that they are known to have had permanent camps in the sections of the Peninsula known as the Tyabb and Footgarook Sands. Whereas in the clay areas, traces of past sites are almost non-existent.

Some very interesting stone implements have been discovered on the property of the Shepherd family, Bramosa Road, Somerville. The orchard, consisting of 175 acres, is situated about two miles along Eramosa Road on the east side of the township. It was first occupied by the family in the eighteen sixties.

Mr. G. Shepherd mentioned to the writer, that many years ago the orchard had an abundance of Melaleucas growing on it; and he remembered his father telling him about the large Eucalypts which used to grow in the area. He kindly supplied for inspection some artifacts which the family had collected over the years. Most of them had been ploughed from a large area on a hilly piece of land on the property. The implements were as follows:

7 Ground-Edge Axes.
1 Broken Ground-Edge Axe.
1 Ground-Edge Chisel.
2 Whetstones (Used for the sharpening of axes).

The axes were in an excellent state of preservation, as was the chisel, which was about two inches long, by one and a half inches wide.

REFERENCES

Flinders, Matthew, 1814, "Voyage to Terra Australis", London.
Howitt, A. W., 1904, "The Native Tribes of South East Australia", London.
Keble, R. A., 1928, "Kitchen Middens on the Mornington Peninsula", *Victorian Naturalist*, XIV.
Le Souef, D., 1902, "On the Shellfish Food Remains of Aboriginals", *Victorian Naturalist*, XIX.
Massola, Aldo, 1959, "History of the Coast Tribe", *Victorian Naturalist* Vol. 76 (7).
Massola, Aldo, 1969, "Journey to Aboriginal Victoria", Melbourne.
Mitchell, S. R., 1948, "A Set of Aboriginal Stone Tools", *Victorian Naturalist*, 64.
Mitchell, S. R., 1949, "Stoneage Craftsmen", Melbourne.
Smyth, R Brough, 1878, "Aborigines of Victoria", Melbourne.

F.N.C.V. PUBLICATIONS AVAILABLE FOR PURCHASE

THE WILD FLOWERS OF THE WILSON'S PROMONTORY NATIONAL PARK, by J. Ros Garnet.
Price $5.25. (discount to members) postage 20c.

THE VEGETATION OF WYPERFELD NATIONAL PARK, by J. R. Garnet.
Coloured frontispiece, 23 half-tone, 100 line drawings of plants and a map. Price $1.50.

Address orders and inquiries to Sales Officer, F.N.C.V., National Herbarium, South Yarra, Victoria.

Payments should include postage (11c on single copy).

Further Notes on Tool-using by Birds

by A. H. Chisholm*

Since writing on the subject of the use by birds of tools and playthings (*Vict. Nat.* 88: 180), I have come upon other notes that supplement the earlier material; and as these are quite distinctive they appear to merit prompt recording.

One note is from my own diary, and therefore should have been recollected sooner. In brief, it concerns observations given me by Mr. and Mrs. Jack Harris of "Conobola", Stuart Town, mid-western N.S.W., on their property in 1967. During a general discussion of bird behaviour Mr. Harris related that Rainbowbirds (*Merops*) sometimes picked up shining pebbles from a garden path and tossed them in the air; and Mrs. Harris stated that White-winged Choughs often wrenched walnuts from a tree in the household garden and, flying to the ground, broke them against stones.

The question arises: Has the Rainbowbirds' curious practice of toying with bright pebbles any relevance to the claim (made, as stated earlier, by another rural resident) that these birds sometimes use "natural lamps" — white bones, mussel-shells, etc.,—in their nest-burrows? As for the Choughs' actions, they obviously strengthen the evidence (cited previously) that these adventurers rank as both primary and secondary tool-users, a fact indicated by their use of "hammers" to open mussels and by their occasional attempts to break a mussel by holding it in the beak and striking it against a hard piece of wood.

Incidentally, the smashing of walnuts as well as mussels, for the same purpose, gives Choughs the distinction of being tool-users in respect of both vegetable and animal foods. And, surely, tribute is due to the birds for their enterprise in having learned somehow, in each case, that edible matter lay beneath the hard shells.

The Egg-Smashers

In the earlier article I quoted a statement made by a Sydney naturalist, the late A. G. Hamilton, in *Bush Rambles* (1937), that "an African eagle" had been known to drop stones on Ostrich eggs. Hamilton did not give any authority for this report and I raised a query on the point. Now, it appears, the source may have been a book termed *Bible Animals*, by J. G. Wood, published in 1877. For, as disclosed in a recent issue of the *Journal of the East African Natural History Society*, (1969, p. 231), the Egyptian Vulture's habit of breaking Ostrich eggs with stones, so heartily proclaimed as a recent "discovery", was in fact recorded in Wood's book some ninety years earlier.

It is, of course, embarrassingly easy to overlook obscure records on a particular subject; and indeed naturalists need to be diligent readers if they are to keep in touch with material that is not obscure but is published in journals outside their own country. (Australian ornithological records of significance have often been neglected by writers else-

*History House, 133 Macquarie St., Sydney

its nest in the hollow of a tree. Lengthy observation, Kilham says, has convinced him that the beetles are not eaten, but are deliberately applied to the bark near the nest-hollow, in a bill-sweeping operation, so that their distasteful exudation may deter squirrels from commandeering the site.

This report — what its author terms his "final hypothesis" on a matter that has long puzzled him — may be regarded with some dubiety. It may appear to credit the little bird with undue "foresight". Reflection suggests, however, that the use of pungent beetles as a repellant has some affinity with the practice of anting: that is, the practice of many birds of rubbing ants and other strong substances (including beetles at times) beneath their wings and under the tails. The ultimate purposes in the two cases differ, of course, but there is similarity in the birds' appreciation of the effects of contact with acidulous substances.

Moreover, with scepticism in mind it may be well to recall that when, in 1934-35, I published in the Melbourne *Argus* basic reports on the remarkable practice that was to become known as "anting", several naturalists flatly rejected the evidence; and yet, since that time, writers in many countries have published scores of observations and discussions on the subject.

Clearly enough, discretion needs to be tempered with flexibility when consideration is being given to reports of extraordinary behaviour on the part of birds.

REFERENCES

Chisholm, A. H. "The History of Anting", *Emu*, 59, pp. 101-30, May, 1959.
Kilham, L. "Use of blister beetle in bill-sweeping by White-breasted Nuthatch", *Auk*, 88, p. 175, Jan., 1971.

Australian Natural History Medallion 1971

by

MARGARET G. CORRICK

Cliff Beauglehole was born at Gorae West, near Portland, in 1920. He was educated at the nearby country school and finally took over his parents' mixed farm, where he lived until moving to Portland in 1968.

His interest in nature was awakened by his upbringing and everyday contacts with the bush around him. By the time he left school he had a good knowledge of nature in his immediate environment. At the age of ten, he knew and could name scientifically, about sixty species of orchids from his district, and began sending specimens to the late W. H. Nicholls. The years 1941 and 1942 were outstanding orchid seasons, during which Cliff discovered two new species, *Prasophyllum beaugleholei* and *P. diversiflorum*. In 1949 he discovered *Pterostylis tenuissima*, another new species.

Soon afer leaving school he began a botanical survey of the Portland area, in which eventually seven hundred and seventy species were listed, an increase of over three hundred to the already known flora of the district. From the very beginning his collections were carefully labelled and preserved, and his herbarium now is among the largest private collections in the country. Over the years its resources have been made freely available to scientists all over Australia, and overseas. He has also made substantial donations from it to the reference sets of several Field Naturalist Clubs. His services as a voluntary collector to the Melbourne Herbarium were soon recognised, and he was accorded the rare privilege of being given a stock of official labels on which to enter data of every specimen donated.

Cliff's interest in entomology, and particularly in bees, seems a logical sequence for a farmer and naturalist already deeply involved in botany. Early in 1950 he began an association with the late Tarlton Rayment, which lasted until the latter's illness and death in 1956. During this period

Cliff worked tirelessly on the collection and study of local bees, resulting finally in the naming of thirty new species. Of these, three were named after the collector, and five after the district.

The years 1951 to 1963 also saw a development in Cliff's already deep interest in birds, when he undertook to organize the collection of beach washed specimens from the wild, and storm lashed shore of Discovery Bay. In all, a total of almost five thousand birds was collected by Cliff and his helpers, including the record total of nine hundred and fifty collected by Cliff and his wife on one day in August, 1959. Anyone inclined to dismiss this work as a pleasant way to spend a day at the seaside, should try carrying the carcases of two athanasses across a mile or two of soft sand. The Beauglehole family have vivid memories of the preparation of skeletal material by natural decomposition, which took place in specially built wire racks in the farm yard. These skeletons were eventually donated to museums in Australia and the United States. This project greatly extended knowledge of the sea birds visiting our shores, and several records of new species were added to the Australian and Victorian lists.

Also during the 1950's Cliff made an extensive survey of the many caves of south-west Victoria, to collect or record bone deposits. He had long been aware of the scientific value of these, and was becoming disturbed at the loss of material caused by rubbish dumping in the caves, excavations for road making materials, and by the indiscriminate scavenger for aboriginal implements. The culmination of this was the discovery of a huge deposit in McEachern's Death Trap Cave. Between January and March, 1963, Cliff and Fred Davies of Portland led a team of local naturalists in the excavation and sieving of several tons of cave sediments. As a result, identifiable remains of some two thousand animals were recovered. From September, 1964, to mid-1965 a second stage in the excavation was carried out under the supervision of Norman Wakefield (*Victorian Nat.* 84, 363-83) which yielded a further several hundredweight of bones. The final assessment of all this material will add immensely to knowledge of the faunal succession of the area.

In spite of the time given to a multitude of special natural history projects — often the collection of material for others engaged in particular research — Cliff continued his botanical exploration, gradually moving into new and wider fields, both botanically and geographically. He branched out into the difficult field of non-vascular plants, and covered the whole south-west area of Victoria in search of mosses, liverworts and lichens. He also studied and collected the fresh water and marine algae of his locality, discovering at least seven species of seaweeds previously unknown to science. His interest in mosses and fungi has been a continuing one, his amazing ability to seek out and recognise new and rare species has added immensely to scientific knowledge. Many of his non-vascular discoveries from all parts of Australia, and Lord Howe Island, particularly of fungi, still await study and classification.

In the 1960's Cliff widened his horizons still further, and undertook two trips to Lord Howe Island, one long trip through Central, North and West Australia, and three shorter trips to Central Australia. On each of these trips he made extensive botanical observations and collections.

Following his return from Central Australia in 1967, he embarked on a full scale botanical survey of the Grampians, believing that in spite of the great amount of work already done by others, there was still much to be discovered, a belief amply justified by his list of new records. This project also led to the first official financial recognition of his work. The interest of the Plant Survey Council of Victoria, and a grant from the Maud Gibson Trust, enabled him to work full time on the project, which became the first full scale research undertaken in Australia on the grid system. Following the Grampians survey he worked for the National Parks Authority in several Western Victorian Parks, and eventually moved on to new terrain in East Gippland.

Cliff's interest in existing, and proposed, National Parks has had high priority in recent years. The comprehensive plant lists now available for many of our Parks are due, in large part, to his efforts. He worked tirelessly in the promotion of the Mt. Richmond and Lower Glenelg National Parks.

Early in 1968 the Beauglehole farm was leased, and the family moved to Portland to enable Cliff to devote more time to his botanical work. He will shortly return to work in western Victoria for the newly formed Land Conservation Council of Victoria, which has commissioned him to do a full scale survey of the Shires of Portland and Glenelg.

Mention has already been made of new species of orchids, fungi, algae and bees discovered by Cliff. To this list of new discoveries must be added twelve species of ants from eastern Australia, believed to be undescribed; four new species of wasp from the Portland area, again one of which was named after Cliff; a new species of fly from Portland; seven new species of marine algae, one bearing Cliff's name; three new species of vascular plants, apart from the orchids previously mentioned, including the tiny trigger plant, *Stylidium beaugleholei*; and a fourth orchid only recently discovered in Gippsland, as yet unnamed.

The following list of species named after Cliff Beauglehole is a remarkable tribute to a modern naturalist's discoveries:—

Bees —
  Exoneura cliffordiella.
  Hylaeus cliffordiellus.
  Megachile cliffordi.

Wasp —
  Serocophorus cliffordi.

Marine algae —
  Helminthocladia beaugleholei.

Orchid —
  Prasophyllum beaugleholei.

Trigger plant —
  Stylidium beaugleholei.

Marsh-flower —
  Villarsia umbricola. var. *beaugleholei.*

There are also massive botanical lists of Cliff's first records for various States, ranging right through from conspicuous shrubs and herbs, to the mosses and fungi.

Cliff's friends and associates share many happy memories of field trips with him. His complete dedication to the task, meticulous attention to detail, and care of collections is an example and inspiration to less experienced field workers. His respect for the natural environment, his boundless energy and enthusiasm are remarkable. A field trip with Cliff, no matter how arduous, is always a joyful experience.

In presenting the Medallion Mr. Dunbavin Butcher described Cliff as "a great amateur in the traditional sense". He belongs to what appears to be a fast vanishing race, the true Field Naturalist.

Rock Fractures Called Joints

by A. W. BEASLEY*

Exposures of rocks on the surface of our Earth generally owe their form to the local joint pattern. A joint is a break in a rock mass where there has been no relative movement of rock on opposite sides of the break. Many rocks are intersected by regular parallel sets of joints which are usually closely spaced. Where several sets of joints are developed, the rocks tend to break into cubic or rectangular-shaped blocks the size of which reflects the spacing of joints in the sets; it can range from a few inches to several yards. The term joint is said to have originated among British coal miners who were reminded of mortar joints that bind bricks or stones together in a wall.

Rocks split along joints and, as well as being very important in the development of our landscape, they are important in many of man's activities. Rock excavations, tunnels, and quarrying operations must take account of the spacing of the joints. Small blocks can be quarried nearly everywhere, but the number of possible sites for quarrying blocks of building stone more than a yard or so in dimension is small. Joints are also useful to the geologist (both professional and amateur) in the field in providing natural smooth surfaces on exposures, generally in several directions, on which the internal structures of the rocks can be observed. In some particularly massive rocks, such as granites, hand specimens would be difficult to collect were joints absent.

Water flowing over a rock surface tends to travel along the joint partings,

*National Museum of Victoria

Sandstone traversed by joints at Eaglehawk Neck, Tasmania.
Photo: R. F. Cane.

with the result that surface streams become aligned along them. Likewise the joint partings direct water moving underground. In limestone, solution proceeds fastest along the joints, and so the general pattern of a cave-system in limestones (such as at Buchan and Jenolan) usually depends on the arrangement of the joints.

Actually, joints play an important part in the decomposition of rocks, as they permit the access of atmospheric and aqueous agents to considerable surfaces which would otherwise have not been exposed. In some rocks joints are inconspicuous, but evidence of their presence is usually discernible. For example, in a reddish, ferruginous sandstone the positions of the joints may be shown by bleached bands. Often a space is developed in the rock between the joint surfaces; this is called an open joint or fissure. This space may be filled later with minerals, such as quartz, calcite and limonite, to form thin veins. Sometimes valuable ore veins are developed along joints.

There seem to be several possible causes for the formation of joints. The fractures are a response to distributed forces (stresses) in rocks,

Columnar jointing in basalt, known as the "Organ Pipes", near Sydenham, Victoria.
Photo: Author

which behaved as a brittle solid when they broke. The stresses may be connected with earth movements and deformation (folding, tilting, distortion and breaking), they can be termal stresses connected with the cooling of igneous rocks, or residual stresses in rocks which were at one time deeply buried beneath other rocks and later brought to the surface by erosion. Many geologists believe that most jointing is associated with a release of pressure from the "unloading" of overlying rocks through uplift and erosion.

In lava flows and minor igneous intrusions such as sills, a pattern of jointing that breaks the rock into hexagonal or pentagonal columns is sometimes present. As the molten material cooled, it necessarily shrank and this gave rise to the regular cracks through the mass. The "Giant's Causeway" in Northern Ireland is one of the world's best-known exposures of columnar jointing in basalt. In Victoria spectacular basalt columns known as the "Organ Pipes" occur in the valley of Jackson's Creek, near Sydenham, some 16 miles N.W. of Melbourne. The columns are up to 70 feet in height and are about 18 inches in diameter. Although they are little known to the general public, they have been visited by generations of Melbourne geology students and field naturalists.

The A. E. G. (Rudd) Campbell Memorial Appeal

Approaching Wyperfeld in August 1971, for the first time since "Rudd's" death, I had a nagging doubt in my mind that I would find it changed, and it would lack its usual dynamic impact.

I feared "Rudd's" sudden demise could have left an unbridgeable gap. The ready smile, the special bushman brand of humour, the ever filled pocket of hottest peppermints, and that wide knowledge would all be lacking.

I had been there less than two days when I realised that my fears had been unfounded. Wyperfeld and "Rudd" Campbell were synonymous. Their long association of half a century made them complementary. He made Wyperfeld Park — just as it helped to form him, and so long as the park exists "Rudd" will still be there. Without realising it newcomers to the park will see him in it.

The memorial appeal opened to remember "Rudd" by building a visitors' centre and museum at the park surely gives us all a wonderful opportunity to commemorate him by contributing to it. If you visit the park the Ranger will be more than pleased to accept your donation. Others may care to gladden Mrs. Campbell's heart by addressing contributions to her at Tresed, via Yaapeet, Vic. 3424.

VICTOR JACOBS

Winter Insect Collecting in the Northern Territory

by J. C. Le Souef

Following a nine weeks collecting trip to the Northern Territory in 1969, my wife and I decided to pay another visit in June last. The main object this time was to check on the life history of the new species of butterfly, *Virachola smilis daleyensis*, which was added to the Australian fauna on the first occasion.

This time we took the road to Adelaide, calling first on Keith Hateley and his wife at Kiata, and spending the first night at Bordertown. Next day we travelled to Port Augusta, taking the road from Murray Bridge through the beautiful Barossa Valley, bypassing Adelaide. The road was good, the traffic very light, and the scenery magnificent. After Port Augusta, a night was spent at each of Kingoonya, Coober Pedy and Kulgera whilst en route to Alice Springs. Here we changed from the relative luxury of motels to the more satisfying life under canvas. After four days at Alice Springs we drove north to Dunmara, west to Timber Creek and on to Kununurra; returning via Katherine to Daly River Crossing. A few days were spent here before we went on to Darwin, coming south to Mataranka Homestead, Alice Springs, Ayers Rock and home again through Melrose and Adelaide.

It seemed a pity to have to hurry through the very interesting country north from Port Augusta, but we made a lunch stop at Lake Hart, not far from Woomera. Here I found the pre-pupal larval shelter of the skipper butterfly, *Telicota ancilla*, an unexpected addition to the South Australian fauna.

During the few days break from driving at Alice Springs, a number of *Ogyris hewitsoni parsonsi* larvae and pupae were taken from the gums in the bed of the Todd River. It is of interest to record here that with little cover on the trunks, the larvae had pupated under the remnants of bark on the trunk at ground level. There was little of interest in the butterfly world here, although there were a number of *Papilio anactus* pupae on the citrus trees in the Wintersun Caravan Park, and several other species were flying in the Richi Pichi Museum. Because of the cold nights, there were few moths or other insects about although several species of grasshopper were taken among the rocks behind the town.

At the CSIRO laboratory there was a fine carpet snake caught locally, considerably lighter in colour than the coastal ones. Several of the central Australian race of the pouched mouse, *Sminthopsis crassicaudatus* were being kept for study by the mammalogist.

During our stay with Constable and Mrs. Taylor at Timber Creek, we learnt much of the history of this very interesting district. I was taken to the site of the old Depot on the Victoria River where all freight was transported by ship before the advent of good roads. It was near here that I had the great satisfaction of accomplishing one of the tasks I had in mind for the trip. I found a set of very old donkey harness, part of the Police Station equipment, discarded by a Constable Pascoe some years ago on a now disused rubbish dump. A number of other fragments of history were found at the Depot.

under shallow water with much gluey mud. From the cast skins seen, there should be a number of other reptiles apart from goannas.

On the escarpment at Middle Spring, one of the "tourist attractions" of Kununurra, I flushed a pair of White Quilled Rock Pigeons, apparently missed by the feral cats which have so altered the ecology of the countryside. This district is noted for its Corella population, which is estimated at 30,000 birds, eating tons of sorghum each day, and providing a hazard for air travellers. The jet service times had to be altered so as to miss the birds as they moved from and to their roosting places.

A sidelight on the mineral potential of this area was given when a local farmer told us that he had checked on the material used on an irrigation barrage and found that it contained fifty per cent. iron ore.

Plate 1
Phasmid, Kununurra, W.A.

While at Kununurra, we spent a day at Wyndham, but there was little of entomological interest there, with only the normal inland dry country insects to be seen. It was, however, of great interest to again find several small patches of the Elephant Eared Acacia (*Acacia dunnei*), growing in the harshest of environments. With the temperature at 120 degrees in the car when we were here, this was a time when we really appreciated the portable refrigerator which formed part of our equipment for this trip.

It was with some anticipation that we arrived at Daly River Crossing, stopping again at the cashew nut tree outside the "hotel", where we had first taken *Virachola smilis daleyensis* on the previous visit. But with the delightful uncertainty of field entomology, there was no sign of the butterfly at all on this occasion, despite hours of searching in the sun. A possible explanation was that the local Aboriginal Mission had gathered the fruits of *Strychnos* in which the larvae feed. However the rare Leopard (*Phalanta phalanta araca*) and the Orange Tiger (*Danaus genutia alexis*) were taken, as well as the Northern Territory race of the Common Tit (*Hypolycaena phorbas ingura*) and Gilbert's Blue (*Holochila gilberti*).

Mr. Bob Judge, on whose cashew nut grove we spent so much time, showed me his "jungle" with a large jungle fowl's mound obviously in use; but he was not able to bring the pittas which usually come to his call. It was here that I saw the only large colony of the butterfly Common Australian Crow (*Euploea core corinna*). There were many hundreds of them occupying quite a large area.

It would seem that this district is one of the few places left in Australia where wallabies are in really large numbers. Here too were a number of Wedge-tailed Eagles flying unmolested, which says something for the ecological outlook of local station owners.

It was with some reluctance that we again headed for Darwin. In contrast to many other parts of the Commonwealth, there is little provision for the camping tourist close to the city. We were lucky in finding a comfortable spot in the Overlander caravan park, our small dog precluding us from the only other park available.

On our first visit to Darwin, Mr.

Plate 2

Fossilised beach section on 700 foot escarpment, Timber Creek, N.T.

Plate 3
Strychnos fruits showing emergence holes of Virachola smilis dalyensis Darwin, N.T.

Norm Byrnes of the Herbarium, was able to direct us to the few suitable collecting areas of pseudo rain forest nearby. He explained that the flora is much the same throughout the State, with the exception of the east coast of Arnhem Land with its much higher rainfall; and the top of the escarpments where there are different plant species, although the country is still dry.

However on this occasion it was the life history of the *Virachola* butterfly, which was the object of our visit. Mr. John Peters of Sydney, who first found the larvae and pupae, told us where he had located the *Strychnos* fruits in which they feed on the Darwin foreshore. Having now identified this tree for the first time, there was no difficulty in finding others. Later at East Point, we took several larvae and a number of pupae after opening over two hundred fruits with emerged pupal cases.

Along the top of the cliff at Lamaroo Beach, well known for its "hippie" colony, is the main insect collecting spot near Darwin with a number of species of butterfly moving along the tops of the trees growing from the shelf above the beach below.

Although at present the authorities seem mainly concerned with providing houses, roads and amenities for the largely civil servant population, there is a band of enthusiasts collecting together what remains of evidence of the pioneers. Housed in an old building in the Botanical Gardens is a fascinating collection of bits and pieces from the early days. It is hoped that eventually they will be housed in a proper setting.

Mataranka Homestead has a threefold attraction. It abounds in wild life in comfortable surroundings, and has a thermal pool being fed by a spring at a constant temperature of some 90 degrees.

It was here that I had hoped to check on the life history of the butterfly *Danaus genutia alexis*, the Orange Tiger, which was flying in some numbers with the Lesser Wanderer (*Danaus chrysippus petilia*), on the first visit. Despite hours spent by day and by torchlight at night, it was not until the last afternoon that I managed to take three specimens of the former.

flying with the Lesser Wanderers. By then it was too late to study them further.

The cold south-easterly winds inhibited night collecting here, as it did on much of the trip, bringing the morning temperature down to forty degrees — unexpected on the Roper even in winter. There were, however, quite a number of insect species attracted to the light during the evenings.

In order to ascertain whether there was any evidence of the brown butterfly, *Geiteneura*, at Ayers Rock, three days were spent there on the return journey. Maggie Springs was gay with flowers and alive with insects when we last paid a visit, but on this occasion there were few flowers and fewer insects. Despite diligent searching, there was no trace of grass being eaten by lepidoptera at all. Again there was no sign of the large blue "fly" seen here in 1969. It had been hoped that specimens could be taken in the light of the recently described *Punops austrae* from Mt. Olga, as Mr. A. Neboiss of the National Museum wanted more material for study.

The Assistant Ranger, Mr. Carwood, told of the virtual disappearance of the Hopping Mouse because of the depredations of feral cats. He stated that he had shot 150 cats recently in the Park.

Travelling in the Northern Territory in the winter nowadays is very different from the earlier days. Beef roads already cover many hundreds of miles, and others are being upgraded in preparation for sealing. The contract has been let for the sealing of the "Broken Spring Highway" south from Alice Springs; and there is even a team now working on the road to Roper Bar, which not many years ago took nearly a week to negotiate. We were able to cover the 9,750 miles in nine weeks quite comfortably without any trouble, in a conventional station wagon.

As there have been over 100 species of butterflies recorded from the Northern Territory, it would seem that there is a considerable drop in the number of species on the wing after the end of the wet season.

F. M. Angel, in his paper on his collecting trip to the Territory with F. E. Parsons (*Trans. Roy. Soc. S Aust.*, 74 (1) March, 1951) lists 59

Plate 4
Larva,
Polyura pyrrhus sempronius,
(Tailed Emperor)
Mataranka,
N.T.

species. They spent five weeks there during April and May, 1948.

While serving with H.M.A.S. Warrega in November, 1945, J. O. Campbell recorded 20 species during visits to Darwin. (*The Australian Zoologist*, Vol. II, pt. 2, June 20, 1947).

As well as the following list of butterflies, some 50 species of moths, 20 species of beetles, and many examples of other orders, were taken.

List of Specimens Taken:

(A: Darwin, B: Mataranka, C: Daly River, D: Kununurra, E: Tennant Creek.)
Telicota ancilla ancilla — C.
Telicota ancilla baudina — B, C.
Telicota colon argeus — A.
Ocybadistes walkeri olivia — B, C.
Cephrenes trichopepla — A.
Pelopidas lyelli lyelli — A, B, C, D, Wyndham.
Papilio anactus — Alice Springs.
Papilio canopus — B, D.
Papilio demoleus sthenelus — widespread.
Cressida cressida cressida — A.
Catopsilia pyranthe crokera, form *luteola* — D, Timber Creek.
Catopsilia pomona pomona, form *pomona* — A, B, Wyndham.
Catopsilia scylla etesia — D.
Eurema hecabe phoebus — B, D, Wyndham.
Eurema smilax — A, B, D, E, Ayers Rock.
Eurema laeta lineata — B.
Eurema herla — A.
Elodina padusa — C.
Elodina perdita walkeri — B.
Delias argenthona fragalactea — A.
Anaphaeis java teutonia — widespread.
Cepora perimale — A, B, C, D, Alligator River.
Appias paulina ega — A.
Danaus genutia alexis — B, C.
Danaus chrysippus petilia — A, B, C, D, Timber Creek.
Danaus affinis affinis — A, B, D, Howard Springs.
Euploea core corinna — widespread.
Euploea sylvester pelor — C.
Euploea darchia darchia — C.
Mycalesis sirius sirius — Howard Springs.
Melanitis leda bankia — A, C.
Hypocysta adiante antirius — Timber Creek.
Polyura pyrrhus sempronius — A, B.
Hypolimnas alimena darwinensis — C.
Hypolimnas bolina nerina — Timber Creek.
Hypolimnas missipus — A, C.
Vanessa cardui kershawi — D, E, Ayers Rock, Wyndham.
Precis hedonia zelima — B, Howard Springs.
Precis villida — B, Alice Springs.
Precis orytha albicincta — A, B, C, D, Howard Springs.
Cethosia penthesilea paksha — A.
Phalanta phalanta araca — A, C.
Acraea andromacha andromacha — A, B, C, D, Timber Creek.
Virachola smilis dalyensis — A, C.
Hypolycaena phorbas ingura — A, C.

Narathura araxes usopus — A, D.
Narathura micale amydon — A, Nourlangie
Ogyris zozhie typhon — Elliott.
Nacaduba kurand felsina — C.
Nacaduba biocellata biocellata — widespread.
Catopyrops florinda estrella — C.
Jamides phaseli — B, C.
Anthene emolus affinis — C.
Anthene lycaenoides godeffroyi — C.
Theclinestes onycha onycha — A, B, C, D,
Lampides boeticus — A, D.
Catochrysops panormeus platissa — B, C, D, Timber Creek, V.R.D.
Euchrysops cnejus cnejus — A.
Zizula hylax attenuata — A.
Zizeeria alsulus alsulus — Victoria River Downs.
Zizeeria otis labradus — Ayers Rock.
Zizeeria knysa karsandra — Victoria River Downs.
Erina erina crina — Howard Springs.
Holochila gilberti — C, D. Howard Springs, Roper Bar.

Flowers and Plants of Victoria in Colour

Copies of this excellent book are still available, and of course would make a wonderful gift. They are obtainable from the F.N.C.V. Treasurer, Mr. D. McInnes.

ENTOMOLOGICAL EQUIPMENT

Butterfly nets, pins, store-boxes, etc.
We are direct importers and manufacturers,
and specialise in Mail Orders
(write for free price list)

**AUSTRALIAN
ENTOMOLOGICAL SUPPLIES**
14 Chisholm St., Greenwich
Sydney 2065   Phone: 43 3972

STEREO BINOCULAR MICROSCOPE

Wide field, three dimensional viewing with long working distance

MATCHED   2X – 4X   OBJECTIVES
PAIRED   10X – 15X   EYEPIECES
INTERPUPILLARY ADJUSTMENT
MAGNIFICATION 20, 30, 40, 60X

Supplied in wooden cabinet
$110.00

STEREO MICROSCOPE

For the examination of whole objects such as insects, flies, organs, parasites etc.

30X and 50X magnification with wide, bright, stereoscopic field
$24.50

GENERY'S SCIENTIFIC EQUIPMENT SUPPLY
183 Little Collins Street
Melbourne: **63 2160**

Victorian Non-Marine Molluscs — 7

by D. C. Long*

The Introduced Zonitids

The pulmonate land snail family Zonitidae is represented in Victoria by a number of introductions; five European species have been identified here and one or two other species may be present in the northern part of the State. The five species are *Oxychilus cellarius* (Muller), *O. draparnaldi* (Beck) and *O. alliarius* (Miller), *Vitrea contracta* (Westerlund) and *Euconulus fulvus* (Müller).

In *Oxychilus* the shell is low spired and the Victorian species have a wide central perforation (umbilicus) at the base; the shell has a glossy surface, is horn to greyish yellow in colour, and is paler below than above. The species are best distinguished by a combination of shell characters and the appearance of the animal.

O. cellarius

This has been verified from five sites in central Victoria including Melbourne, and has a greyish-yellow shell up to 10 mm. in diameter with 5 whorls when fully grown. The animal is grey, somewhat darker on the head and the edge of the mantle is speckled with brown (visible under a lens). Neither of the other two species of *Oxychilus* identified in Victoria has this speckling.

O. draparnaldi (Fig. 1)

This is common in and around Melbourne. It is the largest of the three species and the shell may grow up to 15 mm. in diameter with seven whorls. The shell is horn coloured varying in shade from pale to dark, is often slightly expanded near the mouth and has relatively prominent lines of growth. The animal is blue-grey, including the foot, and has a dark grey mantle edge.

O. alliarius

Practically State-wide in its distribution, it is the smallest of the three species growing to just over 7 mm. in shell diameter, with five whorls, and has a horn coloured shell. The animal is slaty blue-black with a grey foot and emits a smell of garlic on withdrawing into its shell when alarmed.

Vitrea contracta

It has only been identified in Victoria once — at Geelong in 1967. It has a small semi-transparent white shell up to 2.7 mm. in diameter roughly similar to an *Oxychilus* in shape. The animal is greyish white. It may occur elsewhere in settled areas and should be looked for.

Euconulus fulvus

This was found at Cann River in 1928 (C. J. Gabriel), *Victorian Nat.* 46 (6) 130-134); it has a small glossy reddish brown shell up to 3.0 mm. diameter with a conical spire and no umbilicus. The animal is grey-black.

The Zonitidae are omnivorous, mainly nocturnal snails, and do not constitute molluscan pests. They should be encouraged in gardens because they eat other snails and slugs. Nearly all the Victorian Zonitidae records are from settlements or cleared land, but *O. alliarius* has been found twice in semi-open forest.

* 65 Abbeydale Street, Oakleigh, Vic. 3166.

Fig. 1. *Oxychilus draparnaldi*
Drawing: Phyllis Plant

reptiles of victoria – 4

by HANS BESTE

Plate 7

Egernia saxatilis — Black Rock Skink.

A bulky looking skink with a depressed body. Powerfully built.

Length: to almost 12 inches.

A distinct ear-opening. Five fingers and five toes. Long claws, toes well developed. Long tapering tail, slightly depressed. Upper and sides dark grey to brown, often with greenish sheen. Scales carry keels on upper surface. Under paler, merging from light grey on sides to cream or pale salmon pink.

Distribution and Habitat: Found under rocks and among granite outcrops, mainly on coastal mountain ranges. Usually found sunning on large boulders.

Best distinguishing features — dark colouration, whitish throat, habitat.

Plate 8

Egernia whitii — White's Skink.

A slim member of this family.

Length: to over 12 inches.

The head is slim and distinct from the neck. A distinct ear-opening. Five fingers and five toes. Very variable in colour. Two colour forms found in this State. Handsomely marked with brown and black stripes on back. Stripes ornamented with small creamy spots, which are edged with black. Under pale yellow to salmon. Basic colour dark brown-grey or pale orange-brown.

Distribution and Habitat: Found under logs and rocks in forested areas; also in open.

Best distinguishing features — cream spots on sides and back.

Plate 7

Photo: H. Beste

Plate 8

Photo: H. Beste

NATURE CONSERVATION IN VICTORIA

by J. Frankenberg

This book comprises 145 pp. of the results of long investigation by the Victorian National Parks Association to determine requirements for effective Nature Conservation in this State.

Both the achievements and deficiencies in the existing system of Nature Reserves have been noted, and recommendations based on careful and detailed studies in the Victorian scene, provide legislators, authorities, and plain citizens with the guidelines so much needed in the planning of land use, National Parks, and Nature Conservation reserves of the future.

The book is a "must" for every conscientious conservationist.

Price — $2.00 plus 20 cents postage.

Available from:

The Hon. Sec. V.N.P.A. — J Ros Garnet, 23 Cumulus Street, Pascoe Vale 3044; or from the Secretary, National Parks Service, State Public Offices, Treasury Place 3002; or Sales Officer F.N.C.V., 25 Sunhill Avenue, North Ringwood 3134; or from other affiliated Natural History Organisations and Booksellers, who may obtain quantities from the V.N.P.A. or Service.

Field Naturalists Club of Victoria

General Meeting 8 November.

The Herbarium hall was crowded with members who attended the November meeting to see the Natural History Medallion for 1971 presented to Mr Cliff Beauglehole. The presentation of the award was made by Mr. Dunbavin Butcher, the Director of the Fisheries and Wildlife Department of Victoria.

While he did not wish to detract in any way from past winners, Mr. Butcher stated he considered Alexander Clifford Beauglehole to be one of the most worthy recipients of the medallion in its history. Never before have we needed information on which to base conservation so much, he said. Time has been with us for a long time in this country. Now it is against us. Knowledge about our flora and fauna is needed quickly Compared to Australia, which is a boom land legitimately concerned with development, Britain has a great tradition of interest in the study of nature. It is the responsibility of conservationists to put their case unemotionally, with information to back it. Several groups in F.N.C.V., he stated, were contributing to collecting this data. In this field of work Cliff Beauglehole is most eminent.

The dossier submitted to the award committee on the work of Beauglehole is a remarkable record, said Mr. Butcher. He has a private herbarium of 23,000 specimens. His prodigious output of work was rare. He could find no parallel to it. At the age of 10 it is recorded Beauglehole could scientifically name 60 native orchids occurring in his home district. He has collected with skill and perception; collecting not only for himself, but for people all over Australia. His persistence as a collector was shown by his spending 10 years searching an area to find a single species reported to exist there.

Time and time again Beauglehole has helped the Fisheries and Wildlife Department and the National Parks Authority. He has the rare honour of being allowed to use the official Herbarium labels. All this has been done by a man who is only 51. Mr Butcher pointed out that Mrs. Beauglehole figures prominently in Cliff's work, especially his long coastal walks.

Speaking on receiving the award, Mr Beauglehole said he owed a great debt to his parents for encouraging his interests. Also to the Holmes family, who were friends when he was a boy; for his knowledge of orchids he owed to the late Murray Holmes. His wife and two daughters had further encouraged him. Many people had assisted him, he said, including numerous visiting scientists. Discussion with these different specialists led him into covering the wide field he had worked on. There were too many to whom he was indebted to enumerate them all, but one he particularly wished to thank was Jim Willis. He also gave special thanks to those clubs who nominated him for the award; to those who prepared the dossier on his work; and to Mr. Butcher for the description of why he won the award.

Club President Mr. Tom Sault said few realise just what is entailed in giving this award. There are masses of dossiers to be examined. One dossier alone occupied two cubic feet, he said. Therefore he called for a show of appreciation to the Natural History Medallion Award Committee, of which Mr. Butcher is chairman, and to the general committee which prepares the presentations. This was given by a round of applause.

Mr. Beauglehole then returned to the platform to speak about his latest activity — commencing the botanical grid survey of East Gippsland. He has been working on this project for some three years, and has covered an area of 6,000 square miles. At the same time he has made lists of the flora of seven national parks in that area, which cover some 36,000 acres, plus proposed new ones, and extensions to the existing ones that cover a further 430,000 acres. The proposed new ones being Cobberras National Park and the Snowy River National Park; these it is suggested should respectively cover 300,000 acres and 100,000 acres. An article covering the subject of his talk on this work in East Gippsland will be appearing in a later edition. The only thing its readers will miss are the interesting and attractive film slides of areas he visited, and the plants found there, which Mr. Beauglehole showed at the meeting.

Mr. Ros Garnet moved a vote of thanks to Mr. Beauglehole for beginning the botanical grid survey of Victoria.

Mrs. M. Salau was presented with her honorary membership certificate. The President noted she joined the club in 1931, and for many years she has been one of the chief assistants to Mr. McInnes in running the annual Nature Show. She has been responsible for selecting the films shown, writing reports and arranging many exhibits.

In accepting the certificate Mrs. Salau said she thanked her interest in nature to being born in the country, at Heathcote; and recommended the study of things found in the country for filling one's later years.

The award of two other honorary members' certificates awaits a time when the recipients are able to attend a meeting of the club. Three new ordinary members were admitted to membership at the meeting, together with five new joint members, and one new country member. It was announced that Mrs. Webb-Ware has had a serious accident.

Correspondence revealed that the Tasmanian government has no intention of cancelling or altering the Gordon power scheme, which means Lake Pedder will soon be flooded.

Mr. Bill Davis moved that F.N.C.V. should join the Save Westernport Coalition. He said at present it was a loose body involving some 30 organisations. He felt the name of F.N.C.V. would add weight to the coalition, and the inclusion of its name among the members was the chief reason for asking the club to join. The cost would be only a couple of dollars. There was only one speaker against the motion, which was carried.

Mr. Sault announced that a few months ago the Geology Group surveyed the Aboriginal stone axe quarry at Mt. William, a few miles N.E. of Lancefield. A map was prepared from the survey showing the position of all the working sites. Two copies of the map were presented to Romsey Shire Council in whose territory the quarry lies, one to the National Museum, and another is being kept at the Herbarium.

Western Victorian Field Naturalists Clubs
(Continued from last month)

HAMILTON

President: Rev. E. H. Deutscher. Secretary: Mrs. L. Courtney.
Membership 49 Adults, 12 Juniors. Meetings First Monday, Arts Centre.

Guest speakers at meetings included Miss H. Aston, "Aquatic Plants"; Mr. P. Disher, "Reptiles"; Mr. A. Chapman, "Wanderer Butterfly"; Mr. H. Beer, "Control Burning"; Mr. J. Clements, "Eels". Excursions are held monthly. "Photoflora 70" was sponsored and our annual Nature Exhibition during the 'Golden Wattle Festival".

First steps have been taken towards the establishment of a forty acre wild flower sanctuary in Dalyenong Forest and a submission was made to the Senate Select Committee setting out the case for a national park to preserve typical goldfields fauna and flora.

MID-MURRAY
President: Mr R. Curtis.   Secretary: Miss G. Willoughby
Membership: 31 Adults, 14 Juniors.   Meetings Third Friday, Wood Wood.
Excursions monthly (including New Year's Day Picnic). Executive meets monthly also.

An active club with good attendance at activities. Visiting speakers at meetings included Mr. H. Arlt, "Reptiles"; Mr. R. Cooper, "Relationship of Birds to Man"; Prof. Stone, "Mosses", plus one members night, "Acacias".

June long weekend was spent as guests of Sunraysia and another camp-out was held at Wyperfeld.

A slide night was provided by local members for a charitable organisation, and we hosted the Sept. meeting of W.V.F.N.C.A., also sponsored "Photoflora '70", screened two Walt Disney films locally, published a 3rd report and fortnightly Nature Notes column in the local newspaper and purchased club badges.

PORTLAND
President: Miss Ina Watson.   Secretary: Mr. C. Skaer.
Membership 43 (reduced by deaths of three staunch members, and the moving of 6 to 7 others).   Meetings Second Friday, Town Hall.

Speakers at meetings included Alan Berger (Exchange Student from South Africa), Mrs. Jean Aslin, "Snails"; Mr. Gavin Cerini, Mr Bert Curthew, "70 Years of Nature and Photography"; Mr. Murray Gunn, "Bird Banding"; Mr. Leon Farrell, "Nature in Stamps"; Mr. Greg Baker, "Trip to North and Western Australia"; Mr. P. Young, "History and Effects of Controlled Burning", plus a members' night.

During the year the club hosted a W.V.F.N.C.A. meeting and played guide to local attractions for visitors. Several trips further afield were attempted with mixed success.

STAWELL
President: Mr. I. McCann.   Secretary: Mr. N. Bennett.
Membership: 36 Adults, 4 Juniors.   Meetings Fourth Monday, Tech. School.
Excursions One Full Day, one half day per month

The club continues to hold slide nights at Christmas with the immediate aim of fencing the whole of 3 Jacks Sanctuary. One member co-operates with the mollusc survey being carried out by the National Museum

SUNRAYSIA
President: Mr. H. Arlt.   Secretary: Mr. G. Stephenson.
Membership (1969) 107.   Meetings Third Wednesday, Trust Rooms.

This club comprises Botany, Bird and Photography Groups each with their own meeting night and we have an Executive Board. Members have been urged to do all in their power to increase diminishing membership.

Life Membership of the S.N.T. was conferred on Mr. Hal. Thomas

The Trust's rooms were altered by the City Council at a cost of $110 to the Trust. Members painted the rooms and a 16 mm movie projector has been purchased.

"Photoflora '70" was sponsored and a slide exhibition held during the "Back to Mildura" celebrations.

Outings were held to the Raak, Werrimul South, Kulkyne Forest Timberoo and a camp-out to the Sunset Country.

WARRNAMBOOL
President: Mr. R. F. Pierce.   Secretary: Mr. M. J. Yeoman.
Membership: 43 Adults, 9 Juniors.   Meetings Fourth Wednesday, C.W.A. Rooms.

During the year we have continued to pursue all conservation matters. Monthly meetings and outings were well attended and we hope to have delegates at W.V.F.N.C.A. meetings during 1971 as this year.

F.N.C.V. DIARY OF COMING EVENTS
GENERAL MEETINGS

Monday, 13 December — At National Herbarium, The Domain, South Yarra, commencing at 8 p.m.

1. Minutes.

2. Announcements.

3. Subject for evening: —The Future of the F.N.C.V.

4. New Members.

    Ordinary:
    Mr. William Culican, 4 The Outlook, Glen Waverley 3150.
    Mr. W. R. Elliot, Belfast Road, Montrose 3765.
    Miss Sheila Williams, Graduate House, 224 Leicester Street, Carlton 3053.

    Joint:
    Mr. David A. Harbeck and Mrs. Joyce E. Harbeck, 3 Gertrude Street, Burwood East 3151.

5. Correspondence.

6. General Business.

Monday, 10 January, 1972 — Members' Night.

F.N.C.V. EXCURSIONS

nday, 12 December — Point Lonsdale. Subject: Marine Biology; Leaders: Mr. and Mrs. J. Strong. The coach will leave from Batman Avenue at 9.30 a.m. Bring two meals. Fare: $1.70.

unday, 26 December-Monday, 3 January — Fall's Creek. Accommodation has been booked at Spargo Lodge, tariff $55.00 to $60.00 per week, full board. Fare for coach, including day trips $22.00. The fare should be paid to the Excursion Secretary when booking, cheques to be made out to Excursion Trust. The coach will leave from Flinders St. outside the Gas and Fuel Corporation at 8 a.m. Bring a picnic lunch.

GROUP MEETINGS

(8 p.m. at National Herbarium unless otherwise stated.)

hursday, 9 December — Botany Group. Members' night and supper. Please bring a small contribution and your own cup.

o group meetings will be held between the December and January General Meetings.

Field Naturalists Club of Victoria

Established 1880

OBJECTS: To stimulate interest in natural history and to preserve and protect Australian fauna and flora.

Patron:
His Excellency Major-General Sir ROHAN DELACOMBE, K.B.E., C.B., D.S.O

Key Office-Bearers, 1971-1972.

President:
Mr. T. SAULT

Vice-Presidents: Mr. J. H. WILLIS; Mr. P. CURLIS

Hon. Secretary: Mr. R. H. RIORDAN, 15 Regent St., East Brighton, 3187. (92 8579)

Subscription Secretary: Mr. D. E. McINNES, 129 Waverley Road, East Malvern, 3145

Hon. Editor: Mr. G. M. WARD, 54 St. James Road, Heidelberg 3084.

Hon. Librarian: Mr. P. KELLY, c/o National Herbarium, The Domain, South Yarra 3141.

Hon. Excursion Secretary: Miss M. ALLENDER, 19 Hawthorn Avenue, Caulf., 3161. (52 2749).

Magazine Sales Officer: Mr. B. FUHRER, 25 Sunhill Av., North Ringwood, 3134.

Group Secretaries:

Botany: Mr. J. A. BAINES, 45 Eastgate Street, Oakleigh 3166 (57 6206).
Microscopical: Mr. M. H. MEYER, 36 Milroy Street, East Brighton (96 3268).
Mammal Survey: Mr. D. R. PENTON, 43 Duke Street, Richmond, 3121.
Entomology and Marine Biology: Mr. J. W. H. STRONG, Flat 11, "Palm Court", 1160 Dandenong Rd., Murrumbeena 3163 (56 2271).
Geology: Mr. T. SAULT.

MEMBERSHIP

Membership of the F.N.C.V. is open to any person interested in natural history. The *Victorian Naturalist* is distributed free to all members, the club's reference and lending library is available, and other activities are indicated in reports set out in the several preceding pages of this magazine.

Rates of Subscriptions for 1971

Ordinary Members	$7.00
Country Members	$5.00
Joint Members	$2.00
Junior Members	$2.00
Junior Members receiving Vict. Nat.	$4.00
Subscribers to Vict. Nat.	$5.0
Affiliated Societies	$7.00
Life Membership (reducing after 20 years)	$140.00

The cost of individual copies of the Vict. Nat. will be 45 cents.

All subscriptions should be made payable to the Field Naturalists Club of Victoria, and posted to the Subscription Secretary.

JENKIN BUXTON & CO. PTY. LTD., PRINTERS, WEST MELBOURNE

INDEX

Compiled by J. A. Baines, 45 Eastgate St., Oakleigh 3166.

ABORIGINES

Axe-stone Quarry at Mt. William near Lancefield, The Aboriginal (illus.), 273
Grinding Rocks at Stratford (illus.), 58
Lady Julia Percy Island, Victoria, The Aborigines and (illus.), 84
Madimadi Man, The Last (Point Pearce, Yorke Peninsula, S.A.) (illus.), 11
Mornington Peninsula, Aboriginal Relics on the (illus.), 336
Painted Shelter, Boggy Creek (Mt. Difficult Range): A New Locality Record (illus.), 152
Paintings, Aboriginal, at Muline Creek (illus.), 139
Vocabulary of Madimadi (a Kulin language of Murray River, north bank, N.S.W.), 15

AMPHIBIANS

Frog, Marsh, Spotted (photo), 301

AUTHORS

Aitken, Peter F., 103
Anonymous, 73, 201, 238
Baines, J. A., 160
Beadnell, Myfanwy, 306
Beasley, A. W., 291, 347
Beste, Hans, 117, 156, 358
Bird, E. C. F., 189
Blackburn, G., 56
Casey, D. A., 273
Chisholm, A. H., 43, 180, 342
Clyne, Densey, 244
Corrick, Margaret G., 344
Coventry, A. J., 20, 304
Davies, Fred, 39
Dixon, Joan M., 133, 198
Fellows, A. G., 278
Galbraith, Jean, 71, 216
Gasking, W. R., 170, 200
Gill, Edmund D. (with A. L. West), 84, 172, 287
Hampton, J. W. F., 62
Hercus, Luise A. (with I. M. White), 11
Hobbs, R. P., 32
Jacobs, Victor, 254, 349
Johnston, Bob, 116
Jones, David L., 217, 277
Kelly, Peter, 324
Kershaw, Ron C., 4
Kraehenbuehl, Darrell N., 220, 225, 231
Le Souëf, J. C., 23, 350
Lester, Margery J., 120
Long, D. C., 357
Lyndon, Ellen, 61, 212
McEvey, A. R., 172, 202
Massola, Aldo, 58, 139, 152
Parker, S. A., 41
Pollock, R. A., 148
Reid, Hilary, 61
Seebeck, J. H., 310
Smith, Brian J., 45, 155, 272, 325
Spillane, A. E., 336
Timms, B. V., 154
Wakefield, N. A., 46, 48, 92, 158, 203
Ward, G. M., 278
West, Alan L. (with E. D. Gill), 84
Whinray, J. S., 279
White, Isobel M. (with L. A. Hercus), 11
Willis, J. H., 88
Zirkler, Jean, 249

BIRDS

Bass Strait, Birds of (overlooked articles of naturalists), 43
Eagle, Sea-, White-breasted (photo), 241
Kites, Black-shouldered (behaviour of) (note), 61
Lowan, Some Observations on the, at Wychitella, 116
Lyrebirds in Tasmania (Mt. Field National Park), 200
Magpies, Maternal (note), 61
Marsh and Lake, Birds of (report of Joan Forster's talk to F.N.C.V.), 328
Penguins, 306
Tool-using by Birds, Further Notes on, 342
Tools and Playthings, The Use by Birds of (illus.), 180
Tree-creeper, White-throated (at Tyers) (note), 216
Wyperfeld (birds along the Ginap Track) (illus.), 254

CRUSTACEANS

'Australian Crustaceans in Colour', by Anthony Healy and John Yaldwyn (book review), 202

EXCURSIONS

Cann River District, F.N.C.V. Excursion to (with map and botanical species list), 120
Little Desert, F.N.C.V. Camp-out at Broughton's Waterhole (illus.), 165
Mt. Beauty, F.N.C.V., Trip to, 249
Mt. Howitt and the Macalister, Excursion by Bairnsdale and Latrobe Valley F.N.C.'s, 212
Wilson's Promontory, Botany Group Excursion to, 201
Wyperfeld National Park, F.N.C.V. Camp-out in, 160

FIELD NATURALISTS' CLUB OF VICTORIA

Accounts, 76
Affiliated Clubs (reports)
  Ararat, 329
  Ballarat, 329
  Colac, 329
  Creswick, 329
  Donald, 330
  Geelong, 330
  Hamilton, 330
  Hawthorn Junior, 142
  Horsham, 330
  Maryborough, 361
  Mid-Murray, 362
  Montmorency Junior, 173
  Portland, 362
  Stawell, 362
  Sunraysia, 362
  Warrnambool, 362
General Meetings, 25, 74, 173
Group Meetings
  Botany Group, 141, 174, 206, 238
  Marine Biology and Entomology, 50, 75, 298

GEOLOGY AND PALAEONTOLOGY

Fossil Podocarp Root from Telangatuk East, Victoria, 56
Joints, Rock Fractures called (illus.), 347
Plucking of Rocks by the Sea (Von Mueller Creek and Point Sturt, Otway coast, Vic.) (illus.), 287
Safety Beach, Dromana Bay, Port Phillip (illus.), 291

INSECTS

Moths, Fruit, 23
Winter Collection of Insects (North Queensland), 23, (Northern Territory) (illus., and with species list), 350

MAMMALS

Badger, The Life of the (report of talk by Robert Withers), 205
Brisbane Ranges, The Mammals of the (with map and tables), 62
Broughton's Waterhole, Notes on Fauna at (illus.), 170
Burramys parvus (Marsupialia) from Falls Creek, Bogong High Plains, Victoria (illus.), 133
Dingo (*Canis familiaris dingo* (photo), 269
Distribution Data of Victorian Mammals, 48
Echidna (*Tachyglossus aculeatus*), (photo), 29; 53; (diet), 61
Furneaux Group, Bass Strait, Tasmania, The Present Distribution of some Mammals in the (with map and tables), 279
Kangaroo, Tree- (*Dendrolagus lumholtzi*) (photo), 334
Macropus bernardus, Notes on the Small Black Wallaree of Arnhem Land (with map), 41
'Mammals, Native, of Australia, A Guide to the' (book review), 46
Native Cat, Little Northern (*Dasyurus hallucatus*) (photo), 1
Platypus (plural form of) (note), 278
Rat, Broad-toothed (*Mastacomys fuscus*), (photo), 178; (from Falls Creek area, Bogong High Plains) (illus.), 198; (distribution and habitat in Victoria) (illus. and with map and tables), 310
Rat-Kangaroos, Tasmanian, Notes on (illus.), 4
Rattus fuscipes, Southern Bush-rat, Studies of an Island Population of (Greater Glennie Island) (illus.), 32
Sminthopsis psammophilus, Large Desert Sminthopsis or Sandhill Pouched Mouse, Rediscovery of, on Eyre Peninsula, South Australia (illus.), 103
Wallaby, Rock-, Brush-tailed (*Petrogale penicillata*) in Western Victoria (illus.), 92
Wyperfeld National Park, Notes on Fauna of (illus.), 170

MARINE BIOLOGY AND CONCHOLOGY

'Beneath Australian Seas', by Walter Deas and Clarrie Lawler (book review), 202
Marine Biology Group reports, 50, 75, 298
Molluscs, Victorian Non-Marine, No. 3 --*Theba pisana* and *Helicella* spp. (illus.), 45

Ditto, No. 4—*Pygmipanda kershawi* (illus.), 155
Ditto, No. 5—*Ninupala hanleyi* and *Plutiopsis balonnensis* (illus.), 272
Ditto, No. 6—Freshwater Mussels (Family Hyriidae) (illus.), 325
Ditto, No. 7—Introduced Zonitids (illus.), 357

MISCELLANEOUS

Elusive Lakes (letter) (illus.), 154
Ginap, Track, Along the (Wyperfeld National Park) (illus. and with maps), 254
'Merrimu', at Sherbrooke, 203
New Guinea Highlands (report of talk by J. H. Willis), 204
Tides in Bass Strait, A Note on the (with map and tables), 148
Wildflower and Nature Show — 1971, 239

PERSONAL

Beauglehole, Alexander Clifford (Australian Natural History Medallion 1971) (with portrait), 344; (report of presentation), 360
Campbell, A. E. G. (Rudd) (memorial appeal), 349
Churchill, David Maughan, Dr. (new Government Botanist and Director Botanic Gardens), 238
Hooke, Arthur Garnsey (1898-1970), 158
Hunter, William (1893-1971) (Doyen of East Gippsland Botanists) 88
Learmonth, Noel S., A Tribute to, 39

PLANTS

Botany Group meetings, 141, 174, 206, 238
Cann River (species list, F.N.C.V. excursion) 132
Eucalyptus regnans (White Mountain Ash) (photo), 334
Flinders Ranges, A Botanical Bibliography of the (1800-1970), 231
Flinders Ranges, Northern, On a Botanical Collection from the (illus.), 225
Flinders Ranges, Southern, Botanical Exploration in the, 1800-1970 (with map), 220
Gippsland, East (report of Cliff Beauglehole's talk on grid survey), 361
Greenhoods, Two New for Victoria (*Pterostylis lata* and *P. coccinea*), 277
Mangroves as Land-builders (illus.), 189
Northern Strzelecki Heathlands, 71

'Orchids, Australian Native, in Colour', by Leo Cady and E. R. Rotherham (book review), 202
Orchids, Beard- (*Calochilus spp*. at Chiltern) (note), 216
Orchid, Sun-, Veined (*Thelymitra venosa*), A Study of the Self-pollination of, and Some Notes on its Implications (illus.), 217
Proteaceae (report of Miss L. White's talk to Botany Group), 238
Scaevola aemula (Annuello and Hattah Lakes) (correction), 111
Wilson's Promontory, Botany Excursion to, 201

REPTILES

Bluetongue, Western (*Tiliqua occipitalis*) (illus.), 118
Crocodile, Freshwater (*Crocodylus johnstoni*) (photo), 113
Dragon, Bearded (*Amphibolurus barbatus*) (illus.), 156
Dragon, Mountain (*A. diemensis*) (illus.), 296
Dragon, Tree (*A. muricatus*) (illus.), 296
Lizard Genus *Carlia* (Scincidae: Lygosominae), The Discovery of the, in Victoria (illus.), 20
Lizard, Moloch (*Moloch horridus*) (photo), 81
Monitor, Mangrove (*Varanus semiremex*) (photo), 145
Reptiles of Victoria—(Pt. 1), 117; (Pt. 2), 156; (Pt. 3), 296
Skink, Black Rock (*Egernia saxatilis*) (illus.), 358
Skink, Striped (*Ctenotus lesueurii*) (illus.), 118
Skink, White's (*Egernia whitii*) (illus.), 358
Snake, Carpet (photo), 209
Snake, Mitchell's Short-tailed (*Parasuta brevicauda*) (illus.), 156
Snakes, Black-headed (*Denisonia*), Identification of the, within Victoria (illus.), 304
Thorny Devil (*Moloch horridus*) (photo), 81

REVIEWS

'A Guide to the Native Mammals of Australia', by W. D. L. Ride, 46
'Annotated Bibliography of Quaternary Shorelines—Supplement 1965-1969', by H. G. Richards, 172

'Australian Butterflies', by Charles McCubbin, 324

'Australian Crustaceans, in Colour' by Anthony Healy and John Yaldwyn, 202

'Australian Native Orchids, in Colour', by Leo Cady and E. R. Rotherham, 202

'Australian Spiders, in Colour', by Ramon Mascord, 172

'Beneath Australian Seas', by Walter Deas and Clarrie Lawler, 202

'Birds of Victoria 2: The Ranges' (Gould League; illus. by Margo Kröyer-Pedersen), 73

'Nature Conservation in Victoria', by Judith Frankenberg, 360

SPIDERS

'Australian Spiders in Colour', by Ramon Mascord (book review), 172

Huntsman, The Mating of a (*Isopeda vasta*) (illus.), 244

Spider Earthworks (at Charters Towers, Queensland) (note), 278

Lightning Source UK Ltd.
Milton Keynes UK
UKHW02f0807190918
329158UK00007B/153/P